NEW PRODUCT DEVELOPMENT:

From Initial Idea
To
Product Management

NEW PRODUCT DEVELOPMENT:

From Initial Idea To Product Management

Marc A. Annacchino, P.E.

ELSEVIER
BUTTERWORTH
HEINEMANN

An Imprint of Elsevier

Amsterdam Boston Heidelberg London New York Oxford Paris
San Diego San Francisco Singapore Sydney Tokyo

ELSEVIER
BUTTERWORTH
HEINEMANN
An Imprint of Elsevier

Library of Congress Cataloging-in-Publication Data

Annacchino Marc A.
   New product development : from initial idea to product management / by Marc A. Annacchino.
     p. cm.
   Includes bibliographical references and index.
   ISBN 0-7506-7732-5
    1. New products–Planning. 2. Product management. I. Title.

   HF5415.153.A56 2003
   658.5′75–dc21
                                            2003051881

British Library Cataloguing-in-Publication Data
A catalogue record for this book is available from the British Library.

The publisher offers special discounts on bulk orders of this book.
For information, please contact:
Manager of Special Sales
Elsevier
200 Wheeler Road
Tel: 781-313-4700
Fax: 781-221-1615

For information on all Butterworth-Heinemann publications available, contact our World Wide Web home page at: http://www.bh.com

10 9 8 7 6 5 4 3 2 1

Printed in the United States of America

CONTENTS

3 REFINEMENT OF THE PRODUCT CONCEPT INTO A NEW PRODUCT AND BUSINESS, 88

4 THE PRODUCT AND BUSINESS PLAN, 144

10 THE LAUNCH, 437

11 THE PURSUIT, 464

12 NEW PRODUCT DEVELOPMENT RECORDS FORMAT, 502

To Terrie, who allowed the time to write about the future,
To Angel, Ashley, and Alecia, who are our future, and
To my Father, who taught me how to embrace the future.

PREFACE

The purpose of this book is to create understanding of the new product development process in order to maximize the corporate investment. Another purpose of this book is to promote and encourage new product development as a vocation, as well as a means for rounding out the executive. This book can help the reader get an appreciation for the planning execution, timing, and problem-solving skills required to prosecute a program. Another purpose is to create an understanding at the executive level about the process, expectations, pitfalls, and recovery moves to maximize results.

UNDERSTANDING THE NEW PRODUCT DEVELOPMENT PROCESS

This book will attempt to provide the reader with a complete understanding of the new product development process. In order to be an effective influence in this field, the manager's understanding must be significantly wide in scope in order to navigate the new product development process.

From the corporate view, the book's purpose is to maximize the Development results of the investment that the corporation makes.

Interdependency of Factors

In any program there are multiple factors that must be understood. There are technological factors that may remain uncertain initially, there are people dynamics that need to be coalesced, and there are corporation infrastructure issues that need to be addressed. For example, the project may have technological hurdles that must be overcome, with a team that is not yet functioning as a team, working within a system that may not be sufficient to produce the product. Add to this, the need to demonstrate early results to maintain management's enthusiasm! All of these elements must be orchestrated into a functioning program with momentum.

The Executive Must Have a Realistic Approach

The executive must have a good background in new product development in order to be effective in growing the business. Generally, long-term sustainable growth will come from product and market development. It is therefore imperative that the executive have a realistic grasp of the process, the dynamics, and the expectations for results.

Executives Need Good People and Need to Know the Business

There is a common misconception that an "executive" need not know the ins and outs of every job under their domain, they just need to hire "good" people. The misconception exists on two levels. First: even good people need to be integrated into the executive's organization and be coached and counseled through the integration. In new product development the integration process is not one or two months, it is rather one or two product cycles. Consequently to save time and prevent false starts, the executive should have intimate knowledge of the process to ensure their people are kept on track. Secondly: in order to, simply, select "good" people that will integrate into the organization, one must know the process, in order to conduct a performance based interview to a depth sufficient for the organizations needs.

How New Product Development Can Allow You to Chart Your Course in the Organization

One of the most appealing aspects of the new product development arena is that it allows you to effectively chart your course through the corporation. You become the expert of the product and business development and it is in demand. The fundamental aspects and elements of the activity require your cutting across the organization and develop relationships and experience in all areas of the organization.

Like it or not the area of product development can contribute positively or negatively to one's career. Therefore, this book can serve as a guide for keeping your programs on track. It can serve as a means for determining completeness in the investigation, planning and execution phases of a program and serve as checklists for comparison to your individual situation.

Planning, Timing, and Problem-Solving Skills

Unlike the orchestra, however, where one is dealing with little uncertainty in the equation (music), new product development, has a great deal of uncertainty, which must be overcome while still keeping overall timing and a quality result.

Since navigating this course requires a diverse set of skills, many times acquired through numerous product development cycles, this book presents a comprehensive look at those skills, thus enabling the manager not necessarily well versed in all of them to have a resource to refer to.

Bringing the Ship In

As with most elements in business, few people are interested in how many storms a product development team had to transcend; most are concerned only if the "ship" was brought into port. So, too, with product development: executives do not care about the obstacles, just the results. That is why they are giving you the funds to accomplish the task. There is no credit for getting "lost at sea."

Study and Apply Principles

The reader is to study and apply the principles presented, gain an appreciation for the effort required in prosecuting a program, and then establish realistic expectations for new programs.

New Product Development As A Vocation

Another objective for this book is to encourage new product development from a vocational perspective. It can be a very rewarding career for a creative individual by allowing their creativity to assist the growth of a business.

With most vocation, one must always be aware of the danger of falling behind in skill sets, technology, procedures, and the business dynamics. With new product development, one is always at the forefront of each of these. It is required as part of the activity.

Executives of Tomorrow

The executives of tomorrow must know these elements in order to develop sensitivity to the pressure and pinch points of a program. Failure to have an intimate grasp of how the activity interfaces with the management, the dynamics of the organization, and how the product development result should contribute to the business can result in a career oversight.

Geared to the Executive

This book is a mandatory review for the executive newly assigned from a field commission or new to the organization. It will attempt to augment a lack of product development experience by presenting a format and worksheets for the various stages of analysis.

This book encourages the executive to establish realistic expectations in the development arena, and discourage over commitment on the part of the executive to the organization management or the team members to the executive.

The key to any program and most corporate initiatives is to carefully establish the course and stick to it. By establishing the course, and maintaining its direction, one can gain momentum and start to leverage the effort.

THE EXECUTIVE RESPONSIBILITY

The executive's role is one of championing the process through completion with a minimum amount of distraction. When the team's enthusiasm wanes, the executive/manager needs to spark interest and excitement into the team. When the organization becomes disenchanted or looks for greener pastures, the executive/manager must rekindle the interest, excitement and all of the organization's commitment to see the program through.

This book is also helpful when reviewing the executive's relationship with the product people. By creating the understanding and examining the driving forces amongst the team members, the executive can better understand how one must interface with creative people, and establish long term trust. It is imperative to prevent adversarial relationships within the process, while still maintaining control and a sense of urgency.

Depending on organizational management and the technical ability of the executives involved there may be misunderstanding created as to who the customer is. Since the executive holds the funds, sometimes, there is a hazard to yield to opinions or whims that are internally generated. This can cause functional creep and added expense at the best, and at the worst, cause the organization to miss the market in terms of both functionality/price matrix and time. Hopefully, this book will serve as a means for introspection for both the executive and the organization.

As with most things in life, the opportunities exist, how you engage them and respond to them determines general success levels. The executive and management and the product team must respond to the market opportunity together to bring a salable package of values.

HOW THIS BOOK IS STRUCTURED

The information in this book is structured in the general chronological order in which information is required in prosecuting a program. There are breakouts to more detailed explanations on individual topics.

It is also structured to allow random access to specific subjects by use of a detailed outline. Another valuable resource included as part of the CD version is the interactive graphs and charts. These software tools are presented initially for instructional purposes (within the framework of the text proper) and can subsequently be used by the reader for their particular development program. The *Toolbox* is accessed by hyperlinks embedded in the text of the book. After reader cover a specific example in the text, a hyperlink can direct them to the *Toolbox* Where they can apply it to their program.

Marc A. Annacchino

INTRODUCTION

PERSPECTIVES ON NEW PRODUCT DEVELOPMENT

The Importance of Ongoing Activity

New product development is said to be the lifeblood of a company. This is especially more true today than ever before in our history. We have experienced changes in the playing field which are fueled by the availability of information. This has changed the economies of virtually every area of modern day life. Companies that hope to simply exist must now evolve new products at an increasing rate to enhance their competitive posture or even to survive. There are no sacred niches that can be harvested for long periods of time.

We must also be aware of the fact that companies must grow through new product development. To survive and flourish in the long run, the core technology must be grown in the organization. Technology is growing at an ever-increasing rate and must be integrated into product development; otherwise the market niche of today will be obsolete and displaced by companies with the newer technology. The list of technologically displaced products is growing at an increasing rate. Remember the slide rule, or the mimeograph machine? How about mechanical cash registers and nixie tube readouts? There are also transitions taking place every day: 8 mm and super 8 mm transition to video cameras, reel-to-reel tape recorders to cassette, record albums to CD, station wagons to minivans.

Exercise versus Atrophy

New product development in a corporation is similar to exercise for the human condition. Lack of it causes atrophy. If the muscles of innovation are not exercised, the company will loose strength. Companies that lack this activity or have programs that are poorly directed manifest the weakness by loosing financial strength and eventual loss of business. Like training an athlete, the objective of these programs should be to strengthen and tune their performance to its peak. Each development should be like training for another event, each building off of a solid base.

Limberness in the Organization

Likewise you want limberness in the operation of the organization. This manifests itself in the organization's ability to adapt, to change, to react fast to market threats and opportunities. With the knowledge base growing as fast as it is, and the plethora of competitors flocking to the market niches, the ability to adapt and change is a critical factor for survival. The market opportunities are not absolutes but sliding windows of opportunity that must be captured. A delay in exploiting the opportunity means a lost opportunity. Most are not recoverable.

Leadership within the Organization

New product development is the leading edge in the enterprise. It also represents the leadership role in the organization. It must direct, lead, position, and ultimately place the organization on the forefront of a market, an opportunity, or a technology. It's similar to a trip, in that product development picks the vehicle, selects the destination, selects the fuel, determines the speed, navigates the obstacles, and sets the estimated time of arrival. You are in the drivers seat, a significant responsibility!

Historical Perspectives

As you review the roster of companies, there are companies who have traditionally had healthy programs and companies who have had poor programs. 3 M is an example where new product development and commercialization has risen to an art form. Unfortunately, companies with poor programs are either small or non existent.

New product development and commercialization must result in salable products or technologies. If it doesn't or if the hit rate is low on successes, the company suffers significant financial burden. This is because of the investment of hard dollars requiring cash, and lack of revenue to offset the cash expenditure. There is also the drain on company energy that has no payoff. In other words moneys spent now must produce revenues that contain profit as fast as possible to offset the initial expense and to additionally fund the growth of the organization.

Progress is Measured by New Products

Progress is measured by the effectiveness of new products produced by the organization. The lack of an effective program deprives the company of core technology that can be exploited. The objective of this is to leverage the core to exploit opportunities. The new product development team should become a machine generating the means to benefit from the opportunities afforded in the market place. This means that the machine needs to react, be on point and execute the required initiatives. Winning in the marketplace is not the result of a single discreet program, this activity needs to be a continuum with some momentum.

Informed Failure Versus Uninformed Success

The lessons learned from failed product initiatives are invaluable. There generally is the correct amount of analysis and introspection to determine the root causes of the failure. This analysis generates the road map to the future success. Conversely, many times when a program is highly successful, we do not take the requisite time to really understand the critical causes of success. Consequently we generally better understand the failure mechanisms rather than the success mechanisms. This represents only one half of the equation of understanding. In understanding the root causes contributing to success, the organization can leverage these factors to additional sales and profits. Therein lies the true benefit of the activity, leverage.

Learn from the failure to prevent a future failure, and also learn from your success to leverage additional success! It's a simple concept but often not implemented!

Setting the Pace for the Organization

At times the organization can suffer from stagnation. The new product development activity must reset the pace in the organization in order to achieve it's objective. As we will discuss in later chapters, the product development team may integrate the technology into the organization, however the organization must absorb it's output and provide the horsepower to execute the initiative. Therefore, the team must set the pace for the organization. Failure to do so renders the program lethargic and ineffective. Getting people on board, increasing momentum are all necessary to progress. Leveraging the organization is the real payoff.

The Focal Point for Change and New Ideas

Depending on the type of company, your role as a product development member can be as a change agent for the company. Many times, senior management utilizes this activity to institute change in several areas of the company. The degree of success in part lies with the team and the management's commitment to that team during the changes. Full support will reinforce the changes as positive. Lack thereof will result in the team being ineffective as the change agent, as well as fail to integrate the program into the organization.

When entering a company an examination of the product development program is telling. By looking at the number of projects started, the number of successful completion's, the number of successful introductions, you can tell quite a bit about the organization's ability to assimilate new ideas and capitalize on them. Change and integrating that change with agreement is a management skill that is a requirement for new product development teams.

The new product development effort should be the focal point for change in the organization. Furthermore, all change must be consistent with the objectives of the product/business development initiative. The model for this group should be to act as a funnel for new concepts, methods, and ideas and to correlate them to the business objective, and integrate them into the organization for the specific purpose of supporting the new business development effort. The challenge in this endeavor is to move past the biases and preconceived notions of individuals about the new concepts, methods and ideas and evaluate them based on merit, materiality and contribution to success.

Innovation as a Survival Skill

Several years ago, there was a phrase among manufacturing executives: innovate, emigrate, or die. As harsh as the statement may be, it does point out the importance of innovation to an organization. With the increasing rate of change in the marketplace, there are a finite number of competitive weapons that can be used to enhance market share. Innovation is one of the most lucrative of these. It is something not everyone is proficient at. It requires a skill set not necessarily found in most organizations. It is foreign to those who are risk averse. Consequently, it can be a formidable competitive weapon, and those who cultivate it in their organizations will continue to have market advantage.

In examining product evolution, it is apparent that most designs evolve. Each iteration, builds on the previous, adding features, reducing cost, and improving performance. While your product development is doing this, your competitor is doing the same in their product development lab. The best outcome for each is a draw. If you should fall one step behind in this race, you cannot ever hope to regain your position without innovation. Innovation is the key requirement to leapfrog the competitor move during the same time frame thereby placing you in a better position than your competitor.

There are risks associated with innovation, which are offset by the rewards and payoff of a successful program. There is also the understated risk of not innovating. The rewards of innovation far outweigh the risks. As part of the new product development team training, innovation needs to be a key component.

NEW PRODUCT DEVELOPMENT AND THE ECONOMY

This section deals with examining how new product development contributes to the general economy. It starts out with a review of the various means for economic development including manufacturing and how product development plays a role. A discussion of the various types of product development focuses and how they draw on and pay out to the economy is also included. Additionally: a look at our world and what it would be like without some new product developments. Finally is a review of what happens to a company without new product development from a pure financial perspective.

Economic Development

New product development contributes to the economy by generating revenue and profits to a corporation that otherwise would not have been generated. The revenue then is paid out to vendors (other manufacturers). The vendors themselves pay out to their vendors and personnel, or retain as earnings. Salaries are paid to personnel, and they in turn spend funds to purchase goods and services from other profit-making enterprises. The retained earnings fund the long-term growth of the enterprise and increase the value of the business. The profits are taxed and that goes into the pool of funds to govern the community and provide for the common good.

This is the role that private investment plays in the economic development arena. The public sector contributes to the economic development by providing incentives to encourage manufacturers to establish their businesses in their locale. They provide the means for funding business expansion and growth. A collateral activity is their ability to network with other manufacturers on your behalf for future business. Figure 1 illustrates the typical funds flow.

The service sector of the economy has a distinctive funds flow as compared to a manufacturer. A smaller investment in capital equipment, smaller investment in each revenue cycle and a generally faster revenue cycle characterize it. The service business is not characterized by the leverage associated by manufacturers, and as such, generate incremental profits from incremental investment with low fixed costs. This can be shown in Figure 2.

As shown, the service model has relatively constant net profit. This is because there is no major initial investment to absorb, and because there is a small incremental investment with each order. The other costs are somewhat variable and track the incoming order rate and

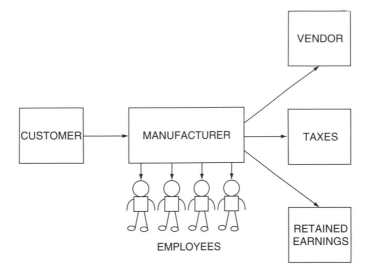

Figure 1. Manufacturer Funds Flow

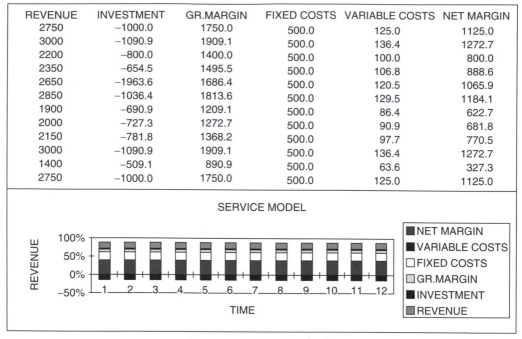

REVENUE	INVESTMENT	GR.MARGIN	FIXED COSTS	VARIABLE COSTS	NET MARGIN
2750	−1000.0	1750.0	500.0	125.0	1125.0
3000	−1090.9	1909.1	500.0	136.4	1272.7
2200	−800.0	1400.0	500.0	100.0	800.0
2350	−654.5	1495.5	500.0	106.8	888.6
2650	−1963.6	1686.4	500.0	120.5	1065.9
2850	−1036.4	1813.6	500.0	129.5	1184.1
1900	−690.9	1209.1	500.0	86.4	622.7
2000	−727.3	1272.7	500.0	90.9	681.8
2150	−781.8	1368.2	500.0	97.7	770.5
3000	−1090.9	1909.1	500.0	136.4	1272.7
1400	−509.1	890.9	500.0	63.6	327.3
2750	−1000.0	1750.0	500.0	125.0	1125.0

Figure 2. Service Funds Flow

incremental profits can track incremental revenue. Service businesses generally have a lower barrier to entry than the manufacturing counterpart.

The manufacturing business, differ markedly from the service business, in that leverage can occur. This leverage is characterized by larger returns for their investment. A larger investment is required at the onset, but revenue is generated through a stream of returns from year to year through the products life cycles. This assumes there is sufficient volume to offset the fixed expenses and the absorption of the initial investment. There is significant investment in capital equipment and processes, along with significant payoff if you're hit on the right opportunity with the right product.

The manufacturing model (Figure 3) shows that net margin increases as at a disproportionate rate as revenue increases. This is because the fixed costs are already absorbed by some base level of business and the incremental revenue does not require incremental investment with each order. Fixed costs decrease as a percent with increasing revenue. However, there is a large initial investment to design the product that can run the length of the diagram in time.

Economic Models for New Product Development's Contribution to Profit

Figure 4 illustrates the interrelationship of how a manufacturing enterprise funds flow as a result of new product development.

REVENUE	COS	GR.MARGIN	FIXED COSTS	VARIABLE COSTS	NET PROFIT
500	1500.0	−1000.0	1500.0	187.5	−2687.5
750	375.0	375.0	1500.0	46.9	−1171.9
1350	875.0	675.0	1500.0	84.4	−909.4
2000	1000.0	1000.0	1500.0	125.0	−625.0
3000	1500.0	1500.0	1500.0	187.5	−187.5
6000	2500.0	2500.0	1500.0	312.5	687.5
7500	3750.0	3750.0	1500.0	468.8	1781.3
10000	5000.0	5000.0	1500.0	625.0	2575.0
12500	6250.0	6250.0	1500.0	781.3	3968.5
16000	7500.0	7500.0	1500.0	937.5	5062.5
18000	9000.0	9000.0	1500.0	1125.0	5375.0
21000	10500.0	10500.0	1500.0	1312.5	7687.5

MANUFACTURING MODEL

Figure 3. Manufacturing Funds Flow

An idea is conceived and qualified in terms of opportunity and overall business sense. There are costs associated with these activities, such as market planning costs, surveys, customer visits, and demographics data analysis. There are also costs in taking the opportunity and the market data fed back and coalescing them into a product opportunity. The scope of the target market is accounted for in this stage, and how the product platform must be laid out to reach the market at a cost-effective price. These costs are generally low compared to the other costs in the product development arena.

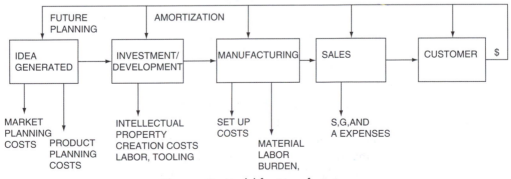

Figure 4. Model for Manufacturing

The next phase is the investment phase and the development phase of the program. This phase takes the product concept and creates the intellectual property required to take a concept and reduce it to bills of material, manufacturing processes and define a manufacturable product that the market will purchase. There are technical, labor, development, tooling, and other capital equipment costs in this phase.

The next phase in the model assumes that all the development is complete and manufacturing takes place. Here, there are set-up costs, material, labor, and overhead costs required to produce the units.

The product moves out of manufacturing and into the sales channel for placement at the customer. In this stage there are a host of S, G, and A (sales, general, and administrative) costs incurred.

Finally, the product is at the customer and funds for the product have changed hands. The manufacturer now can begin to enjoy the profits.

These profits come after the funds are distributed appropriately. Dollars must go back to fund the S, G, and any expenses previously incurred. The manufacturing must be reimbursed for the materials labor, manufacturing expenses and inventory carrying costs.

The development of the product must be amortized so as to be able to fund future developments. In addition: funds must also be channeled back to the product planning, market planning function to allow the proper expenditures to verify product viability, market program viability, and future enhancements and products.

Types of New Product Development and Their Contribution

There are a variety of different types of new product developments in existence today. Each is used for a different reason, and each has its own objectives and dynamics for execution. The following is a common list of the different types and their attributes and contributions.

1. New-to-the-world type products

These are somewhat revolutionary to the marketplace in that the marketplace never had exposure to the product directly, perhaps only as a concept or prediction from a futurist. They generally create entire new markets never before existing. An example would be the cellular telephone. Only predicted vaguely by the Dick Tracy cartoon strip and personified as a "communicator" in Star Trek, the cellular telephone has revolutionized person to person communications in modern day society.

These product development programs generate entire new markets that were not previously there. They enable true growth in the economy by generating revenue to the enterprise. The also have a multiplication effect in the economy by generating requirements for parts and subassemblies that need to be developed and supplied by the vendors. In many cases they generate new channels of sale and new routes to market.

2. New product lines

These new categories of products allow entry into newer markets not previously participated in by the manufacturer. By adding the categories, the manufacturer must be careful to protect the positioning of the existing products, which generate the existing business. Failure to do so will place them in danger of converting loyal customers away from one already successful product to a new one with no net gain in market share. Perhaps a good example of this type of product would be the GM Saturn. Here a large manufacturer with many product lines generated an entire new category of car to serve a more discerning customer base. Careful not to jeopardize their existing base, this "new car company" launched its product in a new dealership with new business practices and also allowed GM to "test drive" innovative marketing programs.

The new product lines generate incremental revenue to the manufacturer by leveraging the markets familiarity with the manufacturer into new categories of products. In many cases, the market's familiarity with the manufacturer paves the way for new categories of products. Sometimes, these products go into new markets, but can also be an alternative to existing ones.

3. Additions to existing product lines

These efforts support existing product lines by creating line completers to extend the influence of the original products' brand to larger audiences or extending range, power, and scope. All are done in the attempt to secure more of a market. An example of this type of product would be the M&M candy extending their product line to M&M peanut and M&M almond and seasonal M&M for Christmas and Easter.

The addition to existing lines has a similar effect to the company's revenue as the new product lines. They generate incremental revenue by leveraging the existing product familiarity rather than the company familiarity. These programs generate incremental improvement in the economy, but generally fall short of the contribution made by the totally new products.

4. Improvements and revisions of existing products

As time marches on, customers have higher expectations do not product, and competition adds features to their offering, it becomes necessary to improve the company's offering to increase market share or to retain it. By redesigning the product or repackaging it, the company can offer a greater value or satisfaction to the customer. It is possible to temporarily affect this by enhancing perceived value; however, an ever more informed customer base will respond to actual value increased in the long run. An example of this type of product development is the automotive companies adding features to the base model each year as standard.

Generally, the improvements to existing products do not generate additional revenue to speak of. They are simply a means to retain the market share or to slightly improve it. They are defensive in nature and in many cases are stopgap measures until a new product program can be prosecuted. These programs do little to generate a vitalized economy in the long run but can provide time and revenue to prosecute the development of the replacement.

5. Repositioning

Another means of increasing or maintaining market share is through repositioning. A repositioning is an exercise in changing the perception in the mind of the consumer. It generally can happen with products that are lower in value (dollar amount) or the consumer spends little time evaluating the actual data. For high dollar decisions, the consumer will generally take the time to evaluate the facts and make their own decision. Repositioning is truly a marketing activity, rather than a development activity. An example of this is a changes of in advertising by focusing the audience on a possible linkage drawn between certain brands of cereal and a high fiber, lower cancer risk, diet.

Repositioning is another stopgap measure for generating revenue from an existing product. It does not generate overall growth in the economy per se, it is similar to the improvement or the revision, except that it doesn't even require a product change necessarily, simply repositioning the product in the mind of the consumer.

6. Cost reductions

These programs are strictly a means for reducing the cost of products to offer similar value. It generally is the result of a competitive initiative either internally generated or from external forces. In many cases, it is simple a means to generate more volume to generate less incremental profit but perhaps more overall profit. Whatever the motive, a cost reduction is generally meant to increase unit volume through the channel. This becomes easier with capital equipment costs and development costs absorbed, and the manufacturer wants to capitalize on the sales channel.

Cost reductions are helpful to the organization by generating additional margin off of the existing product. This margin can absorb development costs, and manufacturing set up costs. In many cases it provides a period of time to continue the product, generated the revenue and allow the organization to position itself with the new product. It does not however generate any real growth in the economy.

Figure 5 summarizes the different types of product developments in terms of time required to prosecute, the revenue to the economy, the revenue to the company, the company's positioning, and the margin impact.

Narrative and Financial Review of a Nondeveloper

Figure 6 illustrates the dynamics of the income statement of a company that has little new product development. As shown in the financial analysis, a company can grind to a slow halt

TYPE OF DEVELOPMENT	TYPE TO PROSECUTE	POTENTIAL REVENUE CONTRIBUTION TO ECONOMY	REVENUE CONTRIBUTION TO COMPANY	COMPANY POSITIONING STRATEGY	POTENTIAL MARGIN IMPACT
NEW TO THE WORLD	LONGEST	HIGHEST POTENTIAL	HIGHEST POTENTIAL	MKT DEVELOP	HIGHEST
NEW PRODUCT LINES	LONG	HIGH POTENTIAL	HIGH POTENTIAL	MKT DEVELOP	HIGH
ADD TO EXISTING	MEDIUM	MEDIUM POTENTIAL	MEDIUM POTENTIAL	LINE COMPLETE	MEDIUM
IMPROVE OR REVISE	SHORT	LITTLE POTENTIAL	MEDIUM POTENTIAL	MARKET SHARE	MEDIUM
REPOSITIONING	SHORTEST	LITTLE POTENTIAL	MEDIUM POTENTIAL	MARKET SHARE	MEDIUM
COST REDUCTIONS	SHORTER	LITTLE POTENTIAL	MEDIUM POTENTIAL	RAISE MARGIN	MEDIUM

Figure 5. Sum of Development Types

	BASE YEAR	YEAR 1	YEAR 2	YEAR 3	YEAR 4	YEAR 5
NET REVENUE	100000	101000	102000	103000	104000	105000
MATERIAL	45000	45950	46400	45850	47300	47750
LABOR	10000	10600	10700	10800	10900	11000
BURDEN	25000	25250	25500	25750	26000	28250
TOTAL COS	80000	81300	82600	83400	84200	85000
CROSS PROFIT	20000	19200	19400	19600	19800	20000
ENGINEERING EXPENSES	4000	4200.0	4410.0	4630.5	4862.0	5105.1
ADMINISTRATIVE EXPENSES	3500	3675.0	3858.8	4051.7	4254.3	4467.0
SALES EXPENSES	3000	3150.0	3307.5	3472.9	3646.5	3828.8
TOTAL EXPENSES	10500.0	11025.0	11576.3	12155.1	12762.8	13401.0
NET PROFIT	9500.0	8175.0	7823.8	7444.9	7037.2	6599.0
PROVISION FOR TAX	4275.0	3678.8	3520.7	3350.2	3166.7	2969.6
NET PROFIT AFTER TAX	5225.0	4496.3	4303.1	4094.7	3870.5	3629.5
PERCENT RETURN ON SALES	5.23	4.45	4.22	3.98	3.72	3.46

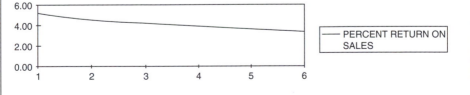

Figure 6. Income Statement Nondeveloper

by not participating in the dynamics of the ever-changing marketplace. Along the top are the years under study. Along the left side are the income statement categories.

1. Cost of sales

The cost of sales has a natural tendency to increase over time. This is a result of the increases of the direct labor costs in manufacturing the unit as well as the effect of vendor increases in pricing. For most manufacturers, the product maintenance function of development has to initiate cost reduction where they can just to stay even. Therefore, to progress down the learning curve, significant initiatives must be made to effect cost reductions as these changes must offset the increases already built in.

2. Profits

Assuming the expenses stay at a constant amount or grow slightly, this increase in materials and labor and manufacturing expenses (burden) create enormous offsets in the net profit of the company. Salaries and benefit expenses increase with time, as the manufacturer must remain competitive with the other more dynamic companies in the industry. Sales expenses will increase as more concessions or incentives or travel must be done to achieve the same result from year to year.

3. Strength

The financial strength of a company degrades with a lack of successful new product development. Unfortunately, the effect is not known immediately. The financial picture initially looks better since the expenditures normally spent on a development are saved. This is however a short term gain, as the market will soon pass the manufacturer by with new products from other firms, leaving the manufacturer with little to sell into the evolved marketplace. In the later analysis, the strength of the enterprise diminishes.

4. Market share

If the manufacturer cannot keep up with the marketplace's appetite, eventually the package of values they offer does not meet the customer's needs, and they begin to loose market share. By loosing the revenue from lost market share, a dangerous downward spiral begins that is difficult to reverse.

5. Response to changes in environment

If the manufacturer cannot keep up the pace established by their industry by evolving product lines on a normal evolutionary path, they are at great risk from the negative effects of changes in the playing field, or legislation changes, or salient attacks from non-traditional competitors. This weakness will manifest itself as a financial problem in time, which will eventually place limitations on the organization's ability to respond in order to survive.

A LOOK AT OUR WORLD WITHOUT SOME NEW PRODUCT DEVELOPMENTS

A World without Apollo

Not all new product development comes from private investment. The US government regularly conducts research in a variety of areas in order to generate understanding, research, and analysis of data. One of the more scrutinized and high profile development programs was the project to land a man on the moon and safely return him to earth in the 1960s.

Aside from the sheer magnitude of the program, the raw objective was to develop methods, produce machinery and hardware, and execute a mission to land and return. The task represented enormous technological hurdles in numerous disciplines. The achievement of this objective in July of 1969 must be historically underscored by the numerous spin-off technologies that have changed our world.

The Apollo program accelerated the development of technology in unprecedented manner. The need drove the course of action. If it existed, use it. If it didn't exist, invent it. Time was literally of the essence. Money was no object, and resources were granted on the basis of how much time it would save, not how much it cost.

There were several technologies that experienced accelerated development during that era. The term accelerated is the operative one in this case. Most technology eventually gets developed as people become interested in the potential and entrepreneurship drives these elements into society, however the Apollo program rocketed development of certain technologies that would have taken many more years to implement without the driving requirement to necessitate it's invention.

Rather than listing and discussing each one, it may be more effective and interesting to look at our technological and societal world without the Apollo program. Let's examine the latter half of the 1990s as we look at A World without Apollo.

You wake up at 6:00 am in the morning. The mechanical radio alarm clock clicks on with the local news. The clock is mechanical because a digital alarm clock is too expensive to build out of discreet transistors. The integrated circuit showed a lot of promise in the 1960s as an idea but was never funded, and so its widespread use is just beginning. However, that is not on your mind.

What is on your mind is the temperature outside and the chill you feel inside. It's a 20° F cold snap outside. Back in the 1970s, oil prices skyrocketed, and so all Americans living in North America need to hold down expensive oil and natural gas usage. Your home is insulated with 1960s technology and cannot hold in the heat very well. The less than optimal insulation used is due to lack of funding to develop the new technology in the 1960s. The Apollo program hardware utilized specialized insulation and polymerized film materials and solved its heat retention and heat rejection problems by the use of advanced insulation systems.

You are planning to go to the football game tonight to see your favorite team play so you hope it warms up, later in the day, upon your return.

Today, your children are going to perform in the school play, and your wife is planning to go food shopping and investigate a new hairstyle at the beauty shop. You are a jet aircraft pilot and have to fly to two cities and return home early this evening.

You are pressed for time as you forgot to buy movie film to film the kid's school play. You can only film 3 minutes of the play and have no means for sound recording of the play to coordinate with the movie film. The video camera has not been invented yet. There was no need to develop the technology for portable television cameras used on the Apollo missions, so there was no spin-off to personal use.

When your wife gets to the food store and is ready to check out, there is a long line at the checkout. Since all groceries are checked out by hand, it takes longer to process the customers. Bar code scanners haven't been invented yet nor is the raw technology in a cost-effective package for use in the commercial sector because integrated circuitry has not progressed far enough down the learning curve.

Now, it's off to the beauty shop to experiment with that new hairstyle. Your wife will be taking a significant risk in experimenting as she will have her hair cut very short for the new hairstyle. In addition, the new style requires a permanent wave treatment. Both items are a one-way commitment to the new style for 2-3 months. The imaging processing technology developed for the Apollo program was never invented, since there was no need. Consequently, no spin-off technology allowed the beauty shops a means for taking your wife's picture and digitizing it to "try out different hairstyles" before cutting or waving.

Your wife now starts to plan the budget for the next month. Unfortunately, she is calculating and adding bills manually and makes a mistake, allowing more expenses than can be covered by the checking account's current balance. The hand-held calculator has not been invented yet. No market need for integrated circuits or cost reductions, so no calculators yet. Eventually you will have one as the world progresses into the acceptance and the implementation of these technologies at the normal nonaccelerated pace.

Meanwhile, you are at the airport and ready to start your day. You climb into the flight deck, and see the traditional stick and rudder used to fly the airplane. There is very little avionics and no inertial navigation system. Everything in terms of flying the plane is done by manual calculation. There are major systems such as navigation or communications systems not in place. The Carousel system for inertial navigation and fly by wire technology has not been invented or implemented in the fleet of planes yet. There was no requirement for inertial navigation for long distances like in the Apollo program, so it never made its way into the commercial marketplace. This system combining computer technology, electromechanical physics and precision machining, inertial navigation algorithms and worldwide infrastructure to implement will be available through natural incremental evolution around the time you plan to retire.

You land in the first city of your journey, and listen to the television in the airport terminal. There is a news story about an entire family who perished in a home fire. How sad and tragic, as you think of your own family. This family, like many others, died because the fire consumed them while they were in their sleep. There was no means for early detection or warning. The smoke detector was never invented for home use. The Skylab program in the early 1970s used mostly Apollo hardware, and NASA needed a means for early detection of smoke and fire for America's first space station.

Your attention shifts back to the weather, and you are thinking about that game tonight and want to know if it will be too cold to go or not. You will not have any way of planning for the weather because modern forecasting methods are limited to predictions only. You see, satellite photos using image enhancement was never invented. There was no requirement for it on Apollo, so no spin off.

Your children go to the school library to do research for a term paper. They comb through the card catalog and cannot find any article or reference material for their paper. The library's focus has not shifted from collection of books to access of information. Your kids will have to wait for the reference material to be sent through the mail because there is no Internet with universal access yet, no lowcost computers affordable enough for the municipal budget, and no network of other users.

You finish your flights, arrive back home, and plan to go to the game. Your wife asks you to anchor a new picture she bought today. You go down in the basement to get your electric drill to drill the hole required for the anchor. Unfortunately, you loaned the extension cord to a neighbor. You have no means for electrical power in the area you will place the picture. Cordless tools have not been invented yet. They are a spin-off from the drill used to core lunar samples on the Apollo missions. You will have to wait till tomorrow for your neighbor to return the extension cord.

You are off to the game at last. It never did warm up nor did you ever get any advanced warning. It would be nice on a night unseasonably cold as this to have a cover over the stadium to retain the heat. Traditional roofs over stadiums are expensive and require huge means for support. Lighter more affordable fabric used in domed roofs for stadiums and airports hasn't been developed yet. (They use spin-off fabric technology from the Apollo Moon suit developed for the astronauts to traverse the lunar surface.) It's so cold in the stadium that you get some hot coffee rather than a cold beer!

Continuing the Product Development Cycle

This short vignette represents only a few technologies developed for the moon program that are in the everyday products we now take for granted.

It has been often said that the program was wasteful in terms of spending the funds on space travel rather than "curing" the problems on earth. By conservative estimates, there are over 30,000 products developed and launched that were the result of the space program. The reality of these results are that the tax on profits alone from the revenues generated from these products and technologies from a have paid for the government's investment in the entire Apollo program many times over each year.

Here are a few examples where continued product development has benefited the customers with better, safer, products.

1. Automobiles

   The Model T progressed to the Lincoln Town Car.

2. Aircraft

   The Ford Tri Motor progressed to the Boeing 777.

3. Computers

   The Eniac progressed to the CRAY supercomputers.

4. Vacuum tubes

   The vacuum tube amplifier progressed through discrete transistors to integrated circuit amplifiers on a chip.

5. Incandescent lights

   The original incandescent light spawned fluorescent tubes and bulbs with numerous multiples of extended life.

The importance of continuing the product life cycle preserves the customer base. It drives competition in the marketplace between manufacturers. Competition is desirable and necessary. It prompts each other to action to improve the products that they offer, so the customer and the manufacturer both benefit.

CHAPTER 1

THE BUSINESS OBJECTIVE

ABSTRACT: This chapter deals with one of the most important steps in the process of new product development. In its essence it is looking at the company, the marketplace, and the competition to determine how the new product opportunity will fit into the overall strategy. There are a variety of elements that contribute to a product's success and the company's success rate in prosecuting new products. These factors go beyond the product specifications and performance. "A better mousetrap does not necessarily make good business." There are other things to consider in addition to the specification items. There are the linkages with the other elements in the organization, the leverage that must occur internally, the predisposition of the organization to new things, the degree of activity in the marketplace, and the aggressiveness of the competitive players. The result of this chapter should be to determine the fit of the new product, the chances of success, and the organization's ability and desire to carry it off.

THE FIT

1. SELECTING THE NEW PRODUCT OPPORTUNITY

Opportunities abound in this world of ours! The challenge is to select and focus on the ones that will yield the desired result. However, what is the desired result for the company? Is it result revenue generation, cash flow, or profit, or does it represent tangible, long-term growth for the company, strategic positioning, and operational requirements? Is it achievable, desirable, and agreed to? All of these questions must be addressed to determine if the opportunity in question is the right one for your company at this time.

Not all opportunities are truly opportunities for a company. Although some of them represent good prospects from a pure market perspective, they may not fit with the company that is evaluating the opportunity. It is very important to factor in to the decision, all of the company attributes that will contribute to or detract from success of the new product. This exercise can be referred to as the "fitability" of the new product into the company.

1

2. UNDERSTANDING THE COMPANIES "FITABILITY" TO THE PRODUCT, CHANNEL, ROUTE TO MARKET, MANUFACTURING, AND TECHNOLOGY

A new product opportunity must easily fit into the operational culture and temperament of the firm. A company evolves with several interlocked disciplines that must work together to effectively assimilate, integrate, and execute a new product program. This must happen from a sales and marketing perspective, an engineering perspective, and manufacturing, service, and quality control perspectives.

The new product opportunity must fit with most of the disciplines to be prosecuted properly. If there is an area that is not initially compatible or leverageable with the opportunity, then steps can be taken to correct or bolster the weakness. If, however, there are too many areas that are inconsistent with the new product, then the chances of developing the product within the market window of opportunity are reduced, especially while effecting changes in the organization and supporting the existing business.

The new product idea must be tested not only in the marketplace, but also within the firm to determine its fitability with the sales organization, the marketing organization, the current and anticipated capabilities in development engineering, and the integratability of the product into the manufacturing organization.

3. HOW DOES THIS OPPORTUNITY FIT INTO THE GOALS AND AGENDAS OF THE COMPANY?

Every company has goals and objectives that generate its agenda. Even if the company doesn't have any stated goals or objectives, there is still an agenda that factors in to the new product analysis. It may be as simple as: The objective is to maintain the status quo. In that case, the agenda is to do nothing in the way of product development. All new opportunities are evaluated, and summarily rejected for one reason or another. Others are evaluated against agendas, such as target growth rates of 15% to 20% per year. Still others are evaluated against requirements demanded by customers, other parts of the organization, or that are purely financially motivated.

There are also the ownership issues, which affect the company's ability to effect a healthy product development program. A publicly held company with significant resources will evaluate an opportunity differently than a privately or closely held company. A subsidiary operating from a remote location will act differently than a division that is part of a larger group within the same venue. A set of poorly prepared objectives for a general manager of a relatively autonomous organization will dictate new product development activity differently than ones given by an onsite senior manager.

The current financial status of the organization will also evaluate the risk and size of the development program. In the case of the subsidiary or the division, the status of the parent company must be examined also.

All of these issues must be factored when evaluating the opportunity. The operative point, simply stated, is that: A new product opportunity must be evaluated not only from the external factors affecting the success, but also from the equally as important internal factors that must contribute to its success.

4. WHO MAKES THE SACRIFICE?

As the saying goes: "No pain, no gain"; so to with new product development. Who in the organization will make the sacrifices? Is it the sales team that will make the extra effort to introduce the product? Is it the marketing team that will take the extra effort to prepare an adequately leading specification that will stand the test of development time? Will it be the development team that will shed the internal conservatism agenda and break new ground in terms of costs, functionality, and manufacturability? Or will it be the service team that will drive the product to success even through initial quality and reliability problems? Each and every part of the organization needs to contribute the extra effort to ensure success of the new product, in spite of the pitfalls it encounters.

If the fit is right the team has a chance at success. The fit alone will not guarantee success; success also requires the commitment of the team members and the company as a whole to prosecute the program.

THE STRATEGY

1. WHAT IS THE COMPANY'S STRATEGY IN PURSUING THIS MARKET— ONE-TIME HIT OR STEADY PURSUIT?

To effectively evaluate the fit of a new product, it is necessary to understand the company's overall product strategy. What has been the track record for embracing new technologies and product concepts? What were the expectations of each, and what were the results? Often many companies do not have a clear objective and as such cannot generate a clear strategy.

The previous section discusses leveraging the organization segments as part of the fit of the product. The product strategy should attempt to leverage the available product opportunities to create a steady stream of products to create a new, sustainable business unit for the company. This forces examination of the interrelationship of the product as well as individual contribution.

Regardless of how lucrative an isolated product opportunity looks, few products can succeed as a one-time hit without subsequent market and product development. It simply puts the organization in a market position that is not defensible. In addition, a single success in an uncontested market will not remain so for very long thus necessitating the need for continued development.

Again, the strategy is dependent on several factors, including the structure of the company. A product or service, which is in its infancy and has a lot of "runway," may be a good choice for a company that will eventually be taken public. It has the longevity and growth potential required for the investor attraction in the initial public offering. A product or developing technology that may be required to complete someone else's product line may be chosen to groom the company as an acquisition target. A privately held company will tend toward market niches that are narrow enough to be defensible and wide enough to be worthwhile.

Consequently, a well-thought-out strategy executable toward an objective is very valuable to the organization.

2. WHAT MAKES THE DEVELOPMENT TEAM THINK THIS OPPORTUNITY WILL TAKE OFF?

New product development is a business venture. It must be evaluated like a banker or venture capitalist evaluates any business venture. Some developments may have very interesting technology and be very fun to work on. However, each must result in tangible value to the organization funding the development activity.

As will be discussed later, the development team consists of several interdisciplinary elements of the organization. Therefore the development team must make the case for the development program. Why will it be successful? What is unique about the product or technology that will be desired and accepted by the marketplace at this particular juncture in time and space? These are the hard questions that must be answered satisfactorily very early in the process.

The strategy, the timing, the functionality, and the company's plans must be imbued into the development team and must be eminently clear to all involved. The team must initiate, champion, and carry the torch of the new product business plan.

3. WHERE IS THE MARKETING HOOK?

At the risk of sounding like P.T. Barnum, I would stress that every product should have a distinct marketing advantage that it brings to the marketplace. Very few products are accepted that are "me to" in nature, so there must be a clear marketing advantage at introduction. This is what starts the momentum that will carry future sales. It is part of the tactics that are generated from the strategy.

The marketing advantage is generally, but not limited to, the integration of features within the product that differentiates it in the marketplace. Longer, lower, wider, less expensive, more functional, easier to use, and more available, are all product design tactics designed to make the product more marketable.

If the "hook" exists, it is because the team placed it there, and it is there to generate sales. If the product is a "new to the world," product its novelty alone goes a long way. However, it is then a sales and training job to gain market acceptance.

4. HOW DO YOU KNOW IT WILL WORK VERSUS A CHANCE AT WORKING IN THE MARKETPLACE? THINK THE SCENARIO THROUGH WITH CONTINGENCY PLANNING

Each new product must have a vision of how it will play out in the marketplace. The scenarios for success and for failure must be thought through, and a recovery procedure for each worked out. This contingency planning is similar to prosecuting battles in a war. When there is competition in attainment of the same businesses, the marketing tactics and strategies are similar to warfare. For example, if a new product is introduced and has existing competition, you are the invader, and the competitor will attempt to defend its territory. If it has uncontested reign of the market, and it is a worthwhile market to go after, the competition will enter the race with an offering of its own. Now the competition is the intruder. Each salient has its own set of measures and countermeasures as well as scenarios that must be developed by each participant to survive.

5. THE BENEFITS OF STRATEGIC PLANNING

The term *strategic planning* has been a very overused term and, in many cases, an underused long-term activity. It cannot be glossed over or generalized in order to be effective. It needs to be specific enough so that its results define the product development and market development direction.

Strategic planning has been traditionally used to prescribe desired market position, dollar amounts of business segments, growth targets, means for justification of acquisitions, and more. It seems to get employed to reset a company's focus and direction initially; however, it soon seems to lose focus in years 3, 4, and 5.

Unfortunately strategic planning has the most value as a learning tool for the organization by setting direction at the beginning of a period, executing a strategy or program, and evaluating its success within the strategic framework initially laid out. Most important, the success or failure of the first year must be factored in when setting second year and subsequent goals.

Consider a strategic plan that calls for accelerated growth by internal development of a steady stream of new products. Also consider that all-important review after that first year, showing the initial target and missing that target by some amount, for example. It is an imperative requirement for the accuracy and legitimacy of the plan that this be factored into the subsequent year's planning. If your company missed the first year, how will the organization make up the difference lost and still meet the subsequent targets?

If the ground rules of strategic planning are well understood and the company carries the discipline to effectively execute the planning process, then the process has several benefits to the organization.

As a review, here are the benefits of the strategic planning process:

A. It improves profitability–by focusing the organization on two levels of profit: gross margin and net profit. It improves profitability regardless of the

measurement criteria: return on investment, return on net assets employed, or returns on total capital employed.

B. It yields a higher growth rate for the enterprise with it than without it. Without a plan for growth, it becomes difficult to determine the best opportunities to pursue. This compromises the overall growth rate. The organization may happen to stumble onto a great opportunity; however, the consistent increase in business and market share is not likely without integrated planning

C. It reduces wasted time, effort, materials, and resources of the organization because of lack of focus. Following on the previous tenet, chasing loose opportunities can prove to consume time and effort.

D. It provides focus on the marketplace and examines the business from the customer's perspective. Every business is about customers. It is they who make the business possible. It is they who comprise the marketplace. Strategic planning, although an internal goal-setting exercise for the organization, forces the management to focus on the customers, their problems, and the opportunities to resolve their problems.

E. It articulates and details future plans to the workforce to allow early buy in to a program. It also serves as an early check for opposition to the future directions of the company. This point is not to be underestimated. Strategic planning is an excellent means to trial balloon issues and initiatives to "test the waters" of the organization. People make up the organization, and people can facilitate or destroy initiatives.

F. It fosters higher personnel commitment to the agreed goals. As an essential part of the management commitment process, a personal commitment must be made and followed through. The strategic planning process allows the participants to contribute and commit to the goals established in the process.

G. It provides for a cohesive management team. One of the most divisive elements in an organization is the fractionalization of the management team. Strategic planning allows for galvanization of this team. If it is not evident that the team can be galvanized in this fashion, it allows for change of the players, if necessary.

H. Strategic planning provides the framework to understand the competition in terms of competitive advantage, workforce, value added, position in the marketplace, and ability to react. This is one of the most lucrative elements of the entire process because it requires the organization to evaluate the competition in a dispassionate objective means and to determine how the organization intends to compete and prevail in the markets it chooses.

I. It provides high visibility on quality from a holistic perspective: quality of the product, process, and entire revenue cycle. The strategic planning process forces quality into the organization. By its very nature of long-term goal and objective setting, the process dictates that the moves must be well considered and execution cannot be sloppy. The philosophy permeates all of the programs generated from the process and drives a higher level of quality onto the organization.

J. **It yields greater customer satisfaction.** Higher quality, an organization that knows where it is going and how to get there, and a realistic customer focus allow a greater level of customer satisfaction.

K. **It provides a framework for continuous improvement.** Since the process is long term and annually reviewed, the organization can continuously improve through introspection, modification, and feedback.

L. **It directs a day-to-day vision that is real and achievable.** Although strategic planning is long term, it affects the organization on a day-to-day basis by reinforcing goals and schedules. It keeps the organization focused and on track, rather than reactionary to every diverse opportunity that comes along.

6. THE FRAMEWORK FOR PLANNING

The planning process has a general framework that must be followed to generate the required results. A well-executed process has three distinct parts. The first is introspective in nature. It encourages examination of the company self to determine strengths, weaknesses, and historical perspectives.

The second part is creative in that the company must determine its future and its desired position in the marketplace.

The third part is pure planning. It defines the means to the end described in the second part. The following is a framework for the strategic planning process that can be used as a general format for initial discussion:

A. **Where we are:**

1. Initiate a search of records to create a 5-year history of the company's performance. This would include financial information and narrative about the company. (See Figure 1-1 for format)

    Analyze and draw initial conclusions from the obtained data. Review trends in sales engineering, manufacturing expenses, in addition to the product-related performance.

2. Use a narrative to describe the general health of the business. Interview a cross section of the employees to find where "the bones are buried" as far as product line performance, results of management attitudes, predisposition to new initiatives, and acceptance criteria. Do not focus on individuals; rather, focus on the results of the organization.

	Year 1	Year 2	Year 3	Year 4	Year 5
Sales					
Gross margin					
Net profit					
Expenses					
Warranty					
Inventory					
Return on investment					
Return on assets employed					
Return on capital					
Accounts receivable					
Cash reserves					

Figure 1-1. Analysis of Business

3. Next, create as detailed as possible a product line analysis to describe the performance of individual products. Examine the costs, features, benefits, trends, and driving forces for each product. It is also important to include the sales, general, and administrative expenses for each because costs may not be allocated on an activity-based cost basis.

Product Maintenance is what absorbs time and slows down new product development, so it is important to examine from a historical perspective how the company handled the products and supported them. In some cases a product appears profitable from a margin analysis, but in actuality may break even and absorb a disproportionate amount of the organization's resources.

Figure 1-2 is a framework for this analysis. Be sure to allow for narrative for some of the intangible elements of the analysis.

Here is a review of the information required for this analysis. The product should be replaced with the description of the product. Consider individual products that are part of families as separate. Next, enter the price that your company is getting in the marketplace for the product. Cite important features and benefits that the customer has given as feedback. Do not use sales and promotional literature for this part; use actual feedback from customers that have used the product.

In the next row, describe the customer base for the individual product segments. Next, discuss the technology employed in your company and describe any differences from the competition. Also include manufacturing costs fully burdened. Look at the current and near-term market driving forces that will affect the

PRODUCT LINE ANALYSIS

PRODUCT	PRODUCT #1	PRODUCT #2	PRODUCT #3
PRICE LEVELS			
FEATURE / BENEFIT			
CUSTOMER BASE			
TECHNOLOGY BASE			
COST			
MARKET DRIVING FORCES			
PERCENT OF SALES			
CONTRIBUTION TO PROFIT			
LIFE CYCLE STAGE			

Figure 1-2. Product Line Analysis

acceptance of your technology, product feature/benefit set, competition, and demographic preferences.

Use accounting data to generate the row of sales data contribution to profit data for each one. Finally, enter the best assessment at the life cycle stage for each product. This may be a numeric or a narrative. It will be used to gauge product development efforts throughout the overall framework.

B. Where are we going?

This section of the strategic planning process identifies where the organization wants to go strategically in the future. This section is one of the most important of the three sections because it is the goal that must be set for the organization. It defines the call to action, the point in the future that the organization wants to attain.

The following are the basic elements of this section and represent the basic requirements for defining the organization's goals.

1. **Mission statement development:** This is an often overused element, and is frequently invoked by senior management as the only means of reinforcement of long-term plans. Simply stated, the mission statement is the short, concise description

of the business of the company and its objectives. It describes the company's identity and role in the marketplace.

2. **Narrative of the dream:** Create a detailed narrative of the "dream" of what you want your company to be in the future. It should discuss the time frame, expected results, financial position, market position, product position, and competitive comparison. It should also outline the ownership structure at that time, if any changes are expected.

3. **Narrative of how the plan and reality will fit together:** Create another narrative that weaves all of the requirements for the future dream and the present and future operational issues together. This is done for the sole purpose to outline plausibility of the plan and goals, and to ensure that they are consistent and compatible.

4. **5-year product line catalog:** This is an interesting and enjoyable exercise because it allows management to sculpt the future offering of the organization by outlining the 5-year product and services catalog. You most likely will be required to go through a detailed product and services evolution flowchart to generate the complete catalog. That process will be outlined later. However, at this time it is sufficient to list the products as they would appear in the catalog.

5. **Product scope definition:** The product scope definition is a means to describe, in a detailed format, the product portfolio in a detailed manner. It should show how the products fit together and leverage off of each another, and where the boundaries are. Also take time to establish low, medium, and high scenarios for each of the products. This will become helpful for the identification of the core requirements of technology that the company must have.

6. **Specific market segments:** Next, detail the specific market segments for each product. Do some market research to determine the trends in the marketplace for these areas and plan scenarios as to how your effort will fare competitively. Management should, as part of this exercise, question why this strategy will work and why this product segment will succeed in the face of the dynamic marketplace.

7. **Industry trends:** Finally the team needs to look at overall industry trends to determine if they will be active through the planning cycle. If basic natural resource industries, such as the energy industry, are not active from a capital spending point for the near term, be aware that you may position the organization for the end of the planning cycle; however, it will be dry through the planning cycle. Strategically it's a good plan; however, from a cash generation point of view, it may not be fiscally allowable.

C. How will we get there?

This section is the pure planning section. It specifically outlines how the organization will move from point A to point B in the planning time frame. The previous narratives described what the organization will look like in the marketplace. This section outlines what specific steps will be taken. Each area of the organization will be examined and planned for in the transition. There will be marketing issues, development issues, manufacturing issues, service issues, and financial issues.

The following outline lists the elements of this section of the plan that must be addressed:

1. **Segment strategy:** This portion of the plan views the market strategy of each segment. It should start with a current market assessment of the segment. The purpose is to find out what is going on in the marketplace and to view the trends, both near and long term. It is also imperative to cite critical factors for success in this section. These two elements then drive the strategy, and how the company will position itself, and determine what role the company will play in the marketplace. The plan also needs to define how the company will play this role. By what means will it prevail in the competitive environment: innovation, price availability, or training? These are the questions that must be answered to complete the strategy.

2. **Implementation:** Since not all growth is internally generated, especially on a 5-year scope, some of the businesses and product lines may be acquired or brand labeled, or be a result of a joint venture development.

   In this part of the plan, each segment is identified and a strategy is developed. In addition, an implementation plan accompanies each one to identify if it is an internal development or some external mechanism. A spreadsheet and Ghant chart illustrates the plan (Figure 1-3).

3. **Investment/Return**

   In order to add financial validity to the plan, it becomes necessary to identify the scope of funds required to carry out the plan. Each investment or market segment that has been identified, each strategy that has been established, and each company organizational requirement that has been outlined must now have a financial picture attached to it. An investment/return money line should be established for each. This is illustrated in Figure 1-4.

   Be sure to include all investments and both development expense and capital equipment expense. Although this is not completely pure from an income statement/balance sheet point of view, you are looking for a scope of funds to be expended. This is most often executed with cash, and therein lies the requirement.

   The return should be calculated as follows: Take each individual segment and determine the number of unit sales. Then generate a projected gross profit based on revenue and costs, and post the product of the gross profit and unit sales to the return stream of payments. It is best not to post revenue only as a percent of

IMPLEMENTATION

	YEAR 0	YEAR 1	YEAR 2	YEAR 3	YEAR 4	YEAR 5
PRODUCT 1						
INTERNAL DEVELOPMENT	X					
JOINT VENTURE						
BRAND LABEL						
PRODUCT 2						
INTERNAL DEVELOPMENT						
JOINT VENTURE		X				
BRAND LABEL						
PRODUCT 3						
INTERNAL DEVELOPMENT			X	X		
JOINT VENTURE						
BRAND LABEL						
PRODUCT 4						
INTERNAL DEVELOPMENT					X	X
JOINT VENTURE						
BRAND LABEL		X				

Figure 1-3. Implementation Matrix

market share obtained or as total dollars, independent of cost and units, because this often lacks substantiation.

Also, the intent here is to evaluate the opportunity of the segment, and as such, you want to compare real dollars invested against real dollars realized.

Next, calculate a growth rate for each and establish a trend measurement for each. In this way the opportunity can be evaluated against a desired financial profile and also against each other.

Finally, aggregate all of the investment and all of the return and post these results to a money line.

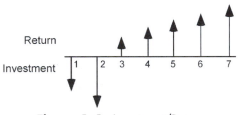

Figure 1-4. Investment/Return

In this way the financial proforma accompanies the narrative of the plan, and senior management will get a financial feel for the plan and the scope of the expenses.

4. **Requirements**

There will be two sets of requirements for the organization and the plan. One will be internally focused and the other will be externally focused. Both sets of requirements must be addressed and cannot be glossed over.

   a. Internal: There may be infrastructure changes as a result of the plan. These should be outlined and specifically listed with timing if they are critical to the success of the plan.

   b. External: There may also be external requirements (e.g., in the sales organization, marketing channel, or vendor links). These need the same treatment as the internal changes to the organization.

5. **Recommendation**

The recommendation is one of the most important parts of the strategic plan. All of the data observations and opportunities must be coalesced into a firm recommendation for the company to follow. It also determines the acceptability of the plan from a management and commitment point of view. There are several elements to the recommendation section that should be included to be complete and to increase chances for acceptance. They are as follows:

   a. Trial balloon the plan with internal people from different elements within the organization. This can be done in stages and/or in parts. By doing this you can get initial feedback from the organization and a measure of the acceptability to certain groups. Next, trial balloon the plan with the sales channel to get their perspective. This will afford you the opportunity to see who in the channel will support the plan and who is not on board with it. This information will be helpful at a later time when evaluating constituent players as the plan is executed. Finally, visit a cross section of the customer base and present it to them. This feedback is the most helpful because it is the customers who will eventually support the plan by continuing to purchase existing and new products from the company.

   b. It now becomes necessary to evaluate the feedback from the internal organization, the channel, and the customer base to factor their input to any modifications that may be required to ensure the plan's success. Any modifications made should be communicated back to the people who suggested them to enhance acceptance and buy in.

   c. Learn the dynamics and the biases of the organization and retreat from confrontation in the recommendation section of the plan. The objective is to create agreement, not to teach or preach to the organization.

d. Next, leave the plan alone for several weeks, if possible, and work on something else. For example, visit some customers, manage other projects, simply distance your-self from the plan if possible.

e. Then reread the plan and examine to see if it's reasonable. Does it still read like a plausible plan? Does it still seem achievable? If so, proceed.

f. Finally, make the recommendation to management and get their agreement.

6. **Action plans**

After you have agreement of the principles, it is necessary to generate an action plan. Without it the plan becomes a conversation piece and none of it gets executed. Worse yet, senior management may assume that action is taking place when it may not be. This will impact credibility and eventual funding. The action plan should be very specific with times, assignments, and completion dates. Even if it is not complete in total, it is important to establish momentum early and set the pace for the organization. If funds are necessary to be appropriated by senior management, put them on the action plan with an assignment and a completion date. You will find out early if this is a real plan or just wishful thinking. The final element in constructing the action plan is to refrain from taking half steps. If the plan is reasonable and the recommendation is well considered and achievable, then do not compromise at this stage by taking halfway measures. Execute the plan as designed.

CONSISTENCY

1. IS THIS A REASONABLE OPPORTUNITY?

As part of determining if the prospect in question is a real opportunity, one needs to conceptualize and visualize the product offering as part of the company's future standard offering. You need to visualize a sales call with your sales team working the channel. Evaluate plausibility of the product. Can you see your team assimilating and effectively promoting it?

Do not assume that the organization will "step up to the plate" to promote the new product.

Take care to strike a delicate balance between a new product stretch for the organization and something that is unachievable given the resources and mix of players.

Given the development cost, resource allocation, and lost opportunity cost incurred in the creation of the product, a certain amount of introspection and due diligence is required to ensure that the organization can be effective.

2. SETTING THE CRITERIA FOR EVALUATION

To provide some consistency in the evaluation process, it is helpful to have a criterion to evaluate these new product prospects. A framework could be as simple as a list of criteria that must be satisfied in order to fit in the organization. This has little to do with the market fit or opportunity for market success; rather, it deals with the internal limitation of the organization.

The benefit of this type of evaluation is that it removes the emotionality in the decision-making process. It also limits the effect of personal preferences and agendas, making the selection more objective and driven by a corporate decision-making process.

To make this even more effective, factor the individual company's time to develop into the criteria, to determine ability to capitalize on the market window of opportunity.

Perhaps the best way to accumulate the criteria is to establish a rating system, which assigns a weight to each of the criterion and allows for an arithmetic means for selection.

In addition, a single opportunity may not have to satisfy all of the criteria. In some instances it may only need to satisfy 70% to 80% to qualify. However, a few of the criteria may be mandatory. The following is an example of criteria for a company that has changed ownership, and is highly leveraged. In this case opportunities taken must be deliberate, and must have large payoffs to offset the risk in valuable capital:

A. Is this a finite opportunity or a new market?

B. Are the projected sales at least $2.5M per year for 4 years?

C. Can our manufacturer's reps sell this in most of the territories?

D. Can we design and build it?

E. Do we have to install?

F. How much after-sale support is required?

G. How many people will we have to add to headcount to support this?

H. Can we sell it to other customers?

I. Can the design be leveraged from another design or to another design?

J. Can we develop it in less than a year?

K. Is the gross margin more than 50%?

As an example, for the highly leveraged organization that must make every dollar spent pay off, criteria B, D, H, and I may be absolutely mandatory, whereas there may be some latitude on the remaining criteria.

3. HOW MUCH OF A STRETCH IS THIS FOR SALES, ENGINEERING, AND MANUFACTURING INFRASTRUCTURE?

It is important to assess how much of a stretch the new product will be for the organization on all of the operational fronts. As discussed earlier, one cannot blaze new trails on every front and expect any leverage to take place. A degree of stretch is desirable in each of the operational elements to keep the organization fresh and competitive. A visual way to look at this would be to generate the the graph in Figure 1-5.

Radially place lines from a center point as shown. Have each line equally placed at the same angle. The length of each line indicates the measure of difficulty to implement the new product. A separate line is used for each of the critical operational areas that contribute to the product's success. There is one line each for sales, engineering, manufacturing, quality, purchasing, service, and repairs. Assess where the organization is with respect to the competition and plot on each the operational. Also plot where the new product will fall with respect to the internal present capabilities of the organization. In this visual way it is easy to assess the degree of difficulty in implementing the new product.

The black line represents the company's capabilities in the various disciplines. The red line represents the present capabilities of the marketplace, and the blue line represents the present requirements of performance for the new product/business.

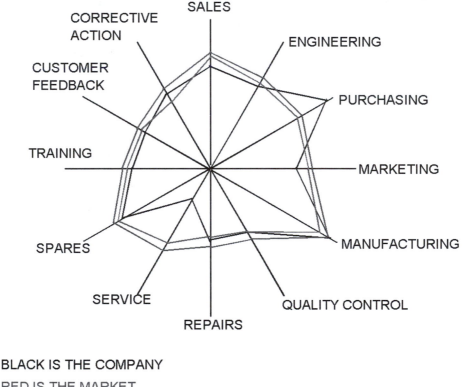

BLACK IS THE COMPANY

RED IS THE MARKET

BLUE IS THE NEW PRODUCT

Figure 1-5. Fitability Graph

By looking at the relative comparison between these various assessments, one can evaluate the company's capabilities to execute the new product development program.

As can be seen, the company is a little short in the sales and marketing discipline and significantly short in service. The diagram also shows that the product requires service capabilities even ahead of the market's expectation for service, therefore the company will have to make some changes.

4. ENSURING ALL OF THE BUSINESS ELEMENTS ARE COMMITTED

As stated earlier, it is crucial to the new product development's success that all essential elements of the organization are in agreement and committed to the execution of the program. It is therefore incumbent on the one entrusted with the leadership role to assess this commitment as early as possible. Interview the key players in the organization to assess their levels of commitment to the program. Correct or alter any attitudes that will stand in the way of success. If you cannot, change the players. The energy needs to be spent in executing the program to position the product favorably with respect to the marketplace. Do not expend the energy trying to convert nonbelievers within in the organization.

5. THE DANGER OF "PROMISED COMMITMENT" OR GETTING COMMITMENT EARLY

There is a danger of getting what seems to be commitment early from several people in the organization in word, but not necessarily in deed. Consequently, it is good practice to involve all the players in the organization at the onset. Require their participation early in the process and observe their attitudes. Don't shy away from confrontation; in fact, invite some of it. In this way the truth comes out, and it can be resolved early on. Also, there is a benefit to some confrontation in that some good ideas and observations may be brought to light (that otherwise would have not been discussed) and integrated into the program. In addition to inviting the confrontation, it is the manager's responsibility to drive it to a conclusion for the team. In this way the group can see how the objections, discussions, and disagreements can be a positive force in bettering the program.

6. TAKING THE "BASAL TEMP" READING

As the program progresses, it becomes necessary to periodically take the basal temperature of the organization with regard to the program. This serves as a reaffirmation of the commitment of the key managers and the key players to ensure a proper sense of urgency to the project. Do not let apathy or negativism begin to permeate the group because this begins to feed on itself, and eventually will slow progress. As the people in the leadership role, your responsibility is to smooth over internal speed bumps and focus on the competitive and market driven issues.

DIFFERENTIATING RESEARCH AND DEVELOPMENT

1. RESEARCH GATHERS AND EXERCISES TECHNOLOGIES AND PROCESSES; DEVELOPMENT ORGANIZES AND APPLIES THESE TECHNOLOGIES AND PROCESSES TO A SPECIFIC END

There is a unique and interesting relationship between research and development in an organization. The two disciplines are diverse and separate, although they are often thrown together in conversation and lumped as one entity. In actuality the two must function as a relay race, with research establishing the lead position and handing off the intellectual property and know-how to the development people to apply and create new products.

Research can be thought of as a strategic element of the organization, whereas development can be thought of as more operational in nature. Many problems in new product development occur when development thinks of itself as strategic. Timelines are not met, costs increase, and they forget that they are an integral part of the revenue-generating organization that must produce results on a specification and a timeline to hit a window of opportunity. For fast-moving technology companies, even research needs to be considered as operational, meaning it needs to have usable results, to be on time, to be well defined, and to be easily transferable to development for industrialization and commercialization.

Another way to think of the difference is that development must have a manufacturing focus: bills of material, specifications, consistency, traceability, costs, performance, and testability in which research is more focused on performance, core technology development, technological capabilities, and eventual cost curves.

There are three basic types of research and development used in one form or another in industry today—incremental, radical, and fundamental. Incremental research is best characterized by a small amount of research and some amount of development. This development is manifested by small, incremental improvements in the product, manufacturing processes, and cost reductions. A lot of research is not required; however, continuous development is. These efforts generally have large aggregated payoffs over time, but can be unsung heroes of the profitability index.

The radical type is best characterized by a large amount of research and a large amount of development. This is the type of development that requires new knowledge to be discovered and understood, as well as development of the means to embody the know-how into a product for a specific purpose. Huge expenses in development follow huge expenses in research. The organization launching into this type of development requires immense amounts of capital to see the programs through to profitability.

The fundamental type of program is best characterized by a large amount of research and little or no development. Many times these programs degenerate into gathering knowledge for the sake of knowledge, rather than gearing the research toward a specific goal. They may be strictly strategic in nature and won't have payoff for many years. The senior management must have a great amount of faith and long-range vision because the rewards of the decision and subsequent sacrifice will accrue to the next generation of management.

2. RESEARCH SPENDS MONEY AND CONTRIBUTES INTELLECTUAL PROPERTY AS AN ASSET ON A FUTURE BALANCE SHEET

Research and development have three major strategic goals in an organization. Their purpose is to contribute to the existing businesses by allowing the organization to defend market and product positions, support sales growth, and expand existing business.

Research fundamentally generates intellectual property for the organization. If the activities of the research element are directed properly, the intellectual property will have tangible value on a balance sheet. This asset will be eventually used to develop new products.

It is therefore incumbent on the senior management to carefully select the programs to fund. They need to select programs that are tied to operational objectives, have a timetable and an end, and result in development generating revenue for the organization through product sales. Management needs to think through the vision of the future in today's time frame and assign research work that will generate timely results for the next product developments.

3. DEVELOPMENT CONTRIBUTES AND MAKES MONEY THROUGH PRODUCT SALES

Development, on the other hand, takes the research and intellectual property and know-how and generates products that can be sold at a profit sufficient to offset the other expenses of the organization. Figure 1-6 is a simplified illustration of the development

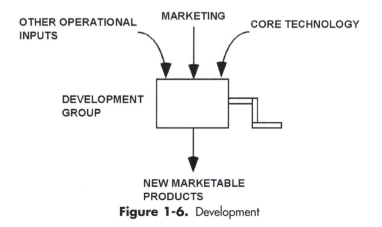

Figure 1-6. Development

role. It can best be characterized by a machine that has marketing input, operational input, and core technology available, and that generates salable products that can be manufactured at a profit.

4. DEVELOPMENT IS MUCH MORE MEASUREABLE

The actual progress of development can be easily measured, since the development team is charged with the responsibility of producing results that are quantifiable, and is committed to a schedule. With the fundamental technology understood by the research function, development must concern itself with items such as bills of material, performance, functionality, manufacturability, and cost. Milestones can be established and tracked with action items, responsibilities, and completion dates. Work content can be managed by manpower hours and loading in general, and results of expenses and time can be evaluated.

5. THE NEED FOR RESEARCH TO BE TIED TO OPERATIONAL GOALS

Since research generates the knowledge and is an integral part of development productizing the technology, it is very important for the research expenditure to be carefully made. Costs to progress from research to development increase dramatically, so diligence is required at the onset. Simply stated, ensure that research function is tied to operational goals. This means that the area of study must directly source the needs of development, and it too must occur on a timeline. If not, development ends up cultivating the core technology and the project becomes irretrievably delayed.

6. THE MODEL FOR R&D

The following is a model for the research function and how it fits into the organization. Although organizationally the reporting relationship may vary, in general, Figure 1-7 represents the flow of information and knowledge.

Figure 1-7 shows that information comes into research from several sources, including cooperative agreements with other companies' shared development in a consortium, from university sources, or from other sources as required. The corporate research department then generates the core technology to lay in to the development programs so products can be developed to hit a market opportunity.

On a strategic basis, Figure 1-8a illustrates how profits are derived from the efforts of research. The goals and objectives identified for the organization's future drives a strategy to achieve them. Research contributes to the development process, and profits are realized through operations. It is a critical link in the chain, and results are required in sequence and on time to be effective.

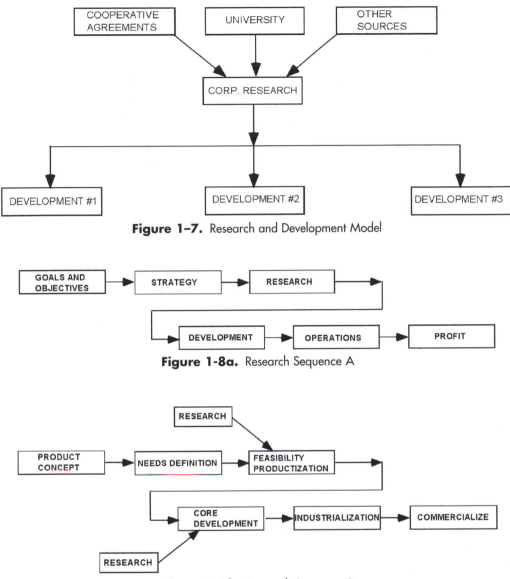

Figure 1–7. Research and Development Model

Figure 1-8a. Research Sequence A

Figure 1-8b. Research Sequence B

On a project basis the research function must contribute knowledge at various points in the planning and development process. Figure 1-8b highlights this contribution of core technology.

The project starts with a concept. The marketing analysis determines the customers' needs definition. Research then contributes in the form of core technology, the necessary knowledge to determine project feasibility. It must answer the question: Can we as a company absorb and cultivate core technology and productize it cost-effectively to create a profit? At the point that the affirmative conclusion is reached, the core development of the technology begins. Research then transfers the technology to development. An integral part of the development is the industrialization of the product, and then commercialization. As shown in Figure 1-8b, research contributes at two critical junctures in the process: at the feasibility stage to determine if the technology can be brought to product status, and at core development when research transfers the technology to development.

LEVERAGE

1. EACH NEW PRODUCT DEVELOPMENT EFFORT SHOULD LEVERAGE VERTICALLY AND HORIZONTALLY WITHIN THE ORGANIZATION

The key to effective new product development is to create leverage in the development arena similar to creating leverage in the manufacturing and sales arenas. It simply is unaffordable for most companies to continuously start from ground zero on every new product development; consequently, the desire is to leverage past development efforts when starting new programs. Figure 1-9 illustrates the issue:

As shown in Figure 1-9, successive development programs can be thought of as developing concentric rings around a base of core technology. As the programs become more complex and involved, the core increases. Each activity then pushes the outer envelope in terms of technology, purchasing, sales and marketing, and manufacturing systems. In this way the company develops an expertise in all areas with each program and experiences real growth in sales, margin, and the infrastructure to support it. A less abstract and more operational way of looking at this can be shown in Figure 1-10.

Figure 1-10 shows that a single product demands company resources in the procurement of materials, the engineering of technology employed, the manufacturing processes, and the sales channel to get it to the marketplace. As new opportunities are evaluated, a portion of the evaluation needs to address the leverageability of the product within the organization for the successive programs. Building a business using like manufacturing processes, engineering and development, and routes to market, all contribute to the leverage discussed.

2. THEIR RESULT SHOULD BE 1,2,4,8 VERSUS 1,2,3,4

By leveraging the sales, general, and administrative elements of the business, a successful series of programs will generate financial leverage in the form of growing revenue and profit. Since the knowledge is being cultivated inhouse and applied to the programs, the leverage shows up as faster program execution times and lower costs. Faster time puts the

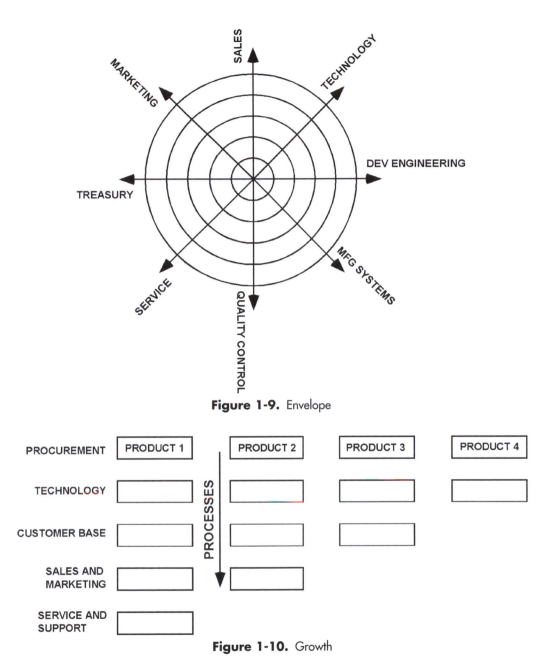

Figure 1-9. Envelope

Figure 1-10. Growth

company in a better market position, and lower costs make the program easier to amortize. Revenue- and profit-generating programs then sum up to a result that generates a profit curve that begins to look exponential, rather than linear. The difference is the leverage that is created. This can be illustrated in Figure 1-11.

	YEAR 1	YEAR 2	YEAR 3	YEAR 4	YEAR 5
PRODUCT 1	1	1.5	2.5	4.375	7
PRODUCT 2		1	1.5	2.5	4.375
PRODUCT 3			1	1.5	2.5
PRODUCT 4				1	1.5
PRODUCT 5					1
TOTAL	1	2.5	5	9.375	16.375

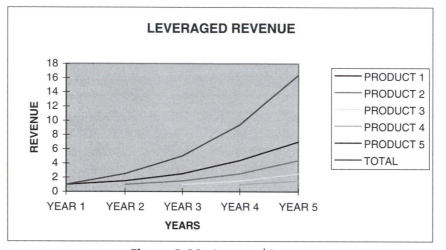

Figure 1-11. Leveraged Revenue

Each product generates growth for the company, starts out small, and builds momentum in the marketplace. As the company then structures the programs in succession (and does this for several product lines), the revenue and profits grow disproportionately. This is the leverage that is desired; however, there is also a collateral benefit, as we shall see in the next section.

3. THIS ALLOWS MORE FLEXIBILITY IF FAILURE SHOULD OCCUR IN ANY ONE AREA

This philosophy will allow the organization to recover from a single failure easier than other strategic philosophies because there is overlap in all areas, and each obstacle does not result in navigating uncharted waters. This philosophy also takes more time and effort to initiate, but also results in a more solid momentum after you get it going.

4. IMPROVING THE CHANCE OF SUCCESS

Consequently, this will better the company's chance of success in a fast-paced, changing marketplace. Most articles on new product development will discuss the percentage of failures in the marketplace and the historical reasons for failure. This philosophy represents tangible, positive steps that can be taken to ensure a higher percentage of successes.

CONTINUITY

1. THE CONTINUITY OF DEVELOPMENT IS AS IMPORTANT AS THE DEVELOPMENT ITSELF

It can be thought that the continuity of development is as important as the development itself. What this statement means is that a company must become proficient at recognizing an opportunity, initiating action to capitalize on it, and prosecuting a program through to success with it. If an organization can achieve this with the determination to weather the obstacles along the way, it will eventually be able to get to anywhere it desires to go.

2. ON AGAIN/OFF AGAIN

Conversely, the developer's nightmare is a management structure that issues on again/off again orders and constantly changes priority on projects. In these cases a bulk of the effort that needs to go into the development of the product is taken with winding down a program, only to later reinitiate it.

3. BUILDING MOMENTUM

Figure 1-12 illustrates this point. If there are interruptions in the development processes, the returns in revenue cannot be additive, according to the model for leverage discussed earlier.

According to the unleveraged revenue model, each successive program builds the revenue and profit curve toward a more exponential shape. Interruptions therefore will alter the time scale and the revenue (missed the market opportunity) and profit generating potential scale away from the desired characteristic shape.

4. GEOMETRIC PROGRESSION OF REWARDS

The objective therefore is to reap a geometric progression of rewards for the invested development. The operative word in the statement is investment, in that a company desires to harvest as much revenue and market share as possible from an investment made in product development. Future revenue dollars are discounted by inflationary pressures, and development dollars are invested presently at full value and generally after tax (outside of R&D tax credits used from time to time). Therefore almost by definition, a certain amount of leverage is mandatory.

	YEAR 1	YEAR 2	YEAR 3	YEAR 4	YEAR 5
PRODUCT 1	DELAYED	1	1.2	1.5	1
PRODUCT 2		DELAYED	1	1.2	1.5
PRODUCT 3			DELAYED	1	1.2
PRODUCT 4				DELAYED	1
PRODUCT 5					
TOTAL	0	1	2.2	3.7	4.7

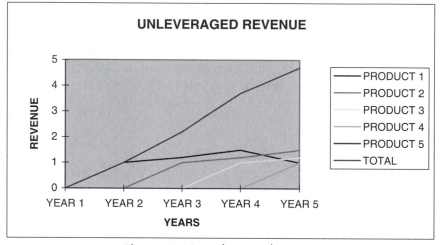

Figure 1-12. Unleveraged Revenue

5. THE PURSUIT IS AS IMPORTANT AS THE ATTACK

The important point of this section is to understand that continuity is critical in new product development. The company must initiate a program, execute it in a timely manner, and harvest the reward of its investment. To that point the pursuit of the business is as important as the original attack. Each program and action must be deliberate and have a payoff. There can be no voids in the continuum because this wastes time and dollars; hence the pursuit is as important as the initial attack on a market opportunity.

FLOW

1. TAKE AND EXECUTE MANAGEABLE CHUNKS OF DEVELOPMENT

To keep the flow of a new product development program smooth and laminar, each step of the process needs to be identified and broken down into component parts for analysis and execution. It is important to refrain from selecting huge chunks of activity to complete without detailing the specific parts. By taking small elements the team can use trial end error in managing uncertainty and essentially "practice" its problem–solving techniques. As will be discussed later, the smaller the component part, the easier to manpower load the team and create a laminar flow of progress.

2. GET GREAT AT EXECUTING SMALL, INCREMENTAL IMPROVEMENTS VERSUS LARGE PROGRAMS THAT ARE LONG AND FROUGHT WITH UNCERTAINTY

The company needs to get highly proficient at executing these small, manageable chunks of development activity, so as to remove the uncertainty associated with the huge project at the onset. By doing these small chunks effectively, repeatedly, and predictably, one can then manage a large program to a budget in terms of time and money. The only way to confront the uncertainty hidden in a large program is to break it down to its component parts for execution. This also has a collateral benefit of ensuring completeness by including all of the component parts required for the program.

3. SMOOTHING THE PROGRESS AND MAKING IT MEASURABLE AND PREDICTABLE

The benefit in terms of project management here is to make these component parts measurable in terms of effort and talent required, and to be in a position to plan and complete activities in a predictable and repeatable fashion. This takes practice and several programs to become proficient. To achieve this means that almost any program can be broken down into its component parts for analysis, and a measurable amount of effort can be applied, and the result can be predicted within a time frame.

SYMBIOSIS

1. EVOLVING A SYMBIOSIS

Within the product development team there should evolve a symbiosis, in which an operational perspective governs the members of the team to work for the project. Should one of the members falter, there is enough strength and cognizance such that the other members fill in the void without affecting the performance of the team in total. In other words, it is similar to a dance between two partners, in which natural choreographing should take place.

2. TAKING THE TIME TO BUILD SYMBIOSIS

This type of symbiosis takes time and several programs to build and cannot occur overnight. However, the program manager should take the time to build this into the temperament of the development group because it removes the specter of prima donna behavior and fosters attitudes that support the overall objective, even in the absence of day-to-day management.

PURSUIT

1. GETTING RELENTLESS AT IMPROVEMENT AND EXECUTION WINS THE BUSINESS

The company should evolve a culture of consistent improvement, incremental product development, and timely execution. The competition does not maintain the status quo, either in isolation or collectively and as such, mandates your firm to initiate consistent, predictable actions that result in increased market share and better competitive posture.

2. KEY VOCABULARY: SMALL, INCREMENTAL, DEFINABLE, MEASURABLE, PREDICTABLE, FAST, AND RELENTLESS

Once the strategic vision is locked in, the company must perform. The watchwords of this performance are small, incremental changes that are definable, measurable, and have leveraged payoff. Any actions that the company takes need to be fairly predictable at the onset, executable in a short period, and relentless in the succession of these steps. The firm needs to maintain the strategic vision and strive to relentlessly execute steps to achieve it.

SINGULARITY VERSUS PLURALITY

1. A NEW PRODUCT SHOULD FIT INTO A CONTINUUM OF NEW PRODUCTS: A FAMILY SUPPORTED BY A STRATEGY

There has been significant discussion regarding strategy, vision, and operational planning, and all of the steps associated with it. The reminder here is to refrain from one time hit type programs. The single-hit type programs are characteristically seductive, lucrative, and have huge upside potential. These programs also have generally longer lead times, significant uncharted development waters to navigate, and often require huge capital outlay. Long-term success of an enterprise relies on the need to stay on strategy, nurture and execute a plan, and focus on solid strategies and the tactical and operational planning to carry them off effectively.

2. PRODUCT DEVELOPMENT SHOULD NOT BE AN ISLAND, TOO EASY TO KNOCK OFF BY COMPETITORS

Take care in the selection process not to select programs that are all alone, like an island. An enterprise must have significant financial resources to prosecute a development program with no previous core technology or background. These firms are at risk when embroiled in these types of programs, as they can consume tremendous amounts of capital and personnel. Progress may be slow, and obstacles require disproportionate effort to overcome. They generate characteristic revenue curves similar to the aforementioned unleveraged revenue example. This makes them easy targets to knock off. They are financially stretched, they have no depth to the core technology, they have not productized it before, and their slow progress allows a competitor or consortium of competitors to leapfrog them technologically with their product offering. The marketplace is very dynamic and fast paced. Loyalty is as strong as your last product advantage. Alliances between companies are fleeting, and your ally today may be your opponent tomorrow.

3. THE FAMILY OF PRODUCTS RETAINS MOMENTUM AND THEREFORE STRENGTH

As discussed previously, a family of related products under development in which the technology employed is one of the core technologies inhouse, serves the best long-term prospects of the organization. It retains the momentum, offers flexibility, and fosters financial and competitive strength for the company.

4. PRACTICING THE CONCEPT OF LEVERAGE

This is the basis for practicing the concept of leverage. Combine efforts to serve a strategic goal or use a single effort to serve related, multiple goals. This philosophy leverages the product development resources in an organization. The managers of several programs should spend their time looking for and creating this leverage. Start with modest goals that are achievable and build on the successes. Along the way, gather ideas and execute selected ones that contribute to an overall portfolio.

MARKET INVESTIGATION

The market investigation and subsequent sections are so important to the long-term growth of a company that they are included here in the following presentation. They are presented here from a "fit" perspective to provoke the strategic thought and planning of new product development early in the process, as it should be. In Chapter 2, and more so in Chapter 3, the mechanics of marketing will be covered in detail. The objective here is to present how the concepts relate to the overall evaluation and selection process, and what one should look for in a new product development program.

1. PRIMARY

There are two basic types of market research in use today—primary and secondary. The primary type of research is one of the most direct and accurate forms available because it deals with information obtained directly from a customer. There are a variety of ways to secure primary market feedback, but most fall into the following categories: surveys, interviews, and demonstrations. Surveys can be administered in a variety of mediums and are structured means for obtaining information. To be conclusive and effective, they need to be administered in the same way each time, with no changes in format. Interviews are a face-to-face means of gathering information directly from the respondent. These are very valuable for securing qualitative-type information, exploratory-type information gathering, and discovery of latent needs among customers. In fact, this type of sampling can be used as research for determining how to structure a survey. Demonstrations are another type of primary market research. This type of research offer information in the form of a demonstration and solicits feedback from the respondent. These are critical to confirming features, benefits, and acceptability of a product to a customer. Each type of primary market research has its advantages, and each needs to be used in different aspects of the information-gathering process.

2. SECONDARY

The second type of research is the secondary market research approach. It is characterized by an indirect means for gathering information from readily available sources. Typically, research of this type consists of, but is not limited to, the following means for obtaining information:

A. Trade shows: The trade show is an excellent means for determining product information as well as determining the scope of companies in any one field of endeavor. In addition to the personal interviewing and observation and literature search, trade shows typically publish a program that list companies by product type or process. These are typically segmented several ways to allow a quick view of the market.

B. Trade literature search: The literature search is an ongoing activity to keep in tune with the marketplace to see who is promoting what to whom. It is an excellent means for gathering information for a future, competitive comparison.

C. Articles: Articles are an excellent means for documenting trends in the marketplace. They also, by virtue of their authors, yield expert sources for further information and perspectives on the marketplace. They also are very useful to the novice who needs to familiarize himself or herself with a product area.

D. News pieces: News pieces are timely information of the actions of the leaders in the marketplace in a given area. They tell you who is active in an area and to what extent.

E. Observation of trends: This is an internally generated activity that looks at what is happening in the marketplace from period to period. By looking at the historical activity, one can extrapolate trends to the future.

F. Financial reports: Financial reports, such as Dun & Bradstreet and Standard & Poor's, and the internet, offer a wealth of information about competitors and, as such, collectively about the market. These can be monitored for information and financial information to get a picture about the finances and their principles and the capitalization of a competitor.

G. Textbooks: Textbooks offer a novice background information on products or services and technology. These sources can be useful to understand elements of the technology.

H. Patent search: Patents are available for search to determine the state-of–the-art in a field of endeavor, as well as to determine your company's position in the state-of-the-art.

I. Lost sales reports: These internally generated documents can be helpful in determining a competitor's ability to use their product to secure business. By evaluating their technology, and your company's performance against them, these reports can be an invaluable source of information.

J. Internet: It almost goes without saying that the Internet is of the most prolific means of information gathering about companies, products, competitors, pricing, trends, and countless other data. It can be an invaluable resource for immediate data gathering, if the data is available.

3. INTERVIEWING

Although previously mentioned as one of the primary means for market investigation, interviewing is such a valuable technique. It is so often overlooked that it may be considered a separate category of its own. If approached properly and under the right conditions, the customer is the most accurate means for assessing your company and its standing in the marketplace. If you listen and attempt to internalize what the customers are telling you in an interview, they will communicate most if not all of the qualitative data you will need.

When conducting an interview, it's best to review the company operations first. This will give you a feel for the operation and how this customer makes money. It will also give you a perspective on how your company and its products will fit into the overall plans, as well as determine the legitimacy of the respondents' comments by substantiation. The interview will allow you to discover latent needs that the customer may not readily know they have. This generates the opportunities for new product ideas.

It's best to have samples, models, examples, and props to enhance the exchange and draw out response. Do not be adverse to broaching issues that may be controversial (pertaining to the subject matter at hand), because this will give perspective on the attention level of the respondent and his or her biases, and will yield perspective on other comments.

4. SURVEYS AND CUSTOMER FEEDBACK

Surveys and customer feedback should not be conducted only at the onset of a program. These need to be an integral part of the process of refinement, product evolution, and subsequent product developments. Never get too far away from the customer because this will lull an organization into complacency, or worse, remove the feedback loop in the process.

5. EXTRACTING THE OPPORTUNITY WITHIN THE COMPLAINT

The customer grants a significant favor by their complaint when you analyze it. The company sold them something that fell short of their expectations, and they thought enough of it to complain. We as market and product development professionals may not like or want to listen to the complaint; however, there is a lot to be mined out of one. In effect the customer is telling you what you did wrong. If one evaluated the math of the situation, a single complaint is worth quite a bit to the organization if you are going to do something about it. If you view it as a nuisance, then it has no value; however, if you plan on changing to meet the customer's expectations, then it has great value. In other words, how many other customers who didn't complain would have to stop buying from you for you to lose enough market share to notice? How many customers would you have to survey to find out what this one complaining customer is telling you directly? Clearly it is a valuable means of feedback.

6. ENGAGING THE CUSTOMER

As part of the visit you will want to engage the customer in a frank discussion about their business, its driving forces, its vulnerability, and how they navigate the competitive threats to find success. This discussion should begin to have symbiosis as an eventual tone. To have a real appreciation for the customer and their needs, it becomes necessary to understand their business. The following list of questions is of a market exploratory nature and will give a flavor for the type of questions you may want to use to engage the customer. These discussion interrogatories are open-ended and require thought to create mutual understanding.

A. What are the areas in your organization that currently fall short of skill set and performance index required to achieve your strategic vision? Who, what, where, how, and why? *This allows for an open-ended forum between you and the respondent. It allows you to draw out complaints initially, and then to get specific ideas on the table for discussion. Share your company's experiences also.*

B. From a historical perspective, what are you doing about them? *This determines the commitment level of the management and the organization to change, and to the strategic vision.*

C. What is your strategic vision? *This is a good test to determine if the organization has established a vision and if it has filtered down through the organization.*

D. What operational or product changes are you making to achieve this result? *Again, this determines the effectiveness of the management structure to effect change toward a specific goal.*

E. What do you need to make your job easier, more efficient, to increase output, and to exercise better control? *This open-ended discussion allows the respondent to identify any latent needs in the organization.*

F. If you were asked to double output with the same resources, what would you expect to be required to achieve this? *This removes the specter of the status quo and forces the respondent to push performance of the organization. It helps the respondent to identify future needs.*

G. What is currently considered impossible or impractical, but if you had access to it, you could increase effectiveness? *This question allows the respondent to dream of a future workplace and identify latent needs to achieve it.*

H. What do you look for in a supplier? *This identifies the customer's expectations*

I. How can we become a preferred supplier to your company? *This determines if you have a chance at the business and the account.*

J. What are your critical competitive pressures? *This allows you to determine a way to partner with them.*

The key to identifying the potential new product is to synergistically combine all of the ideas and requirements into the tangible improvement of the next-generation product.

7. EXCHANGE OF KNOWLEDGE AND EXPERIENCES

During your discussion, it is critical to establish an open and free exchange of ideas, problems, and opportunities between the two companies. Establish a spirit of cooperation at the management level and drive it through the organization. Create the atmosphere of going after a market opportunity together, with each contributing a portion of the talent and effort to get there. Constantly look for how the two companies can both profit from the market segment, since very few companies are absolute leaders in a market with a position strong enough to dominate and effect control of the segment by themselves.

8. ESTABLISHING THE RAPPORT, EXPERTISE, ASSISTANCE, IMPROVEMENT, AND AUTHORITY

This exchange should establish a lifelong rapport between the personnel of the businesses that will transcend any minor market fluctuation, upset, or competitive thrust. Your actions need to position you as an indispensable partner integral to your customer's plans. Make sure to plug the customer (now your partner) into your organization. Balance the sacrifice you will make in availing the customer of your resources with the positioning of indispensability within their organization.

9. COALESCING THE DATA

Examine the responses carefully and group the data into some meaningful way to draw product conclusions from it. Look for product configuration patterns in the summary of data so that product planning can serve the market with the proper product.

10. CORRELATING THE DATA

Does the data received make sense with other information gathered? Has the data been corroborated in some way to substantiate the results and accurately draw conclusions? Did you notice any conflicting data that would require more research? Summarize it in some kind of a spreadsheet to allow viewing of the data in a holistic fashion. Organize it for eventual presentation to management. The results should tell the story of the opportunity to management.

11. TESTING THE DATA

Finally, test the results and the data supporting the conclusions by visualizing the product available for sale today. Evaluate how it will sell in the marketplace today. Would it sell? Why or why not? Visualize the product selling in the future, project the future market needs, and (after development time has elapsed) determine if it will still sell or if you would have lost the window of opportunity. Do the results of this reinforce that you are targeting the right generation of product?

UNDERSTANDING THE MARKETPLACE

1. IT'S A COMPLEX WORLD

The world is becoming more homogeneous as time progresses. Actions taken overseas affect transactions and business here. Changes in currency affect global trade and individual buying decisions. The marketplace is simply becoming more complex, and initiatives and actions need to be multidimensionally evaluated before implementation, or they may be ineffective or disastrous in other areas.

As an example, let's assume your company supplies electronics to an Original Equipment Manufacturer (OEM) in North America for sale to end users in North America. A sister company to the OEM (half ownership) resides in the Tuscany region of Italy and sells their equipment to the European continent. The Italian company is the primary mechanical manufacturer for the domestic OEM and has electronics supplied by an Italian manufacturer of electronics, supplying the same package near your cost. Do you have a domestic threat from the Italian electronics company here? What are the critical factors to secure continued domestic market share? Can you maintain domestic loyalty here even if the Italian mechanical manufacturer prefers its own supplier? Can you be functionally displaced by the Italian electronics manufacturer who is working in close conjunction with the Italian mechanical manufacturer on a new platform that will be brought to North America?

All of these factors must be evaluated. You will find that it begins to mimic a chess game with an international flavor. The operative point being: the world is a complex place.

2. WHAT ARE THE CUSTOMER PROBLEMS TO FOCUS ON?

The effective way to navigate the marketplace is to focus on the customers' needs. Their problems are the ones that need to be solved in the embodiment of your products and services. By keeping close to the customers' wants and needs by resolving their problems, you position yourself as a problem solver and a consultant to be trusted and relied on, an aspiring position that every supplier wants.

3. UNDERSTANDING THE DYNAMICS

Since the marketplace is a fast–paced, dynamic medium, it is most important to understand the dynamics of the market. Your company is attempting to do one of two things. The first is that you are trying to develop a new market that currently doesn't exist. The other is that you are trying to increase market share by taking it from others. Both actions cause reactions in the market. The first will draw attention to the lucrative opportunity. The second is basically a catfight to secure the business. Knowing the dynamics will assist in developing the strategy for each. Failing to understand the dynamics can be fatal to the effort.

4. UNDERSTANDING WHAT THE CUSTOMER VALUES AND IS GOING TO BUY (i.e., INSURANCE POLICY CLAUSES AND "PEACE OF MIND")

What does the customer purchase? Do they purchase product, satisfaction to a need, or peace of mind? This question and its answer are the essence of the marketing and sales of a new product. The products and services are fundamentally the embodiments of the satisfied need the customer has. Fail to satisfy the need and no compensation exists, that will offset what is lacking. Meet the need well and there is a lot that will be overlooked, all else being equal. Remembering that the product you sell and get profit from is simply the embodiment of a satisfied need is crucial to formulating any program. Failure to understand and accept this fact may cause wasted time and effort for the company. A basic example of this is an insurance policy: Is the customer purchasing a document with legalese and clauses to satisfy a need, or are they purchasing peace of mind from the insurance company?

5. POSITIONING THE PRODUCT

This is why it is so important to position the product in the mind of the consumer. This positioning must tell the story in a convincing manner so that the buyer's perception of the product satisfies the need. It is important to understand that at the onset, it is the customer's perception that is making the decision. Subsequent buy decisions after the first sale will be made on product performance in meeting the need. In other words, no matter how good any product is, fundamentally, the customer must have the perception that the product will satisfy their need at first sale or you will never get the opportunity to let performance satisfy the need. Positioning the product creates that initial perception.

Positioning the product means different levels of difficulty for different industries and products. For example, repositioning a product by repackaging the same basic elements is on the easier end of the spectrum. Redesigning a product platform for an added or different feature set is quite more involved.

6. WHAT TRIGGERS THE PURCHASE ORDER

A study of Business Transactions would indicate that, at some time in the exchange between the company representative and the customer, there is a point in which the customer has all or most of the perception needs met and decides that this is the best deal I can reasonably get. This is the precise point where a closing needs to take place. This is what will trigger the purchase order. Asking for one before this point is reached is pointless, and waiting past this point is dangerous; it becomes a critical success factor in a sales transaction to seize the opportunity to create this point and act on it.

7. PRACTICING

They say that practice makes perfect, and that is generally true, because inherent in practice is the scientific methodology. Try, evaluate, learn, modify, try, evaluate, etc. This is the value in the lost sales report in that it is a key element in the analysis of the transaction and is integral to any improvement change to become ultimately successful.

8. TESTING THE EFFECTIVITY IN DIFFERENT SALES STYLES

There are different sales strategies and styles for different products, and it is important to match the sales style to the product type. If, for example, you have a technological leadership position, the sales style would be different than if you were supplying a "me to" type product, the difference being a value-based style versus a price-based style. This is an area that must be identified and coordinated with the product launch and that is integral to the marketing plan. If it is not, you may be wasting the sales personnel's time and risking their disillusionment.

9. MODIFYING THE STORY FOR EFFECTIVENESS

Marketing and selling a product is similar to telling a story to the customer, at least from the promotional aspect. Ideally the interaction needs to have the symbiosis discussed earlier, but for purposes of this discussion, the marketing "story" needs to be consistent with the launch objective. If it is observed that for whatever reason, the presentation needs to be modified; it should be reevaluated in conjunction with the entire product plan and product launch.

10. TIERED MARKET STRUCTURES: HI, MED, LOW

Also, different tiers in the market demand use of different sales strategies and tactics. If the product is a vehicle that traditionally had status and utility associated with it; there are several strategies to use to be successful with each tier of luxury, value, and low-cost transportation. Parts of the promotion need to appeal to the different wants and needs of the buyer. The operative point is that the multiple-tiered market must match up to the multiple product line to be effective.

11. HOW CAN IT BE SOLD?

Product marketing personnel need to constantly evaluate the effectiveness of the promotional programs and continuously ask how the product can be sold. Since the landscape is constantly changing, the positioning must be modified to keep pace.

12. WHERE IS THE VALUE CREATED AND EXCHANGED?

To effectively market and sell products, the company must determine where the value is created and where it will be exchanged. Fundamentally this is the essence of the transaction. The customer is looking for the solution to a need at a good price. The manufacturer must add value at a profit to continue operations and to fund future developments. The value must be created and exchanged for revenue so that both parties come out to the good. Anything less than this is temporary and fleeting.

GLOBAL PRODUCT AND BUSINESS DEVELOPMENT

SMALL WORLD, ISN'T IT?

Disney Corporation dedicated an entire Magic Kingdom attraction to it, the Internet has made it an electronic reality, and Boeing's 777 made it easy. It is indeed a small world and getting smaller every day. What does this mean for new product development? It simply means that isolationism in one country or one continent is no longer possible. What your company does, to a degree, affects the global marketplace for that particular market, and what the rest of the world does may indeed affect your company. Product development and market development no longer occur in a vacuum. The downside is more and more intense competition. The upside is instant access to information and data.

2. TYPICAL FOREIGN CONTENT OF TODAY'S PRODUCTS: AUTO, VCR, CAMERA, CELLULAR

If it is not obvious that the world is getting smaller from a communications aspect, consider the foreign content of many of today's products. The foreign content has become so commonplace that several products publish it in terms of a percentage. Consider the Honda automobiles manufactured in the Marysville, Ohio plant, in that they are manufactured by Americans for a Japanese company. It has American content and Japanese content in materials and labor on both sides. It would be an exercise just to document in detail the funds flow in terms of raw, manufactured cost and intellectual property amortization. The same split content occurs in cameras, video cassette recorders, cellular telephones, and computers. Another interesting exercise would be to document the mileage on parts in a subassembly through manufacturing to its final destination. Consider the following route a power transistor takes in getting to its final destination: The power transistor will be used in the final output stage of an electronic drive for controlling the speed of an electric motor, which is part of an elevator drive system in Algeria. The power transistor is manufactured in Japan and is shipped to a distributor in North America. That distributor ships it to a drive manufacturer in the Midwest for integration into a drive. The drive is shipped to the drive manufacturer's subsidiary in the United Kingdom as part of a stocking order. The drive is pulled from stock and shipped to Bologna, Italy for integration into the elevator control panel. The elevator control panel containing the drive is tested and shipped to the end user, a building in Algeria!

3. ANALYSIS OF EXCHANGE RATES' SENSITIVITY ON FOREIGN CONTENT

The presence of foreign content in products also adds a dimension to product costing. The actual value paid for foreign content components is subject to the changes in currency from one country to another. This can place pressure points on a product design from a cost sensitivity point of view. In addition, the supply of these products can be affected by global competition, export compliance, and political turmoil.

4. VENUES FOR PROCUREMENT

This means that the venue for procurement of the component must be carefully selected, in addition to the specification and the cost. Failure to consider ramification may cause loss of supply midstream or intense cost pressure.

5. VENUES FOR MANUFACTURING

The same can be said for manufacturing; however, an additional dimension occurs here. What happens if the local lawmakers pass legislation that affects the plant. Can your capital equipment be idled by changes in law, political pressure, or local labor initiatives? In this case your company is at greater risk than the procurement case because your manufacturing capacity is affected.

6. LOCAL LAWS

Local laws can place greater demands on new product development, since some require adherence to international standards. These may be stricter than U.S. laws and standards; consequently, the more complex and costly the development may have to be.

7. EXPORT COMPLIANCE

The United States requires manufacturers to adhere to export compliance legislation. This is mostly a marketing effort to be selective in targeting certain locations to exclude from sales and shipment. Failure to comply may result in revocation of an international license or worse. The export compliance guidelines exist to limit shipment of U.S. technology to foreign countries that may be a security threat. It can significantly affect the marketing effort.

8. TECHNOLOGY TRANSFER

Technology transfer is generally a company decision, although the U.S. government can play a role on certain technologies. Some governments mandate a transfer of technology to their country as part of a sale. This can adversely affect your company's market position and strength elsewhere, and the decision to transfer technology must be carefully evaluated.

9. JOINT VENTURES WITH PARTNERSHIPS, CULTURES, PROFIT, MOTIVES, AND TAXES—KNOWING YOUR PARTNER'S AGENDA

Partnerships such as those mentioned can be very successful or very costly. Watch where the money changes hands to determine allegiances, agendas, and motives. This will be discussed in detail in Chapter 2.

10. COST ROLL-UP SPREADSHEET WITH SENSITIVITY

Figure 1-13 summarizes the effect on product factory cost impact from foreign content. As can be seen, the effect can be favorable or unfavorable depending on current market

FOREIGN CONTENT SENSITIVITY ANALYSIS W/ LABOR ADJUSTMENT

BILL OF MATERIAL

ITEM	DESCRIPTION	P/N	QTY	COST EA	EXT. COST	DOMESTIC	FOREIGN	CURRENCY FAV/UNFAV @STD COST	COST IMPACT
1	PART 1	654321	2	43.25	86.5	YES	NO	1	86.5
2	PART 1	654320	2	0.23	0.46	YES	NO	1	0.46
3	PART 1	654319	2	0.48	0.96	YES	NO	1	0.96
4	PART 1	654318	6	4.3	25.8	YES	NO	1	25.8
5	PART 1	654317	1	6.2	6.2	YES	NO	1	6.2
6	PART 1	654316	4	7.9	31.6	YES	NO	1	31.6
7	PART 1	654315	8	12.3	98.4	YES	NO	1	98.4
8	PART 1	654314	10	8.75	87.5	YES	NO	1	87.5
9	PART 1	654313	3	3.4	10.2	NO	YES	0.99	10.098
10	PART 1	654312	7	1.25	8.75	NO	YES	0.98	8.575
11	PART 1	654311	5	1.65	8.25	NO	YES	1.12	9.24
12	PART 1	654310	2	2.4	4.8	NO	YES	1.12	5.376
13	PART 1	654309	1	2.75	2.75	NO	YES	1.12	3.08
14	PART 1	654308	3	3.45	10.35	NO	YES	1.12	11.592
15	PART 1	654307	4	6.45	25.8	NO	YES	1.12	28.896
16	PART 1	654306	5	4.12	20.6	NO	YES	1.12	23.072
17	PART 1	654305	7	2.34	16.38	NO	YES	1.12	18.3456
18	PART 1	654304	8	45.7	365.6	NO	YES	1.12	409.472
19									
20	LABOR		12	35.5	426	YES		1.05	447.3
21									
22	BURDEN		1.75		745.5	YES		1	782.775
23									
24	MISC EXPENSES		100		100	YES		1	100
25									
26									
27									
28									
29									
30									

TOTAL COST 2082.4

ADJUSTED COST 2195.2416

IMPACT 112.8416

%IMPACT 5.42

Figure 1-13. Foreign Content Sensitivity Analysis

conditions. The object point is that your company may enter an agreement with the conditions being favorable, and lock into terms that will force cost pressures and eventual market pressure, if the conditions go unfavorable. This puts your effort in a precarious position. In addition, the manufacturing venue can cause greater problems depending on the percent of labor, local laws, and relative numerical difference in wages. This also assumes that the hours required to produce are uniform and homogeneous across the globe, and they are not because different venues have different skill force levels.

As can be seen in Figure 1-13, many factors contribute to the product factory cost, and as such, it is the responsibility of the manager to account for these issues. The architecture of the product transcends the bill of material to all of these contributing issues. The product needs to be planned out from all aspects to ensure continuity in sales and manufacturing.

DEFINING THE MARKETPLACE

1. WHO IS THE CUSTOMER?

Always understand and know who the customer is. They are the ones generating the revenue through acceptance of your solution to their need. Continue to solicit their opinion, their wants, their changing markets, and their pressure points. This will ensure that as changes in the landscape take place, your company will be positioned to respond appropriately.

2. DEMOGRAPHICS

Do not be lulled into a false sense of security by demographic summaries. Since so much of our lives are governed by lumping people together and treating their issues as common, keep in mind when serving an OEM market, the customer is trying to differentiate themself to create a competitive advantage. Consequently, they need your company to customize your solution for them. In trying to serve a consumer market, keep in mind that there are different classes of customers with different needs that have to be met by the product or service. Products cannot necessarily be lumped together and mass marketed with success like the demographic summary may indicate.

3. COLLECTION OF USERS, OEMS RESELLERS, AGENTS, SIMILAR APPLICATIONS, ETC.

Remember that a market is a collection of customers, not a uniform homogeneous purchasing machine that can be pushed, prodded, and calculated with decimal place accuracy every time. There are intermediaries in the process who have their own problems, issues, and agendas. Leveraging the initial product development and the launch requires real-time monitoring and action to build and preserve momentum.

4. WHAT ARE THE CUSTOMER ALTERNATIVES VERSUS YOUR SOLUTION?

Always remember that the customer has alternatives. They may be competitive offering at a lower cost, they may be competitive offering at a higher cost, they may be functional displacements, and in some cases, there may be personal apathy on the part of the customer. The operative point here is that the customer your company is approaching to sell may *not* be predisposed to the solution you are offering.

5. HOW IS YOUR COMPETITIVE ADVANTAGE EMBODIED IN THE PRODUCT?

Often this is evident in examining competitor's solutions to customer needs. Many times a unique or lower-cost method may be used to accomplish the same function. The difference may be that the development engineers have biases and mental paradigms, as we all do in addressing and solving problems. The innovative engineers, designs, and companies break these paradigms to offer alternative approaches embodied in their products for satisfying customer needs. That is the business objective in new product development.

SUMMARY

This chapter had several objectives for the reader to achieve. The thematic thrust of the chapter was to draw the relationship between new product development as the business objective. It also was designed to give a qualitative feel of the different steps in the process of new product development. As the chapter unfolded the reader learned the various aspects involved with the process. There was a discussion on the new product opportunity's "fit" in the organization, and how it tied into the strategy of the organization.

There was also a discussion on the directed purpose of research and the function of development, how to leverage efforts, and create continuity. Product development programs should each contribute toward a corporate goal, and the market investigation is used to refine the goals and the opportunities.

The chapter should also give the reader an appreciation of the interrelationship of the various disciplines within the enterprise that must function in concert to successfully prosecute a program. Some of the material presented in broad overview will be discussed in a more-detailed manner in later chapters. However, they were presented here for continuity in outlining the process of taking an idea from inception to completion. The subsequent chapters will detail the process, starting with Chapter 2: The Market Opportunity.

CHAPTER 2

THE MARKET
OPPORTUNITY

ABSTRACT: Identifying and evaluating the market opportunity is key to new product success. The opportunity must be clear and tangible, not obscure and indirect. There are many factors and issues that impact a development throughout the gestational period, so a fleeting opportunity is not real, nor will it have any longevity on which to build a business. Test and retest the strategy and the opportunity throughout the process. Examine things from a multidimensional perspective as they relate to the market and the company. Establish contingency plans for longer developments because conditions may change, forcing you to regroup and fall back on these plans. These precepts are important because all else follows from what is established in this chapter.

THE BUSINESS CONCEPT EMBODIED IN THE
NEW PRODUCT IDEA

1. THE BETTER MOUSETRAP

Design a better mousetrap and the world will beat a path to your door. This has been the battle cry for new product developers for many decades. However, there are increasing requirements for success today, such as: Who else makes mousetraps? What must our mousetrap do to induce people to buy it over others? Are we a low-cost mousetrap producer or are we a niche mousetrap player? Have we complied with the new mousetrap standards? What steps have we documented to ensure that humans will not get injured using our mousetraps? Is there a significant influx of foreign mousetrap manufacturers encroaching on our market? Are there new ways to functionally displace mousetraps, such as high-frequency deterrents? All of these issues make the age-old adage now multidimensional.

2. MOST SYSTEMS GENERATE THE IDEA AND FAIL TO ENCOMPASS A PLAN

Most systems for generating new product ideas focus on the idea, the market opportunity, and the timing. The concept for a new product must encompass the plan to carry it off effectively. The business concept must be entwined with the product concept. Each will draw on the other in the development phase. Each contributes to each other's progress.

3. ENGINEERS CREATE BILLS OF MATERIAL; MANUFACTURING DESIGNS THE PROCESS

There has been a lot of discussion about interdisciplinary project teams and group effort. Within this group mentality the engineers need to design a bill of material and a manufacturing process that will bring the product concept to a business reality. The team needs to build a business around these products. Anything less will fail to achieve the desired result.

4. TIE THE BUSINESS OPERATIONS TO THE IDEA

The manager needs to visualize the business operations as the product is being outlined. This visualization will assist in defining product features and configuration that will aid in future business operations. For example, the demographic data may indicate a number of versions of a product; however, the business operations—manufacturing, procurement, and inventory control—require a finite, practical number of versions that are more manageable.

5. MUST LOOK AT THE BUSINESS ASPECT OF THE PRODUCT (i.e., WHO, WHAT, WHERE, WHY, AND HOW)

The field of new product development can be very exciting, and intoxicating at the same time. Consequently, the manager must look at the business aspect of the program to complete the analysis and plan. How will the company do business with this new product? Where will the money change hands? Why will the customer be willing to part with money for the product? How will the transaction take place? Provisions and steps must be taken to address these issues to have an effective launch and program.

As an example, the electric car is an age-old idea that with the advent of computerization, modern batteries, and electronic drive technology, it can begin to become a market reality. If an electric car was designed, produced, and introduced in a mass market, there would be significant infrastructure issues to deal with. The issues such as: safe disposal, recharge locations, battery remanufacturing, financing and supply/demand changes could affect the price of conventional fuel and thus could affect operating costs. Operating cost changes could affect market acceptance. In addition, all of the "off peak" charging of batteries at "night" at reduced rates would now create new peak requirements in power generation. It is not just a great product that makes great businesses. It is also the infrastructure that supports the product that must be integrated.

6. TIE IT TOGETHER

The saying goes: "To the victor goes the spoils"; so too with the business of new product development. The program that covers all or most of the bases to ensure success will most likely be successful. This is because the manager asked those hard questions of the program rather than ignoring the issues altogether, only to find the marketplace asking the same questions and displaying little patience for lack of planning.

SOLVING THE CUSTOMER'S PROBLEM IN THE PRODUCT

1. THEY BUY THE PERCEIVED VALUE OF THE SOLUTION TO A CUSTOMER'S PROBLEM

The buying decision is made based in part on the perceived value of the solution to the customer's problem. The customer buys a solution; the manufacturer constructs a bill of material and processes to assemble it into a product. It is therefore a requirement for successful marketing, design, and sales to embody the solution into the product. The value placed on it is generally governed by the customer. They will determine the amount paid based on this initial perception, and subsequent follow-up sales by experience factors that may weigh the value differently.

2. HOW DOES THIS NEW PRODUCT DEVELOPMENT SOLVE A CUSTOMER'S NEED/PROBLEM?

This is the basis for a marketing plan. Effectively answering this question will directly tie the customer's need into the product concept. The pathway is the sales presentation. It relates to the customer and how the solution to their problem or need is embodied in the product your company is selling them. As we know, different issues affect classes of customers differently. What is a critical issue for a buying decision for one customer is not necessarily all that important for another. The market must be broken down to effectively create the product's architecture so that the lion's share of the available market can be addressed with a basic platform design. This platform design allows a wide variety of features and a feature gradient to be incorporated in a basic unit, with options to scale up performance or packaging of values for the customer.

3. UNDER WHAT CONDITIONS DOES THE CUSTOMER COMMIT?

What is the universal model for customer acceptance? The answer to this question is as complex as the number of different customers there are in the marketplace. The mental arithmetic that takes place in the mind of the purchaser varies with situation, need, time pressure, alternatives, and many other factors.

For example, under normal circumstances, a cool glass of water is worth little more than an act of kindness is worth a generous tip in a restaurant. Under conditions that are much starker and manifested by shortage, a cool glass of water can be worth considerably more, if the need is great enough. Certainly, in a desperate situation, one would be willing to pay considerably large sums of money or earning power for a less-than cool, less-than clear glass of water, than under normal circumstances. What would be the quality/price gradient in this example?

The operative lesson in the example is to understand and know the conditions and the frame of mind the customer is in when they make that buying decision. Failure to understand and plan for it properly may position your company with product inventory, and wondering why customers aren't buying.

An additional means for gathering data on the customer buying-decision dynamics is to monitor competition. How are they selling products? What is their pitch, and is it effective? Gather and analyze brochures, data sheets, marketing literature, lost sales reports, and field sales reports to obtain a "feel" of their strategy and tactics.

4. HOW CAN THIS NEW PRODUCT DEVELOPMENT FIT THE CONDITIONS DEMOGRAPHICALLY?

The marketing assessment outlines the market segments (i.e., that group of customers who have identified features, needs, and preferences that are similar). The manager needs to break down the market opportunities into the component parts so they may be reassembled into product versions, to be used to address the market. What version of the basic platform will be used where, and why? Will it offer a competitive advantage? In a more macro sense, what will the product mix consist of? What will it cost, and how will it be priced? These decisions need to be made correctly early in the process. A sensitivity analysis needs to be conducted on the cost and the price. If it's expected that the price will lower after introduction and competitive response, this new price needs to be factored in to see if there are any latent pressure points in profitability due to a skewed mix.

THE PRODUCT AS A BUSINESS

1. HOW DOES IT USE AND GENERATE CASH, PROFIT, AND VALUE?

All products use and generate cash. It becomes an important factor as product development programs become larger and more of a factor in the overall finances of a company. Depending on the type of business and the labor content and material content, every dollar of added sales requires additional dollars of operating cash to fund it. This cash finds itself as inventory either in raw goods or completed assemblies, as well as accounts receivable.

The development requires cash for the basic intellectual property development (technology) and also the capital equipment needed to manufacture the products. The organization gets a cash pressure on several fronts: expenses, capital, inventory, and accounts receivable. This can be shown in Figure 2-1.

As illustrated in Figure 2-1, the negative cash flow exists for the duration of the development program. Depending on the cost of capital employed, and the inflation rate and the time to develop, this negative cash flow can impact the organization significantly, if it is not recovered rapidly through revenue.

2. PRODUCT AS A SERVANT TO THE BUSINESS, NOT CONVERSELY

This is why the product and the technology must be a servant to the strategic plan of the business, and not visa versa. The product requires enormous amounts of real time cash that could have been retained earnings. The impact manifests itself in the value of the stock and shareholder equity. Consequently, timing is of the essence in recovering the initial investment.

3. THE END IS DOLLAR AND VALUE CONTRIBUTION TO MARGIN DOLLARS

To use a British colloquialism, "at the end of the day," the objective is to have the development program pay off in the form of margin contribution and value contribution to the organization financially. It may be a long road to get there, but if the opportunity is real and the plan achievable, the organization must have the faith to see it through. The following example illustrates this point.

In 1948 the Halloid Company paid $25,000 for the right to develop xerography. Their original target for introducing a Xerox machine was in 1950, some 2 years later. The model,

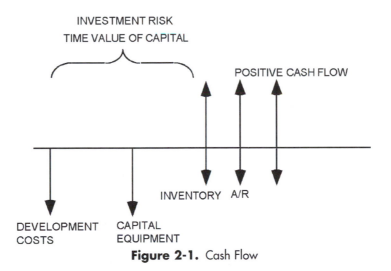

Figure 2-1. Cash Flow

a copier, had 1260 separate components. Fraught with problems, the first real marketable unit wouldn't be introduced until 1960. The financial drain on the company to develop and market the product was tremendous. The total development cost was $75 million, which amounted to more than twice the operating earnings of the company during that time frame!

The first unit sold for $29,500 at introduction in 1960. The financial impact of the revenue was staggering. By 1970 Xerox's stock grew in value 66-fold: a fine example of big risk, big payoff.

4. PLUG THE DEVELOPMENT TEAM INTO THE CONCEPT OF BUSINESS

One of the most positive things a manager can do to improve the efficiency of the development team is to "plug" them into the business. Most members of a development team know something about business. They are generally familiar with the cost structures, the material content, the overhead and burden rates, and have a general product-costing orientation. Very few, however, have an appreciation for the more strategic and tactical aspects of the business. In addition, very few have an appreciation for the sense of urgency required in the product development playing field.

Consequently, a very synergistic move is to train the team as a team in these aspects. The manager needs to outline the impact of slipped schedules and creeping factory costs, and what that means to the revenue line, the recovery of investment, and the loss of competitive advantage. Selected elements of accounting and finance should be reviewed with the team in total and reinforced on an individual basis in those situations that lend themselves so.

THE COMPETITIVE GAME

1. UPSETTING THE PLAYING FIELD TO YOUR ADVANTAGE

There are several marketing strategies in use today in terms of leadership roles: guerrilla tactics for niche players, defensive positions, and offensive positions. However, all of these strategies become all that more valuable when the new product initiative upsets the current playing field. By finding a way to upset the playing field, you improve your position twofold. The first way is to compete with the existing players with a new product. The second way catches them off guard as to the approach used. Depending on the uniqueness of the approach, it may be all that more difficult for the competition to respond.

The following example illustrates the concept from two perspectives, long-term and short-term. In the 1970s U.S. automobile manufacturers lacked certain quality features that the American public was looking for in the product. Several foreign manufacturers embarked on a program to improve the quality and service of their products to the American consumer. By the 1980s the perception had shifted for the foreign manufacturer, from a shoddy goods perception to a quality, consistent product with exceptional service. Their battle cry: American consumers are entitled to exceptional quality! This changed the playing field for Detroit and ultimately, severely impacted its market share. This was a long-term change in the playing field that led to a distinct market advantage. Tremendous amounts of time and money were required for Detroit to place the competition at bay.

In more recent history the computer industry exhibited short-term changes in the playing field to create market advantage. Different architectures in operating systems, software, and license issues affect access and market position every day.

2. FASTER, MORE INFORMED SOLVING OF THE CUSTOMERS PROBLEM

A key competitive advantage is to mobilize the operation to excel at determining business and product opportunities by engaging the customer, and becoming proficient at designing and introducing products with customer solutions embodied in them. The organization needs to bring formidable factors internal and external to the organization to bear on these opportunities. The better one is at this, the better the chances for growth.

3. QUALITY PRODUCTS/ON-TIME DELIVERY/DOC; ALL IS ASSUMED, THE PLAYING FIELD IS IN APPLICATION TO SPECIFIC NEEDS

The marketplace is a different one from that of 10 or 20 years ago. We have tasted well-made products at ever-lowering costs. We have experienced better service and more availability of product. Out-of-the-box failure rates are increasingly more scarce among different industries. Competitive edge won't come from these areas; they are assumed. The competitive advantage will come from mass customization, applying basic platforms to a diverse set of customers, availability, application of specific products, and from low-cost, feature-rich innovative producers.

IDEA EVALUATION WITHIN THE FRAMEWORK OF THE BUSINESS

1. SYSTEM FOR EVALUATION

To objectively evaluate a wide variety of diverse new product ideas, it becomes desirable to have criteria. This removes the emotions from the evaluation and allows different personnel in the organization to do the evaluation. Each idea then must be tested against the criteria the same way with the same questions.

2. SETTING UP THE CRITERIA

One of the challenges is in setting up the criteria. What questions are important? What are the pressure points of the company? What are things to avoid? For example, is the opportunity dependent on the development of technology that requires a few select people?

Are they available within the time frame you desire? If they leave the project, what means for recovery do you have? This is an example of a pressure point as it relates to personnel. There may be a host of other pressure points that affect selection. As a rigorous part of the strategic plan, it is important to develop these criteria in a customized manner for your specific business.

3. TESTING EACH IDEA

Each idea should be tested against the criteria, as previously stated. Figure 2-2 is a format that can be used to evaluate an idea. It is structured with a weighted sum arrangement of criteria. Each criterion is listed as a separate item. Next to each criteria is a preference assessment of that specific criterion. The preference assessment directly relates back to the strategic initiative. Does the business desire that attribute in a new product development? The range is from −1 to +1. Next to this preference value is the assessment of how the product relates to that specific criteria. The columns are multiplied and summed to get the weighted value.

By observing the individual values posted for each criterion or attribute, one can learn a fair amount about how the organization will absorb and prosecute the program. As can be seen, the organization has a high desire to select an opportunity that has sales of $2.5M for 4 years. They are highly technical and are not concerned about absorbing the technology.

They do, however, have an aversion to field service and support, perhaps a weak field organization. This criteria shows up as a negative value. Consequently, if the product being

EVALUATION CRITERIA FOR NEW PRODUCT DEVELOPMENT

ITEM	EVALUATION CRITERIA	CO. DESIRE WEIGHT (-1-0-1.00)	ACTUAL SCORE (.1-1.0)	PRODUCT # 1 WEIGHTED SCORE	ACTUAL SCORE (.1-1.0)	PRODUCT #2 WEIGHTED SCORE
1	SALES > 500k PER YEAR FOR 4 YEARS	1	0.9	0.9	0.9	0.9
2	EXISTING SALES CHANNEL SECURE ORDERS	0.75	0.85	0.6375	0.6	0.45
3	ABSORB THE TECH AND DESIGN THE UNIT	0.85	0.45	0.3825	0.75	0.6375
4	TRAIN, SERVICE & SUPPORT IN THE FIELD	−0.25	0.2	−0.05	0.75	−0.1875
5	SOLD TO A VARIETY OF CUSTOMERS	0.1	0.6	0.06	0.8	0.08
6	ALREADY OUR CUSTOMER BASE	0.9	0.7	0.63	0.5	0.45
7	LEVERAGE TO ANOTHER PRODUCT LINE	0.5	0.35	0.175	0.2	0.1
8	PROJECTED MARGIN OVER 50 PERCENT	0.65	0.4	0.26	0.6	0.39
9	INSTALLATION IS REQUIRED	−0.5	0.95	−0.475	0.85	−0.425
	TOTAL			2.520		2.395

Figure 2-2. Evaluation Criteria

evaluated has a high dependence on this for success multiplying the preference and product assessment will be negative. When summed up with the other criteria, these will detract from the overall weighted value. This analysis simply serves as a numerical evaluation, which is objective, rather than talking yourself into the assumption; the product or business won't need that much field organization support.

Another aspect is to admit that the criterion of field support is not one your company is good at, and to reexamine your commitment in that area. Additionally, perhaps the product and the sales channel could be designed such that a field support organization won't be required.

As with any weighted summation, the absolute value of the sum does not hold a lot of meaning. However, when two programs are set side by side, the sums can be compared on a numerical basis to select the best alternative. This can be a valuable tool, under two conditions:

1. Be very careful when setting up the company criteria. Make sure it is an accurate reflection of the desire and capacity of the company.

2. Be accurate and honest when evaluating an opportunity against these criteria.

If you are diligent about these two items, this will serve as a useful tool. The table presented here has fixed values and fixed criteria. In the Tool Box an interactive chart exists where you can configure your own criteria and preference weights and assessments.

4. CONSISTENCY

Consistency is one of the key benefits in using this objective type of evaluation. As times change and personnel changes, it becomes a challenge to establish and maintain consistency in the organization. This exists in all aspects of the organization, not just in the identification of new product opportunities. It is most important to be consistent in the new product area, however, since this area drives all future movements. Wild swings in strategy and tactics will rod the organization around thus removing whatever shred of leverage it may have had, and rendering it ineffective.

5. MEASURE OF ONGOING EFFECTIVENESS

As the system is put in place and used for some time, it is important to measure its effectiveness. What ideas were brought in, how were they evaluated, and what were the results in selection? What degree of success was obtained in pursuing the market? Are the evaluation criteria still the right ones? Are they consistent with the strategic plan? These questions need to be periodically reviewed. Figure 2-3 is an illustration of how the process should work.

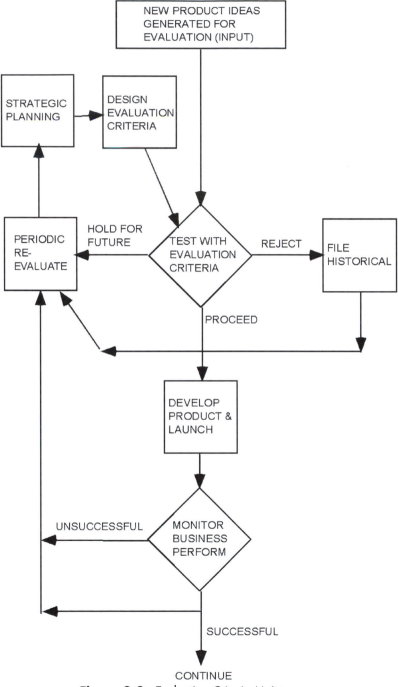

Figure 2-3. Evaluation Criteria Maintenance

6. OBJECTIVE CRITERIA VERSUS AREAS OF COMFORT/AGENDAS

The final and summary thought in this section is that new product opportunities must be evaluated with two methodologies in mind. The first is that they need to be evaluated in a consistent manner. The second is that the consistent manner be tied back to the strategic plan and the goals of the company. With these two prerequisites for an evaluation system, new product ideas can be brought in and evaluated in a systematic manner (free from personal agenda and immune to organizational areas of comfort), and can be initiated within the strategic framework.

PLAYING THE GAME WELL

1. IDENTIFY THOSE OPPORTUNITIES IN WHICH SMALL INCREMENTAL CHANGES MEAN BIG ADVANTAGES

If you loosely define the competition of new product development as a game, then consider it to be in the organization's best interest to learn to play the game well. It has been previously discussed to become proficient at incremental changes. An adjunctive activity is to become proficient at identifying those items that are small and can be accomplished with a minimum of effort, and execute them for incremental gains. A series of these will mean big payoffs in the near and long term. They serve as practice for larger programs and also keep competition at bay, by constantly raising the barrier to expectation level in the specific market place.

2. DETERMINE LATENT OPPORTUNITIES FOR COMPETITIVE ADVANTAGE AND INCORPORATE IT WITHIN THE PRODUCT

Here is where the "engagement" of the customer becomes so important. It is at this interface, that if you listen carefully, the latent needs of the customer will become evident. It is in that latent need that the identification of the product features and/or configuration will be defined and value posted by the customer. By incorporating these features, the company can gain a competitive advantage with small, incremental effort and protect margin against erosion as a result of competitive pricing actions.

IT'S A FASTER, CHANGING WORLD

1. NO COMPANY OR MARKET IS AN ISLAND

To give some perspective to the concept of the faster, changing world, consider that by some estimates, the sum total of mankind's knowledge base doubles every 5 years. In addition, the speed at which this knowledge diffuses into the hands of more and more people is increasing at an astounding rate. This makes more knowledge available to more people every day.

It would be more straightforward if the plans your company makes could be made in isolation. They could be well considered and linearly planned out. However, no company really operates in a vacuum, either on the purchasing or the selling side. Each side is interconnected by a network of suppliers, vendors, customers, and employees. Since there is this interconnection, information travels fast and fairly pervasively, albeit not always so accurately. Consequently, advantage is short lived and information must be acted on rapidly and decisively.

2. IT'S AN INTERCONNECTED RACE; OTHERS ARE AFTER THE PRIZE ALSO OR SOON WILL BE

Business is an interconnected race between companies to secure additional business. What one supplier loses in business, another secures. It is a zero sum game. Therefore an individual firm must move rapidly to arrive at the opportunity before their peer businesses. Many compete for the same incremental business. If there currently is no competition in an uncontested area, there soon will be. Undiscovered or newly discovered niches and segments will attract competitive activity. Lucrative niches developed by one firm will be plundered by another. It is the manager's challenge to carve out their place in the market and hold on to it amidst all the activity.

3. SIX DEGREES OF CONNECTION, NETWORKING, ETC.

There is an AT&T commercial that talks about the connectivity that exists in our world. It describes a situation in which we are all interconnected by a vast network of people that essentially can link us with any other person, within six or less links. Figure 2-4 illustrates this point.

This is an important fact in that it describes a sphere of influence that surrounds us all. This being the case we can gain access through each other and daisy chain to otherwise unreachable people. The same can be said of organizations. They each have a sphere of influence that extends to other companies, which can allow access to otherwise unreachable contacts, customers, corporations, and suppliers. This is best illustrated in Figure 2-5.

Company initiatives are communicated through your sphere of influence to suppliers, and customers and competitors via customers and suppliers. It is therefore understandable that care must be taken in dealing with this fact in initiating plans and product developments to prevent inadvertent communication of sensitive material. In other words, if you can gain access to competitive data via your network, they can gain access to your data. It is a faster, changing world indeed.

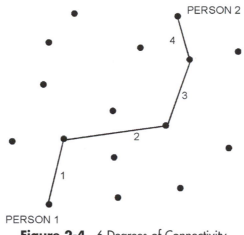

Figure 2-4. 6 Degrees of Connectivity

SPHERES OF INFLUENCE

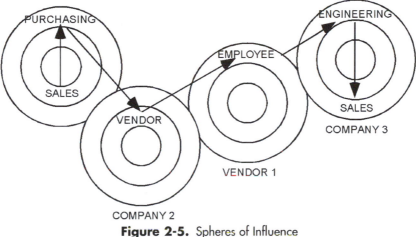

Figure 2-5. Spheres of Influence

WINNING THE GAME

1. DEMONSTRATE SPEED, ACCURACY, AND AGILITY TO A CUSTOMER BEFORE YOUR COMPETITOR

How does your company win the competitive game? Everyone seems to be pursuing the same business that you are. How can you differentiate your company from all the others? The perspectives in answering these questions lie in your company's ability and desire to demonstrate speed and agility to the changing market conditions.

It was previously discussed that no company or market is an island. It also follows that conditions of advantage and marketability do not remain fixed. Responding to these changing conditions in a timely manner will ensure success.

It would seem from the tone generated in this and previous sections that winning in the new product arena is a formidable, almost impossible task. It could be, for firms that do not continually strive to improve their products, their marketability, and their position in the marketplace. If, however, your company does execute these continual steps in new product and market development, you will be well positioned to compete. If you have a strong foothold among the customer base, it will be that much easier to defend and grow.

2. CONSULTIVE SELLING

"Our customers come to us because they view us as the leaders and trust our expertise to resolve their needs." This statement exemplifies the result of continued consultative selling. The company has positioned itself as the resource that the customer will use as a normal course of their business. It also creates a dependence on the company that results in a competitive advantage. In addition to the normal advantage in a bidding or competitive situation, it also positions the company to be on the customer's cutting edge of new developments, placing them there first. This type of selling allows the customer to become a partner in their future success. Such a relationship requires mutual trust and involvement.

3. CREATING THE DEPENDENCE (THEY MUST OFFLOAD OR REDEPLOY HEADCOUNT)

How can you create this dependence between your company and the customer? Your solution needs to provide features and services that the customer would ordinarily need to provide. By your solution of providing them, you can offload the customer of having to provide this service. If they redeploy or reduce headcount, then you have created the dependence. You need to be selective in this area, because if not, you run the risk of attempting to be all things to all people. Create the dependence, where it does not cost your company too much in time and effort, or better yet, in an area where you are exceptionally proficient.

4. USING MORE ENERGY TO SELL, OFFSET BY ORGANIZING AND EXECUTING EFFICIENTLY

This type of selling, although very successful in the long run, does require a significant amount of energy and funds to develop and cultivate. Your company itself needs to make a commitment to select a customer and see it through to completion. By effective partnering with the customer's future plans, links between the two companies can be cemented, which will transcend competitive thrusts.

NEW PRODUCT DEVELOPMENT AS A COMPETITIVE WEAPON

1. NEW PRODUCT DEVELOPMENT AS A SURVIVAL MEANS IN A MARKET

Companies do not have to compete on price and delivery alone. In fact, it is quite a boring existence to do so. New product development can be used as a competitive weapon in the marketplace. With most companies, their history, patterns, and areas of comfort become known. New product development is a means for a company to gain advantage, secure a position, or win a new customer. It is the ultimate weapon in changing the playing field and should be used as such.

2. LEAPFROGGING THE ACCOUNT AND THE INDUSTRY

Examine the industry and the marketplace. Determine the products and the technology employed to solve customer problems. Cite advantages and disadvantages of each of these implementations, then forecast the future of their technologies. This is where the competition will be at your next product introduction. Will you exceed customer requirements, or be at parity with your competitor? Your program needs to leap ahead of where the others will be at that future point. Aim further as a starting point so that at the time of introduction, you will be in a superior position.

3. ATTACKING YOUR ENTRENCHED POSITION

So things are going fairly well. No major problems, volume is reasonable, quality is acceptable, and costs permit a good margin. Time to relax and reap the benefit of all the work right? Wrong. Now is the time to attack yourself. Do the proper amount of introspection to determine how long you will have the advantage, what the vulnerable points of the business are, what steps need to be taken to secure the future business levels and more. The competition will be doing this, so it is in the company's best interest to do the same, and to do it before anyone else. This activity is healthy as it reduces complacency and rekindles a sense of urgency in continually improving the processes, the products, and the relationships with customers. Now is a good time to revisit that loyal customer and solicit their feedback.

4. SALIENT IN THE LATENT SITUATION

Are there latent opportunities among your customer base? What are the specific requirements that they need to compete better? Find that specific requirement and seize the opportunity to capitalize on it. These are the actions that bind you to a loyal customer. The difficulty is in finding out what the latency is, as it is not always readily obvious. Often the customer needs to be led through the discussion to obtain this information. However, since it is somewhat difficult to obtain for your company, it will be difficult for your competitor also.

5. RAISING THE BARRIER TO ENTRY

In any situation, it benefits the company to keep the barriers to entry as high as possible. Surround and imbue the company with expertise and equipment that gives market specific advantage. Investing in capital equipment financed from profits raises the barrier thus discouraging start-ups from encroaching on your territory. Investment in the business will not only raise the barrier, but will also lower costs and improve repeatability and quality. These have a disproportionate effect in terms of competitive advantage and position. It makes someone chasing your accounts and business all that much more difficult to capture.

6. RAISING THE EXPECTATIONS OF THE CUSTOMER AND WHAT THEY EXPECT FROM THE MARKETPLACE, THEREBY ENHANCING YOUR POSITION

Your company's actions should raise the customer's expectations of provided products and services. By setting the standard for performance in your marketplace, you make the competition dance to the standards you set, not visa versa. This continual resetting of the customer's expectation level will also raise the barrier to entry from a features and services perspective, another difficult target to reach for others. This can also catch others off guard because they won't understand the schedules that the partnership establishes for implementation. Your schedules will chart the pathway and will be known only to you and the customer/partner, making it a guessing game for anyone else.

THE STRATEGIC DIFFERENCE BETWEEN LARGE AND SMALL COMPANIES

1. LARGE COMPANIES: BUY, DOMINATE, INTIMIDATE, STEAMROLL, AND COVER THE MARKET

In a homogeneous market condition, a large company can have an advantage. By virtue of its size, its capabilities, and its staying power, a large company can dominate a market. If it is a dominant player in the specific market segment, it can influence the future of the segment and its general progress. An entrenched player has the capability to eliminate competition by pricing actions, availability, supply logistics, and produced costs.

2. SMALL COMPANIES, NICHE PLAYERS, AND GUERILLA WARFARE

Small competitors have advantages of their own, but in different ways. A small company can react to changing market conditions, and capitalize on them. The smaller company will enter a market to exploit an opportunity prepared to leave it at a moment's notice. They do not have the staying power or temperament for a long drawn-out market battle. By virtue of their culture, they will abdicate certain markets under contest and look for new opportunities. Fast to enter and fast to leave is often their operational perspective.

3. WHILE LARGE PLANS EVALUATE AND STRATEGIZE, SMALL COMPANIES EXECUTE TACTICS

In contrast, large firms, plan, strategize, evaluate, and then act. The result can be summarized by the product life cycle chart in Figure 2-6, which illustrates a review of their life cycle and how large and small companies differ in their approach, execution, and longevity in the specific segment. In addition, examination is given to the financial profile generated in each of the examples and how they differ.

4. LARGE TOP-DOWN STRATEGY VERSUS SMALL BOTTOM-UP TACTICS

The difference in the size of the company also determines the differences in implementation. As can be seen from the Figure 2-6, larger companies may be slower at the initial implementation but achieve significant volume once the massive selling and manufacturing machine is mobilized. The small company may be lightning quick at the first article implementation; however, it will quickly lose momentum compared to the larger company as a result of lack of critical mass. The consequences are clear when the following financial analysis is evaluated, concurrent with the chart in Figure 2-6.

Figure 2-7 shows the traditional product life cycle. There are three distinct phases of the product life cycle (i.e., emergence, growth, and decline). The emergence phase is the start of the new product's business. The volume is low, and the product is embryonic in its market share. The second phase is the growth phase, where planned volumes are achieved. The third phase is the decline if left untended, the company will lose market share and volume will drop off. Generally at this point the company is no longer competitive in that specific market segment.

Ideally, at the time when volume is stabilized and the company is enjoying good profit margin and good volume, the project to replace the product with an updated or next-generation

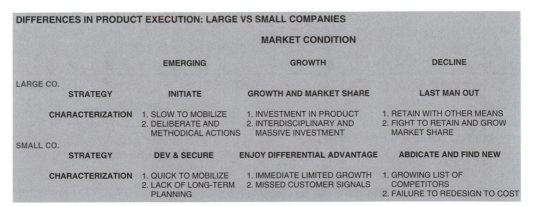

Figure 2-6. Large Vs. Small Companies

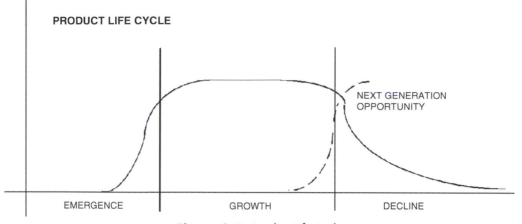

Figure 2-7. Product Life Cycle

product is initiated. This is timed so that when this development is complete, it will coincide with the reduction in volume of the first-generation product. It is through these follow on programs and product improvements and cost reductions that the company will retain and grow its market share. At this time the company may want to reexamine the market and segment the product offering to capture different tiers in the market thus enhancing market share as time passes.

As this is the ideal, homogeneous method of new product implementation, an examination of how a small firm and a large firm differ in the implementation of new product opportunities follows. The financial model in Figure 2-8 best characterizes the small company's implementation:

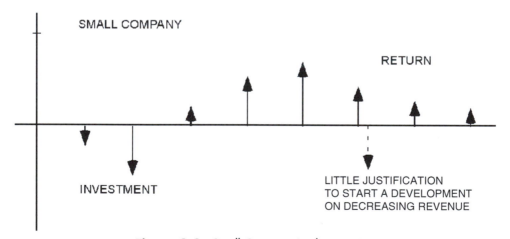

Figure 2-8. Small Company Implementation

The small company characterization in Figure 2-8 reinforces the assumption that as a product opportunity is discovered, a rapid implementation is generated, and put into the marketplace for consumption. The firm makes a small investment to "test" the marketability. Investment in capital equipment is minimal, and the product is introduced. Initial volume looks promising as it begins to build up each successive year.

The problem that now results is twofold. The first is lack of sufficient market share initially established to gain hold of the market. The second is cost pressure due to lack of low-cost product platforms. This results from lack of sufficient capital expenditures to keep the costs low. In addition, there is not an adequate barrier to entry in this market from a capital perspective. The development commitment should not be to "test" the market. It should have been to secure the market.

At the time when the next generation product should be developed, the outlook for the first generation product is in decline. Consequently, the company decides that it is not worth pursuing the opportunity, in light of all the other competitors, and seeks another uncontested niche.

In contrast, let's examine a large company's implementation. This is best characterized in Figure 2-9.

In this example, the large company researches the opportunity and commits to exploiting it. This involves an interdisciplinary effort within the corporation that focuses on developing the product and the business for the long term.

A significant investment is made in development in year 1 and 2. This investment is in developing the product and the manufacturing infrastructure to gain market share. The result is a rapid growth to a favorable market share figure.

As time progresses, the market share and volume increase to a peak. As the volume begins to taper off, in subsequent years, the second-generation product has been scheduled

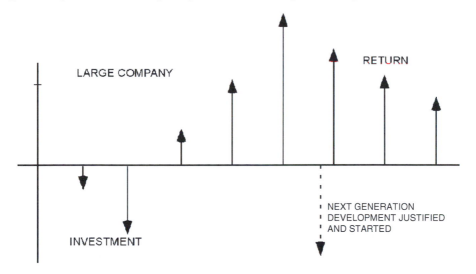

Figure 2-9. Large Company Implementation

for development and is introduced to keep market share stable and growing. The result is a much stronger position from the onset through the product life cycle thereby generating revenue for profit and subsequent development. This is how a solid commitment in a well-researched program can contribute to long-term growth that can be retained and built on.

THE PRODUCT EVOLUTION FLOWCHART (THE ROADMAP TO SUCCESS)

1. IDENTIFYING THE OPPORTUNITIES THAT CONTRIBUTE TO THE VISION

The product evolution is one of the most valuable documents a forward-thinking organization can make. It is the essence of the road map to success. It defines where the company plans to go, in what time frame, and how it will get there. In the case of planning the content of a well-considered flowchart can be worth more than the strategic plan itself, since this defines the product path to be taken to the future. The strategic plan then defines the financials and the standing of the firm against the competitors. The first step therefore is to identify the opportunities that contribute to the vision of the future and focus on only them. If the vision of the future is agreed to and locked on to, then reject those opportunities that cannot contribute to it, regardless of how lucrative they may first appear.

2. MIX OF PRODUCT AND SERVICE OFFERINGS

The flowchart should not be limited to hard products. They should also include the service aspect of the business. In addition, the mix of products and services should also be projected, as this will define infrastructure requirements, finances, and personnel.

3. CONTRIBUTION TO REVENUE AND PROFIT

Where will the revenue and profit come from in the next 5 years? What percentage will come from new products and new services, and what will the margin structure look like? These questions, although not having to be answered, need to be reflected on to prepare a well-considered flowchart. Figure 2-10 illustrates the concept.

The chart in Figure 2-10 represents an attempt to factor profit contribution centers into the product evolution flowchart. The chart is split into two areas, namely products and services. The products can be existing, and new products to be added in the future. Services, likewise, can be existing and new. The material, labor, and burden components, together with the price and quantity, drive the contribution to margin analysis. By extending this chart, trends can be drawn by loading in historical data, and future mix analysis can occur by forecasting and planning new products and services. This will assist in assembling the flowchart as well as adding financial credibility to it.

PRODUCT AND SERVICES MIX

ITEM	PRODUCTS	MAT	LAB	BUR	COST	PRICE	MARGIN	QTY	CONTRIBUTION TO PROFIT	% CONT	% OF TOTAL
1	PRODUCTS										
2	#1	1	23	4	28	55	0.49	10	270	8.9	5.4
3	#2	2	24	4	30	67	0.55	10	370	12.2	7.4
4	#3	2	23	5	30	68	0.56	10	380	12.5	7.6
5	#4	2	25	6	33	89	0.63	10	560	18.5	11.3
6	#5	2	56	7	65	125	0.48	10	600	19.8	12.1
7	#6	2	45	8	55	100	0.45	10	450	14.9	9.1
8	#7	2	6	8	16	34	0.53	10	180	5.9	3.6
9	#8	2	7	9	18	40	0.55	10	220	7.3	4.4
10						% SUB TOTAL			3030	100	61.0
11											
12											
13	SERVICES										
14											
15	#1	0.5	39		39.5	75	0.47	10	355	18.3	7.1
16	#2	0.5	32		32.5	75	0.57	10	425	21.9	8.6
17	#3	0.5	24		24.5	75	0.67	10	505	26.0	10.2
18	#4	0.5	76		76.5	75	-0.02	10	-15	-0.8	-0.3
19	#5	0.5	87		87.5	75	-0.17	10	-125	-6.4	-2.5
20	#6	0.5	98		98.5	75	-0.31	10	-235	-12.1	-4.7
21	#7	0.5	12		12.5	75	0.83	10	625	32.2	12.6
22	#8	0.5	34		34.5	75	0.54	10	405	20.9	8.1
23						% SUB TOTAL			1940	100.0	39.0
						TOTAL			4970		

Figure 2-10. Product Services Mix Analysis

In The Tool Box a more comprehensive chart is available for your use in planning.

It extends the chart through 5 years and displays trends analysis also. This spreadsheet details the 5-year projection and/or history, summary, and trend analysis. Sheets 1-5 are details for each year, sheet 6 is a summary sheet, and sheet 7 is the trend analysis.

4. PRODUCT MIX ANALYSIS, NEW VERSUS EXISTING, AND TECHNOLOGY PATTERNS

The real value in the process is to get a "feel" for the trends, and the degree of sensitivity each item has, and how it affects the company as a whole. In a historical analysis, look for patterns or performance behaviors in products and businesses. In a future-planning exercise, decide which products and businesses will be the big contributors and which will create large variations in margin and profit numbers. This is a more objective way of evaluating risk factors associated with new technologies and the products that are generated from them.

5. MECHANICS OF THE PRODUCT EVOLUTION FLOWCHART

It is time to begin the product evolution flowchart development process. As discussed earlier, this chart will serve as a road map for the products that will be developed, acquired, or brand labeled. The product evolution flowchart, if prepared properly, can be a helpful tool in charting the future course of the company. By its very nature of specificity; it forces identification of the products and services that will be required to support the narrative of the strategic plan. It also forces product planning to the point of determining contribution to the overall financials of the strategic plan. Figure 2-11 is a typical illustration of a chart. It shows several things: the pathway that products will take in evolving, how they combine with others in establishing product platforms, and differentiation of products serving different market segments. It also establishes a time base for when all of these things will happen.

As shown in Figure 2-11, products 1-5 exist in basic form in year 1. The product evolution flowchart shows how these products can be combined from a features and functional-

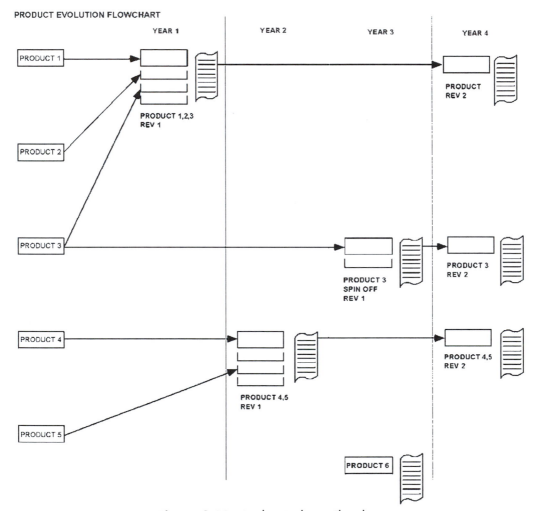

Figure 2-11. Product Evolution Flowchart

ity perspective. For example, products 1, 2, and 3 combine as a result of a development program started in year 1 and introduced in year 2. All of the features in all three products are combined into one product configuration or platform. The basic platform can be configured in one of three versions. This product implementation then runs as a product for years 2 and 3. In year 4 the next-generation product, which combines all the functionality and features of the different configurations, is launched. Newer technology and cost reductions make this condensing of products cost-effective.

In a similar manner, product three will spin off a series of product configurations in year 3 off of a basic platform. In year 4 a refined product will also be introduced.

To make the flowchart more meaningful, a narrative should accompany each product implementation, which explains the actions taken and for what reason. Also, the objective of the implementation should be included. The narrative should contain the following components: each should have a description of the product, a definition of the markets served, the degree to which the competitive stance is improved, the product configuration, and the expected level of technology employed.

In this example of a flowchart, product six could be added to the portfolio as a result of an acquisition or a brand label arrangement.

As an additional perspective to this product-planning exercise, it may be desirable to prepare several alternative flowcharts showing alternative options and pathways. In this way the best approach can be selected for implementation.

6. INTERNAL REQUIREMENTS, EXTERNAL REQUIREMENTS

Up until this point, we have examined the product-planning process as an external market driven exercise only. There may be instances in which a strategic requirement for a new product would be a cost reduction or a quality improvement. In these cases these internally generated reasons for new product development must require a narrative in the flowchart. From a general management perspective, each product must generate cash and profit. Existing programs that are in trouble, either from a prime cost issue or a cost of quality issue, need to be addressed.

7. FUTURE CATALOG OF PRODUCTS AND SERVICES, FEATURES, AND TECHNOLOGY

What will the future product catalog look like? What types of technology will be employed? What level of complexity will it involve. Will training be a marketable commodity or will it be assumed as part of a sales transaction?

This process should actually result in a future company catalog with the complete offering included. The catalog will also drive a thematic narrative of company operations and services in the future. By assembling this catalog, one can see if the strategic story holds together when the products are attached to it.

8. PROJECTION OF COMPETITION, COMPANY, FRAMEWORK, PRODUCTS, TECHNOLOGY, AND POSITIONING

With all this great planning, your company is positioned to capitalize on the future. The competition is not standing still, however, or at least not all of them. Consequently, it is a good exercise to project what they will do based on what you currently know about them. This should be done on the product evolution flowchart and the 5-year future catalog. If for no other reason, it will force consideration of their plans and allow you to evaluate the risk of your plans resulting in future product offering that are only at parity with the competition, or worse yet, nearly obsolete by comparison.

9. WHO, WHAT, WHERE, WHY, AND HOW?

A good plan requires specifics of the mobilization to be clearly articulated. What are the specifics of the product evolution flowchart? This product was developed and introduced by such a date... This technology was absorbed into the organization by this date... Additional channels of sale was established, in place, and functional by this date. These are the specific statements that actions must follow to accomplish results. The flowchart, the narrative, and the 5-year future catalog must answer the five questions of specificity.

10. ORGANIZE STAGES

Finally, organize the action items and the stages of the plan so that it all ties together into a cohesive set of specific objectives and their supporting action items. Look for commonality of purpose and resources to optimize the work needed.

11. OPTIMIZE AND MAXIMIZE BENEFIT FOR DOLLARS SPENT

Resources for development are always in demand. In addition, the resources for development represent only a small part of the overall investment in a new business opportunity. It is therefore in the company's best financial interest to maximize the results for each dollar spent. Financially sound, well-run, small companies consistently deliver the maximum benefit for every dollar spent. They must in order to survive. Larger companies run the risk of failing to do so, because of politics, lack of consistent information, and communication flow. In either case the cost of lost opportunity and the hard cost of research and development demand that the development team maximize the results of dollars spent.

BUSINESS CAPABILITIES

1. YOU HAVE THE HEART; DO YOU HAVE THE TOOLS?

Now that you have defined the vision, the pathway, and the products and services to get your company to the future, the next step is to determine if you are equipped to execute the plan. What areas of the organization need remedial assistance or additional infrastructure to be a contributing factor to success? There are specific tools required to execute the program, and an introspective look at the strengths and weaknesses is in order. The corporate capabilities need to be integrated and coordinated at near full value. If one or a set of these required capabilities is weak or missing, it must be addressed before the program begins. Failure to do this will result in a resource-starved program, dooming it to failure of timing, performance, or momentum. Examine each area of the business to ensure it can support and contribute the necessary resources to effect success of the program. If the organization is lacking, fix it now!

2. CAN YOU GET THEM OR AT LEAST HAVE ACCESS TO THEM? THE OBJECTIVE IS TO HAVE A CUT SHEET OF LUMBER, NOT TO OWN THE SAW!

Is it possible to obtain the tools for the organization from other sources? In the next section, a variety of ways will be discussed to allow companies to compete in nontraditional markets by partnering for capabilities. As the title states, the objective is akin to obtaining a cut sheet of lumber, not to necessarily own a sawmill. If this is the case, don't go looking to buy a sawmill, unless you use so much cut lumber that it is in your firm's best interest to invest in capital to become a low-cost producer, and have control of the operation. This would only take place when usage drives the make or buy decision.

Obtain the tools that make most sense to the operation and the new product development.

PARTNERING FOR CAPABILITIES

1. DON'T BE PROUD, NOT INVENTED HERE, BE PRACTICAL

Corporations are not inanimate entities; they are comprised of people. People have needs, emotions, and biases. People feel threatened and react to those feelings. Many a logical, forward thinking, and beneficial idea has been sidestepped rather than embraced because of these factors. Consequently, the notion of securing assistance or know-how from outside of the organization often does not receive wholehearted support. It is the manager's responsibility to ensure the organization rises above these factors to embrace the tools required for success.

2. GETTING THE TOOLS AND REMEMBERING THE OBJECTIVE

There is a corporate trust in new product development that is unwritten, and that exists between senior management and the new product development champion. It is a covenant whereby management will accept the developer's leadership in exchange for a certain amount of due diligence in prosecuting their program. Consequently, when obtaining the tools needed for a program, it is necessary to exercise care in their procurement. Get only the tools needed. Get the tools as a means to the new product end. It is a vocational hazard to quickly get in the habit of procurement on a grand strategic scale, and bypassing the critical mass just required for the program. The operative point is that these tools must be absorbed and amortized by the program, and so restraint is best practiced at this time.

In the course of procurement, it is wise to procure in a forward-thinking, forward-compatible manner, and to that end, where it is a small, incremental cost to prevent obsolescence of equipment, expenditures should be made freely.

When getting the tools, remember the objective. In the same way you engage the customer to assess where they are, it is a requirement to engage the partner in a more detailed way. Find out their plans, aspirations, and desires. Only in this way can you determine the best deal to strike for both parties.

3. BACKUP PLANS, EVALUATING WITH A PARTNER, CHECKING THE PARTNER'S FUTURE INHOUSE CAPABILITY

"The best laid plans of mice and men often go awry." This saying gives an implicit definition of the scope of new product development, but especially in the framework of partnering. Knowing this is an accepted fact, make sure to plan contingencies within the partnering arrangement.

An unfortunate fact of life is that unforeseen driving forces can alter a partnering relationship for reasons unrelated to the development, its progress, or its potential. A financial hardship may present itself to the partner of your company and force abandonment of the program and their contribution.

There are also instances in which management focus may change within the partner's management structure. A different agenda may be embraced, or a separate alliance that may potentially overshadow or conflict with the partnership may have been reached. Therefore it is best to identify alternative and contingency plans at each stage of the process of partnering. As the deal is negotiated, constantly ask yourself, "What if this doesn't work out, what will be my options to achieve the new product development objective?"

One possible alternative may be to plan on absorbing the technology that the partner has, into your organization after the term of the partnership. In many cases a relationship where this is identified at the onset, eventually may be the best for both companies.

The initial interplay is therefore very telling in how the relationships will progress. Carefully watch the partner's behavior in several situations. Assess their loyalty to your company, the program and the team of personnel involved. Determine if they act in a long-term, decisive manner or if they operate in a day-to-day tactical mode with loose focus. It is imperative to assess this quickly in certain programs replete with uncertainty, as you will need to evaluate the difference between time lost in overcoming problems or chasing other opportunities before completing tasks at hand.

Take care to set up a communications infrastructure that preserves secrecy where necessary but encourages a free exchange of knowledge where it enhances the effort, relationship, or the program.

Also set up communications at different levels between the two firms as shown in Figure 2-12.

As the figure illustrates, your firm and the partnering firm should set up a communication protocol that is multilevel. Patterned after a hierarchy, bilateral communications can take place at the operative level. For most issues, if nonconcurrence occurs at the lower level, it can be arbitrated at the next level. Senior management preserves and orchestrates the relationship between the two firms. Avoid single-point communications, as it is only as strong as the individual at that point. Multiple communications can also allow information to be brought to light in a non-biased way, whereas the single-point entry can be prone to one individual's biases.

4. 100% OF NOTHING VERSUS 50% OF SOMETHING, FOCUS, TENACITY

The basic concept behind partnering is that together, two entities can achieve more than either one could alone. The partnership fills in the gaps possessed by the individual players. Reiterating again, remember this objective: 100% of nothing is nowhere as beneficial to your company as 50% of something.

The very structure and operational dynamics of a relationship between two companies contributes to diffusion of focus. It will soon be apparent that many interesting things

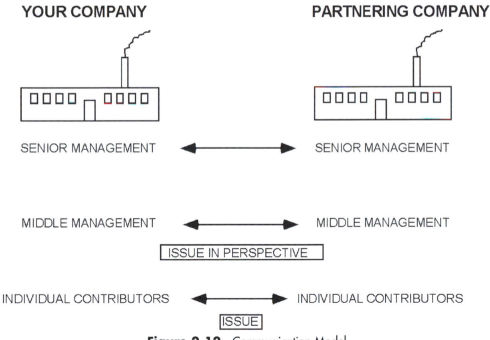

Figure 2-12. Communication Model

surface from the collaboration, not necessarily related to the program. It is the manager's responsibility to refocus the companies on task.

In addition, the partnering relationship may not work out for every company. There has to be a certain corporate appreciation for the differences between the players, their history, and very importantly, relative importance of the project to each firm.

As you architect an arrangement, ensure the players selected have the patience for the partnership.

5. CONCEPTS ON CORPORATE STRATEGY

There are differences in various companies' corporate strategies. Although the following model could have been presented in the strategic planning section in Chapter 1, it has more reference value here where it can be recognized in a potential partner during negotiations. Presenting a "hard and fast format" for corporate strategy in the strategic planning section would tend to preempt creative thinking by encouraging comparison and locking onto a specific type prematurely. When contemplating or planning a joint venture, brand label arrangement, or an acquisition, it is critical to determine the seller's corporate strategy. Although most companies espouse some sort of strategy, some do not practice it effectively. In some cases this depends on the size of the company, as well as the ownership structure. For example, a small privately owned company will execute a strategy differently than a large firm. This is because of three factors:

1. Influence of top management in the process
2. Sensitivity to changing market conditions
3. Inertia of the organization

A large company sets it pathway and starts execution. It wields its inertia to start in the chosen direction. Changing market conditions generally do not affect this. This is the appeal of the smaller firm, which can react to changing market conditions. This resiliency, however, is a danger to executing a plan because management may change direction more often thereby negating any momentum. For the past 20 years larger firms have acquired smaller firms. It is incumbent on the acquirer to understand these dynamics.

It would be tempting to create a chart of the various scenarios of acquisitions, joint ventures, and brand label arrangements to summarize the corporate strategies of each, and how to react to them; however, each case is its own set of circumstances that directly affect the dynamics of a deal. Consequently, each must be evaluated and a pathway must be charted through the circumstances.

A large company with healthy long-term debt may sell a division off at some price not necessarily reflective of value, and recognize the sale on financial statement. A smaller firm would execute the transaction totally different. It is important to investigate the driving forces behind the deal as you are creating the deal. This can have the most payoff when negotiating.

Some of the factors that come into play are as follows:

- Large versus small
- Public versus private
- Technology developer versus pure manufacturer
- Product offering, which is commodity versus service
- Profitable versus not profitable
- Management under fire versus not under fire
- Availability of cash
- Competitive nature of the "deal"
- Perception in the marketplace
- Perception of weakness
- Limiting factors on one firm in the deal
- Recoverability index for both companies should a deal go bad

These frameworks also factor into the success of a launch of a new product.

This list outlines the various concepts and practices of corporate strategy. It is desirable to study the philosophy of each of the types and to become familiar with them as you negotiate the partnership arrangement. In this way it will be easier to evaluate the partner's motives, pressure points, and actions.

MODEL FOR BRAND LABELS, JOINT VENTURES, AND ACQUISITIONS

1. PERSPECTIVES ON PARTNERING

A brand label, joint venture, or acquisition for purposes of new product development can be a successful venture. These relationships allow the company to gain access to markets quickly and with products that are relatively on target.

Partnerships, however, do not last forever. They are simply a means to an end. Consequently, initial plans in negotiating a venture should include plans for how the company will absorb the technology, buy the company, or continue the partnership past the term. Failure to adequately plan this at the onset will result in consequences at the end of the term, with minimized alternatives and options. After all the effort to initiate a development and create a business, it can be devastating to lose control or lose the business because of a lack of transition-planning at the beginning of the negotiation.

This section will present a model for brand labels, joint ventures, and acquisitions, from a product development perspective. It can be used as a planning guide for initial discussions and as a check-off list for implementation.

2. LEVELS OF INVESTIGATION AND DISCUSSION

There are four basic levels of investigation required to understand another partner's business. Depending on the framework of the partnership (i.e., brand label, joint venture, or acquisition), the more in-depth the investigation should be. An acquisition will require more information than a brand label arrangement, for example.

The levels of investigation are arranged therefore in increasing depth into the partner organization. Level 1 is cursory, whereas level 4 delves into the core of the company, and finances and structure. The following is a summary of the levels:

LEVEL 1

A. GENERAL

The general concept behind level one is to investigate if there is the feasibility of creating a deal between the two companies. This section allows a summary of the opportunity to be documented.

B. BACKGROUND

The background outlines the market driving forces and events that lead up to the identified opportunity.

C. MARKET INFORMATION (SECONDARY)

The market information cited here is secondary and is obtained from trade litera-ture, periodicals, and industry-published data. It is differentiated from primary mar-ket data by the fact that it is information obtained indirectly rather than through primary customer contact. It is useful as trend data and initial substantiation, but cannot be used as verifiable market data. That must come from primary customer contact.

1. SIZE, DEMOGRAPHICS

This allows the company to quantify the opportunity and breakdown location and segments.

2. MARKET POSITION

The market position outlines the partnering company's present position in the market. It will allow your company to evaluate the market's perception of the value of the technology.

3. MARKET PLAN

The market plan outlines the steps that will be taken to capitalize on the market opportunity.

D. CONTACTS

Who are the company contacts?

1. HIERARCHY

What is the organizational hierarchy?

KEY PLAYERS

Who are the key players in both companies and who are the dealmakers and breakers?

E. STRUCTURE OF THE DEAL

What is the basic structure of the deal? This should be constrained to a single-page explanation that is understandable to all concerned.

1. DIAGRAMATIC

Diagram the relationship between the two companies and show what goes where, and where everything comes from. If this cannot be done easily and understandably, recheck the thinking of the opportunity and the partnering relationship. Something is not right.

2. PRODUCT FLOW

Establish where the product comes from and how it gets through the infrastructure to the customer. Place notations on the diagram to note key points, areas of concern, and potential failure areas.

3. SUPPORT

Show how the product will be supported during the relationship. If you can complete this in a relatively short amount of time without major conflicts, the basic tenets of a deal are in place to build on.

LEVEL 2

A. RELATION TO OVERALL PLAN (STRATEGIC)

Establish how the proposed partnership arrangement firms into your company's overall strategic plan. Is this a means to gain expertise and immediate entry to a marketplace, and if so, what is the plan for after the term of the agreement?

B. COMMERCIAL

Take care in outlining the commercial implications of the arrangement. If a careless analysis is performed, the entire objective of the program could be missed.

1. COSTS AND PRICING

What are the projected costs and prices? Is there enough margin for each of the players? What will happen if the marketplace puts pressure on prices, exacerbating the margin pressure?

2. FEATURE/BENEFIT MIX

Carefully outline the feature and benefit matrix and tie it to price. Do you have the properly configured platform to take to market with your partner? Is the product properly positioned in the marketplace at the time of product introduction?

3. RAMP UP PLAN, VOLUMES (REASONABLE AND ACHIEVABLE?)

Check to make sure that the two firms are capable of the market driven ramp up. Is there a manufacturing infrastructure to support the projected business levels? Are the business-generated unit volumes required to justify the program reasonable and achievable from a manufacturing perspective?

4. PRIMARY MARKET INFORMATION TO CHECK SECONDARY

Be diligent in performing primary market research to verify feature acceptance, pricing, value, and trends in the marketplace. Does the product need to be changed to fit the updated market criteria?

5. BEST-CASE SCENARIO AND EFFECTS

One of the best ways to determine feasibility and reasonableness is to outline a best-case scenario and the assumptions needed to achieve it. Also project this best-case success onto the organizations (both yours and the partner organization) to determine the effect it may have. Will success change the philosophy of the agreement and the principles? Will the contributing organization feel well served in the transaction, or will they tend to regroup and exercise an alternative strategy?

6. WORST-CASE SCENARIO AND EFFECTS

Likewise, create a worst-case analysis to determine downside effects. Also include the effect of missed opportunities because of time pressure. This can be used to create a sense of urgency in situations that may need arbitration between the parties. Also take time to examine effects of a worst-case analysis in terms of actions that may be taken by both companies. If the contributing company senses that the venture will not work out to plan, they may have a tendency to regroup and refocus effort elsewhere, which will further jeopardize the venture.

7. MOST LIKELY SCENARIO AND EFFECT

Finally, determine the most likely scenario that could occur. Hopefully this scenario will generate the best-desired result for both parties. Take time to also project your plans and proposals for after the term of the agreement, based on the success. This can provide the backdrop for future negotiations.

8. PRODUCT LINE ANALYSIS, SECOND GENERATION

It is also helpful to project the second-generation product analysis at this time. It forces a realistic, critical examination of the features and benefits to determine market fit. If a fit is lacking certain features at this time, the second-generation analysis will highlight it and allow a judgment to be made.

9. ORDER FLOW

At this time it is also beneficial to generate a flowchart of product and order flow. Design the process by which product orders will be entered and prosecuted within the organizations and delivered to the customer. If there is partnership beyond design and manufacturing through to the sales channel, then it becomes necessary to outline the protocol and flow of product here.

C. TECHNICAL

This area deals with the technical aspect of the arrangement. There are a host of elements that must be orchestrated between the two firms to effectively meet the customer's requirements, and this section will outline those requirements.

1. MEETS FUNCTIONAL DESCRIPTION

Does the product, as designed, meet the functional requirements of the marketplace? Does it pass the basic features and performance criteria at a deliverable cost?

2. MEETS RELEVANT STANDARDS

Does the product, as designed, meet the relevant standards that govern the marketplace, or is redesign necessary to incorporate measures to meet these standards?

3. MEETS BRAND LABELED DOCUMENTATION

Has the required documentation been effectively prepared to be acceptable to the marketplace? Is it consistent with the corporate requirements of the brand labeler?

4. MANUFACTURABILITY STUDY AND COST REDUCTION

How about the manufacturability of the product? Are the manufacturing strategy and product platform open ended or close ended from an enhancement and cost reduction point of view?

5. QUALITY SYSTEM ASSESSMENT

What is the level of quality system in the contributing and utilizing corporation? Are the two systems consistent, as this is required if one corporation is to design and the other is to produce? Are the systems capable of continuous improvement and tracking each other? This is important to resolve at this level because it is virtually impossible to reconstruct one in the midst of a product problem or recall situation.

D. RISK ANALYSIS

Finally, as part of a level 2 investigation, it is desirable to perform some numerical risk assessment in terms of finances. Also, what is the qualitative risk in terms of reputation and positioning in the marketplace?

LEVEL 3

A. REVIEW BUSINESS PLAN IN RELATION TO ORIGINAL ESTIMATES AND MANAGEMENT TRACK RECORD

A level 3 assessment will evaluate the partner company's ability to set plans in motion and execute them to completion. It is important to know this perspective, since the proposed venture will depend on the ability to be decisive in their action. As an individual manager, it is also a good idea to do a self-assessment at this time in order to evaluate your own company's ability to carry out a plan.

B. DETAILED REVIEW OF FINANCIALS AND PERFORMANCE DATA

An assessment of level 3 will involve a detailed analysis of the financial performance of the company. This is the level where a joint venture would require a form of due diligence investigation. Since a joint venture would require significant deployment of intellectual and hard assets, your company must perform enough of an investigation to protect these assets.

1. INCOME STATEMENTS

Look at the income statement to see trends and cyclical business cycles. Cyclical activity generates cyclical performance and rearrangement of priorities over time. This reprioritization can affect progress on a project.

2. BALANCE SHEETS

The balance sheet can indicate the structure and financial strength of the potential partner. Lack of financial strength may cause delays, and possibly require additional contribution of cash on your company's part to complete the program.

3. FUNDS FLOW

If possible, have a funds flow statement drawn up for your review. The funds flow indicates how the money is spent by the organization. Determine how it comes in and where it goes out. This will go a long way toward determination of waste and degree of financial efficiency of the partner organization.

4. STRUCTURE OF ORGANIZATION, NUMBER OF SHAREHOLDERS AND WHO THEY ARE, INTERRELATIONSHIPS, LENGTH OF INVESTMENTS, INTERVIEWS, AND PROFILE OF INVESTORS

Get a feel for the ownership structure of the firm. Who are the key decision-makers on finances and policy? Determine if there will be any planned change of ownership within the term of the deal. As with much of this information, it may not be located in a single package, neatly wrapped up for your review; however, by listening carefully and gathering secondary information through third-party services (i.e., D&B, S&P, etc.), you can amass a fair amount of information.

5. REVIEW ACCOUNTING AND SALES ORDER ENTRY SYSTEMS FOR ACCURACY, COMPATIBILITY, CHANGES

Review the accounting system for accuracy and consistency. Are the expenses recognized similarly from year to year? Is the accounting system accurate? Does the inventory system reasonably track the value of the inventory without adjustments every year-end? Can the sales order entry system be aligned with your system with some means of internal controls? Are the two systems effectively compatible to prosecute the deal and keep track of the required data? What is the measurement of the cost of quality, and is it consistent with your system? The thought here is that the two systems need to be compatible enough to measure performance of the venture to the satisfaction of both parties.

6. INVESTIGATE COST-REPORTING SYSTEM AND COST ROLLUPS (TARGET VERSUS ACTUAL)

Finally, investigate the specifics of the cost rollup system so that product costs are kept accurate from year to year. This will ensure that there will be no disagreement on profits or margins in case the agreement calls for shared margin or percent of profit. If possible, it would also be of value to your company to assess the original and actual product costs historically within the partner company, as this will demonstrate their track record in achieving a design-to-cost regimen.

LEVEL 4

A. COMPANY FITABILITY: YOURS, THEIRS, AND OVERALL PLANS OF BOTH

A level 4 investigation is a more comprehensive analysis of the partner organization from an acquisition perspective. It is part and parcel the valuation of the company that will be purchased. It is the measurement of hard assets, know-how, and the ability to please customer and widen its presence in the marketplace.

However, a company is not a thing to be weighed and bought by the pound, so to speak. It is a living, dynamic organization comprised of nonlinear, emotionally driven and often unpredictable people, whose diligence and effort contribute to the numbers being evaluated dispassionately. As such, people have hopes and dreams and goals. How are these transferred into the organization? What are the goals of the organization and how do they compare with your collection of contributors? Knowing this is essential to effecting a good marriage between the two companies. This effort must transcend the new product development portion of the agreement and permeate the relationship.

B. VALUATION OF COMPANY

The decision has been made to purchase the company, and now it becomes necessary to evaluate its worth. In many ways worth is based in the eye of the purchaser, and desired value on behalf of the seller is based on circumstance. In simplest terms, there are a minimum of five ways to value a going concern: owner's investment value, a balance sheet analysis, a rate of return, income statement analysis, buyer investment valuation, a liquidation alternative, or finally, a replacement value calculation. Each of these circumstantial ways may affect the perceived value of the following items; however, it is still necessary to have a starting point for each of these.

1. ASSETS, CASH PROPERTY, PLANTS, AND EQUIPMENT

These hard assets comprise the visible portion of the balance sheet where a buyer can set their hand on tangible things. Although these items are procured and depreciated with time to a very low value, they still represent a significant portion of the business's value at the time of acquisition. Watch out for overstated nontangible assets in a deal.

2. LIABILITIES

These represent the obligations of the business. They are the bills and expenses that your company will incur after the transition.

3. GOODWILL

This intangible represents the reputation of the organization in the marketplace, like brand names and longevity of relationships with customers, and is often used to accumulate the dollars beyond hard identifiable assets in an acquisition. This is true where the purchase price exceeds hard asset value alone.

4. ORDERS

This is valuable from a study point of view to see where the orders are coming from and what the orders consist of. By examining the orders of an organization, several useful things can be learned. Items such as number of orders, line items per order, purchasing authorization, and pricing and discount classifications, all tell a story about the business in question.

5. BACKLOG

The backlog is useful in determining the inertia of the company. It provides the stabilizing influence during the transition. A large backlog gives the acquiring company some breathing room in making changes and absorbing the acquisition.

6. RECEIVABLES

This represents the monies owed to the company. A close examination of the accounts receivable records will indicate how the company does business. Are terms extended to customers as an inducement to offset product shortcomings? Are late payers overlooked as an appeasement for product issues? Are the receivable days out of step with the industry norm?

7. INVENTORY

By and large, the inventory is one of the most uncertain aspects of an acquisition. Of all of the tangible assets, the inventory is the only one that is recoverable out of operations. Inventory is of very low value on the street or to anyone else other than the company. It must be called out by a clean set of product documentation and be useable as delivered without significant addition of labor to recover its cost.

8. HEADCOUNT

How does the headcount compare to industry norms? What is the dollar contribution for both direct and indirect employees? What are the trends in headcount, past and projected? What is the average longevity of the employees? A lot of information is available by researching this area thoroughly.

9. TRENDS ANALYSIS AND PROFORMA PROJECTION

Finally, what does the trend analysis say about each of these measurements? Does it foretell a growing, vital organization or a stagnant one? What has the organization been through recently, and how are they positioned for change?

C. STRUCTURE THE VALUATIONS WITH RISK ANALYSIS

As a purchasing decision is being made, a summary of the valuations and options, and their substantiating data, needs to be reviewed. With each one a risk analysis needs to be incorporated as part of the review. This risk analysis will shade the valuation with a bit of realism in pricing each alternative.

D. COMPARATIVE ANALYSIS WITH INDUSTRY AVERAGES

Every company is different and must be evaluated based on its own merit and future contribution to your company. It is therefore only a measurement and not a valuation to compare industry averages between companies. In the acquisition arena, when the reason for the acquisition is to generate a new product, the comparative analysis and industry averages assessment can only be used as a guide.

E. SET PRICE FIVE WAYS

As mentioned earlier, there are five basic ways to value a business. They are as follows:

1. Owner's investment value

2. Rate of return

3. Buyer investment value

4. Liquidation value

5. Replacement value

Each of these has a justification for the valuation and each can be used to modify a valuation; however, there is an additional consideration that must be given when acquiring a business, when the primary driving force is to obtain intellectual property, trade secrets, or know-how.

If additional new business is to be built on the existing business, then the complexion of the original will change. Investments may have to be made, and profit structures and cost structures may be different. In these cases, it becomes necessary to perform a more in-depth analysis of the business and the proforma investment in it, not only from an acquisition perspective, but also from the incremental investment perspective.

F. SET TERMS

Finally, set the terms of the deal. If the partner company must, as part of the deal, generate a new product, then the sales price and terms of inflow of cash will alter accordingly, as the intellectual value does not exist until these programs are well on their way to completion. When constructing deals in these areas of new product development, linkages and demonstrable results take on new importance as moneys exchange hands.

1. MISCELLANEOUS TOPICS TO AGREE ON

There are several miscellaneous topics that should form the basis for negotiations as a deal is put together. Although each is not a deal maker or a deal breaker, they should be addressed from the perspective of completeness, in that resolving the issues will make the deal go that much more smoothly. Each of the topics along with a short explanation is included in the Tool Box for your reference and should serve as a check off list in negotiating a deal.

2. MATRIX OF ANALYSIS

Figure 2-13 represents a matrix of analysis for the different product partnership arrangements discussed in this chapter. It serves as a matrix of essentials for investigation in negotiating a brand label, joint venture, or acquisition arrangement. Although this list is not presented in its entirety, it does serve as a useful tool in determining scope of investigation in an arrangement.

3. HOW TO HANDLE A FORCED ARRANGEMENT

There are times when the new product manager may not be part of a brand label, joint venture, or acquisition decision, but may be required to prosecute a new product development as part of it.

Often decisions are arrived at rather quickly without a lot of thought given to the details. If this is the case, there are several questions that you as the new product manager need to clarify as your company proceeds down this pathway.

There are five major areas of concern when a deal is forced into the organization. As a manager, it is in your best interest to resolve the issues defined in the five areas as part of the overall plan. The five areas are outlined in Figure 2-14a and a diligence checklist is presented in Figure 2-14b.

A separate Brand Label Checklist is available in the Tool Box for your reference. Additionally, a Diligence Checklist is also available.

MATRIX OF ESSENTIALS IN A BRAND LABEL, JOINT VENTURE, OR ACQUISITION ARRANGEMENT

	BRAND LABEL	JOINT VENTURE	ACQUISITION
LEVEL 1	X	X	X
LEVEL 2	X	X	X
LEVEL 3	N/A	X	X
LEVEL 4	N/A	N/A	X
MISC TOPICS	X	X	X

Figure 2-13. Matrix of Essentials

ISSUES TO RESOLVE IN A BRAND LABEL ARRANGEMENT

MARKETING

1. HOW WAS THE MARKETING STUDY DONE?
   PRIMARY, SECONDARY
2. WHAT IS THE PRODUCT COST STRUCTURE
3. PRICING, FEATURE, AND COMPETITIVE COMPARISON
4. COMMERCIAL FEASIBILITY
5. FIT WITH OVERALL STRATEGY

DEVELOPMENT

1. PROFIT AND COST SENSITIVITY
2. FUNCTIONAL TESTING
3. QUALITY TRACK RECORD
4. TECHNICAL FEASIBILITY/ANALYSIS
5. STANDARDS TESTING AND CONFORMANCE
6. DESIGN VALIDATION

CHANNEL ISSUES

1. EXCLUSIVE ARRANGEMENTS AND NON-COMPETE
2. RESELLER REVENUE STRUCTURES
3. NEW MARKET IDENTIFICATION, DEVELOPMENT,
   AND PROMOTION
4. INTENDED ROUTE TO MARKET

PRODUCTION

1. MANUFACTURING PLAN
2. MANUFACTURING VENUE
3. CAPITAL EQUIPMENT
4. MANUFACTURING DOCUMENTATION
5. CHANGE NOTICES AND SUPPORT
6. VENDOR LINKAGES
7. PRODUCTION TESTING

QUALITY

1. QUALITY SYSTEM AND REPORTING
2. CORRECTIVE ACTION SYSTEM
3. TRACK RECORD FOR SIMILAR PRODUCTS
4. WARRANTY COSTS
5. FIELD FEEDBACK SYSTEM FOR IMPROVEMENTS
   QUALITY EVENT CHART

Figure 2-14a. Brand Label Checklist

4. KEEPING BALANCE INTERNALLY AND EXTERNALLY

Managing a joint development program, or even a brand label program, is an exercise in pleasing many principles. The marketplace has requirements that need to be met, the partner

DILIGENCE CHECKLIST					
ITEM	DESCRIPTION	RESP	SUBSTANTIATION	DATE 1	DATE 2
1	PRODUCT OPPORTUNITY FITS SCOPE OF CORPORATE STRATEGY				
2	CONSISTENT ORGANIZATIONAL SYSTEM FOR EVALUATION				
3	CONSISTENT CRITERIA				
4	DOCUMENTED AND FOLLOWED PROCEDURES				
5	LINKING DEPARTMENT OBJECTIVES				
6	SOUND MARKET PLANNING				
7	POSITIONING				
8	SEGMENTATION				
9	CAPITALIZATION				
10	PRICING				
11	PACED DEVELOPMENT TO CAPTURE OPPORTUNITY				
12	PRODUCT DISTINCTIVENESS				
13	SUPPORTING EVIDENCE OF OPPORTUNITY				
14	SIZABLE MARKET WITH RESEARCH AND FORECAST				
15	PRECISE PRODUCT DEFINITION				
16	SOLID DESIGN CRITERIA AND REVIEW				
17	TECHNICAL FEASIBILITY IN THE COMPANY				
18	ACHIEVING TARGET FACTORY COST OBJECTIVES				
19	PROJECTED COMPETITIVE RESPONSE				
20	STABILITY OF THE TEAM/COMPANY				

Figure 2-14b.

company has needs, and your company has needs. At times the demands of each may conflict with internal pressures, and the manager needs to arbitrate issues.

Keeping balance of these requirements is essential to the program's success. The interrelationship between the issues and the individual factions needs to be identified at the onset and managed throughout the term of the deal. If one area fails to make its own share of the sacrifice, the relationship eventually will be in jeopardy.

Since you're charged with the responsibility of acting in your company's best interest, and the partner company is a participant, your role is one of chief negotiator, and each issue must be resolved to the company's and market's satisfaction in a timely and decisive manner. Don't waste time to market arbitrating and trying to please everyone. Execute the deal and bring the product to market in a form that will be accepted by the market.

5. DEALING WITH ENTREPRENEURS, LICENSING, AND TECHNOLOGY TRANSFER

Dealing with technology in a raw, unprocessed form, as in the case of entrepreneurs, can be difficult. The challenge occurs when taking raw technology and imbuing it into a manufacturing organization. There are differences between the entrepreneur and the manufacturer that cloud the issue, in the perception of completeness and manufacturability. Inventors "solve a customer's problem" by creating "designs to suit your needs." The manufacturer serves a cross section of a market, builds non-changing product in volume, and resists change by its desire for uniformity demanded by its structure.

Both the entrepreneur and the manufacturer have similarities that must be preserved, such as desire for quality designs, commitment to customer satisfaction, and commitment to the venture. There are differences, however, that must be resolved, such as designs must be frozen and manufacturable, problems must be addressed swiftly together, and both must diligently work toward equitable participation in the venture.

BENEFITS OF SHORT-TERM AND LONG-TERM TRADEOFFS

1. DOUBLE-EDGED SWORD

When sculpting a new company by executing a long-term growth plan, there is a balance that must be struck between short-term and long-term tradeoffs. The long-term view requires investment and time not recognizable in the short term. This means that the manager and the organization must exercise patience in achieving the long-term objective and be decisive in preserving the investment during those times when it is financially tempting to terminate the expenses.

2. STREET SMARTS

Negotiating and prosecuting a brand label or joint venture agreement will require a certain amount of street smarts in navigating both companies through the relationship. It will require cognizance of each company's allegiances, pressure points, sensitivities, and goals out of the deal.

If you watch where the monies exchange hands, it will be easier to understand the dynamics of the relationship.

3. NOT GETTING HELD HOSTAGE

Administering a venture between two companies can leave your company vulnerable, in that a commitment has been made to see the venture to completion. Alternatives are no longer as beneficial or available after you have locked on to the partner company. Consequently, you run a risk of being held hostage with their technology and know-how. Being held hostage can have long-lasting negative effects that transcend a single obstacle or crisis.

The only way to counteract an attempt to be held hostage by an employee or a partner company is to have a good alternative waiting in the background to take over. In fact, you may want to provide for this at the onset in the contract clauses.

In any event, being held hostage must be mitigated by having alternative means to prosecute the program, and demonstrating the resolve to exercise it.

4. GOOD DEAL; WIN-WIN NOT TO TAKE ADVANTAGE

It is said over and over that a good deal is the result of a win-win situation for both parties. It is fundamentally based on the fact that if both parties do not benefit from the venture, sooner or later, the deal will fall apart. Neither party will knowingly continue a relationship indefinitely without tangible benefit.

5. MAKING IT LAST

Given that win-win is the preferred result, the best way to preserve a deal, or to continue a deal that has come to term, is to structure additional benefits for both parties. It takes a significant effort to make a deal work in the first place, and if one can be extended, additional work can be saved; leveraging the ventures now adds to the worth of the company by continuing to achieve results without significant additional effort. Additionally, the individuals involved begin to know each other and overall communication improves.

6. BENEFITS: QUICK ACCESS, EXPERIENCE, WILLING PARTNERS, AND INSTANT CRITICAL MASS

To reiterate the benefits of a venture: your company gains quick access to a market and benefits from the partner's experience in the market and/or technology. With a willing and cooperative partner throughout the term of the venture, you can add critical mass to your company for a fraction of the investment cost.

7. DANGERS: "CAMEL IN THE TENT SYNDROME": EXPOSING YOUR POSITION, BRIEFING DEVELOPMENT PEOPLE

If the partner company is bringing a product technology to the deal, and if your company is providing the marketing and channel, remember the "camel in the tent" syndrome. This describes a competitive situation whereby your company exposes the customer base to the partner company, who has the means to take the customers away from you.

It is not unheard of to have the partner company secure customers directly after a time and effectively remove you from the transaction, thus allowing the "camel in your tent". It is most dangerous when your company is not the manufacturer.

In this case the manufacturer has a built-in advantage financially, by the minimum margin your company would use in selling the product. By bypassing your channel, they can afford to sell at your cost and still secure the same margin as if they were selling to you for resale. Figure 2-15 illustrates the issue.

As can be seen in Figure 2-15, the partner can bypass your company to get to the customer and effectively deliver the same goods at an advantage over your company, which

CAMEL IN THE TENT SYNDROME						
CHANNEL	PARTNER COST	YOUR COST	YOUR MARK UP	END USERS PRICE	PARTNERS MARGIN	YOUR MARGIN
PARTNER	100.00	200.00	285.71	285.71	100.00	85.71
DIRECT	100.00	BYPASS	BYPASS	200.00	100.00	0
		" PARTNERS " ADVANTAGE		–85.71		

Figure 2-15. Channel Contention

preserves their manufactured margin. When negotiating a venture with a partner company, legislate steps in the agreement to ensure they remain "partners."

8. PLAN FOR THE FUTURE

A final note on the subject of brand labels and joint ventures. It is important to keep in mind that with the exception of the acquisition case, a partnership deal doesn't last forever. Two companies will eventually go their separate ways after a venture, because each will develop different strategies, needs, and goals. The two companies originally partnered as a result of expediency, and that expediency will not continue forever. Therefore it is wise to plan for the future by absorbing the technology and manufacturing know-how (or whatever the partner company brought to the deal) to position yourself for after the term of the venture. Participate in the venture at 100% effort and make it work for both parties, then plan for the future. Remember the objective is to create long-term growth and value for your company, so it's incumbent on the venture manager to provide this continuity.

SUMMARY

This chapter's objective was to familiarize the reader with the basic concept of market opportunity. A market opportunity was related to the solving of a customer's problem and to making a profitable business out of the opportunity while satisfying the customer need.

Evaluation of ideas within the strategic framework of the business and competitive activities was presented. Tactics and strategy differences between large and small firms drive different results, and means for accomplishing those results through partnerships was presented as an alternative to executing a long-range plan.

Finally, perspectives on partnering within the framework of new product development was discussed.

The reader should now have a basic, practical understanding of how a new product idea relates to a business, and how it must relate to sustain the business in the long-term. They should also have a working knowledge of partnership arrangements, their caveats, and benefits.

This serves as the basis for the next section, where the product opportunity and market opportunity will be refined into a product.

CHAPTER 3

REFINEMENT OF THE PRODUCT CONCEPT INTO A NEW PRODUCT AND BUSINESS

ABSTRACT: New product success is in the details. Consideration to the details throughout the project, and the new product program, will generate the desired results. This chapter will outline the steps and requirements needed to chart the course of new product definition. Throughout the process of new product development, the new product developer breaks ground on several fronts: pricing, costs, functionality, features, route to market, and manufacturing venue. Each of these items will divert and fall short of expectations as a natural part of the process. It is the new product manager's job to keep pressure on all of these fronts and manage them to an optimal amount. Through this type of multidimensional due diligence the new product can carve out and retain a business and build on it.

THE IDEA

1. GENERATING THE IDEA

How does the new product idea get generated? There are a variety of ways this is accomplished. As will be seen, listening to the customer is a very important part of generating the idea and screening is a very important part of taking action.

There are six generally accepted venues for new product idea generation. They are:

A. Customer needs and wants

The customer needs and wants are the result of several actions:

1. **Direct customer surveys**. Surveys are a valuable tool to engage a customer to solicit feedback and to obtain information in a directed manner.

2. **Focused group discussions**. This method allows multiple-person input and prompted discussion to generate ideas.

3. **Suggestion systems and communication from customers**. These are the results of unsolicited feedback from users and relatively anonymous input from stakeholders of the company.

4. **Customer complaints**. This is an excellent feedback system for those who truly listen to the customer complaint for what it actually is and take the additional step to correct it through a product implementation.

B. Scientific research

This method occurs generally as a result of scientific breakthrough of totally new technology, or more often from applied research in the improvement in products or processes. Scientific research is responsible for breakthrough products or processes.

C. Competitors

Watch what the competitors are doing in the way of new product development. They are a yardstick measurement of the marketplace activity and an indication of how your company measures up to the marketplace. In the previous chapter, we discussed how spheres of influence of individual companies extend beyond their boundaries due to employees, suppliers, and customers. In the same manner you can monitor competitive product development activities.

D. Company dealers and representatives

These sales channels can feedback customer ideas and also generate ideas of their own. However, this medium must be evaluated in light of any individual agenda that may exist. The sales function is market-share driven, with a top-line orientation, and driven by orders. This is not necessarily the best source of new product ideas because feedback may be slanted to make their sales job easier, rather than what may be in the best interest of the firm in total. Watch where the money changes hands and under what circumstances, and you won't get confused.

E. Top management

This can be a valuable source or a destructive influence in product development. Top management has the largest bat to drive home a new product development idea through the organization. A program can be initiated and prosecuted in short order and with amazing results. However, less-than constructive results can occur in cases in which supporting data, primary customer feedback, and due diligence are not practiced and are glossed in favor of exercising executive privilege.

F. Miscellaneous sources

Finally, there are a wide variety of miscellaneous sources that can generate new product ideas. The important fact to remember is that it is not as important to remember where the new product idea came from, as it is to evaluate and screen suitable ideas, lock onto them, and execute the best choices. Some of the miscellaneous sources are inventors, patent counsel, university laboratories, government licensing offices, industrial consultants, marketing research firms, and periodical digests relating to an industry.

There are several techniques for generating new product ideas, including the following:

1. Attribute listing: This method requires generating a detailed list of the major attributes of the existing product and envisioning modifying each attribute by maximizing, minimizing, substituting or rearranging, and generating new combinations to improve the product or secure a new niche. As an example, the electric screwdriver was a result of studying a basic screwdriver and modifying the attribute of manual rotary motion to extend to motorized rotary motion. A next logical step would be to enhance the "feel" the traditional device user has through the hand and set torque limits in some numerical fashion.

2. Forced relationships: This method considers several products simultaneously, examines the interrelationship of these products with each other in the group, and generates alternatives that combine functionality into a single new product. For example, it could be said that the personal digital assistants on the market today are the result of combined functionality of a calendar, calculator, and address book.

3. Morphological analysis: This technique involves creative problem solving by examining a problem or objective, then singling out the most important dimensions of the problem and analyzing the interrelationships among them. For example:

Objective: Getting from point A to point B

Type of vehicle	Medium of travel	Power source
Cart, chair, wagon	air, water, ground rail	electric, magnetic

Now generate combinations that will be appealing, lower cost, more desirable, and/or novel. This is an example of morphological analysis.

4. Problem analysis: This method is different than the previously mentioned three methods in that it does not start at the company and work outward to the customer and marketplace; it starts with the customer and traverses back into the organization. The customer is interviewed to surface problems with the usage of a specific

product or category of products. The manufacturer then determines the product enhancement required to resolve the customer issue or complaint. Obviously, every issue cannot be addressed by adding enhancements, so the manufacturer needs to categorize the feedback into categories, such as seriousness, incidence frequency, and cost to implement a remedy.

In effect this is just a feedback method of engaging the customer to determine satisfaction level. The theme repeatedly surfaces: Stay close to the customer to get accurate assessment of product acceptance and business status.

G. Brainstorming

This method is oriented to generating ideas primarily in a free-flow fashion. It is specifically structured to generate ideas by removing mental paradigms in the participants. Generally limited to six people and 1 hour, a session consists of the leader defining the problem to be solved specifically as practical and encouraging idea generation by observing the following rules: criticism is not allowed, freewheeling thinking is encouraged, quantity of ideas is desired, and combination and/or improvement is sought out.

H. Directed brainstorming techniques

This is a subset of the previous methods, in which issues are not defined specifically in the beginning of the exercise. The problem is presented in the broadest frame of reference possible, and as the group exhausts the combinations and the creative suggestions, the moderator interjects facts that further define the issue, driving the group to a specific solution for a specific problem. For example, if the objective is to resolve a problem regarding an airtight body suit as a safety device, the moderator would spark responses by discussing the overall issue of protective coverings. As the group exhausts different protective coverings, the moderator interjects more specificity into the discussion. This technique can be very helpful as it renders preconceived notions about how to solve a problem.

2. A NEW PRODUCT CONCEPT MUST BE TESTED TO BE REFINED

It is said that Rome wasn't built in a day; so too with new product ideas and their translation into new business. An idea doesn't become an overnight, long-lasting success without a significant amount of consideration and reconsideration during its gestational period. Ideas become refined with three major factors influencing their refinement: the passage of time, the venue of consideration, and the frequency of review.

With respect to the passage of time, some concepts seem sure winners initially, but the excitement wanes. This is because the concept's authenticity and appeal are fleeting and cannot support a business to be built with.

Venue plays an equally important role when evaluating ideas because ideas will be received differently in different venues. Consequently, testing an idea on a traveling road show technique has enormous value in generating responses representative of a cross section of people.

Frequency of review also has value in testing an idea. Every day the mind receives new inputs and stimulation. Tomorrow's opinion of today's idea will be more informed than today's, so it's a good technique to revisit those concepts and ideas to see if they still have as much merit tomorrow as they do today.

3. THIRD PARTY INFLUENCE AND FEEDBACK IS KEY

The key to valued feedback is to get informed responses as distant as possible from the source of ideas generated. This will result in an unbiased assessment of the concept, and also allow the new product manager to draft the promotional material and present it to see if the story works, in effect. This third-party detached opinion can be your worst critic or your best supporter. In either event their feedback will drive you one way or the other. Either the feedback is good and you proceed, or it's negative. There can be no shame in negative feedback or rejection of the product concept at this point. In fact, it is a favor to know as early as possible in the game because it will save time and money not to pursue something that will be unacceptable to the marketplace.

4. PRIMARY CUSTOMER CONTACT HAS BIG PAYOFFS

The use of primary customer feedback cannot be stressed enough. It serves two major purposes: (1) it provides direct product feedback in terms of features, and pricing, and configuration; and (2) it provides feedback on how the customer perceives your firm in designing, manufacturing, and marketing this product. This positioning feedback is very important to determine customer and market acceptance of your firm's association with the product and business.

The use of primary customer influence ensures specificity in defining the product. Consider a case in which the product is not positioned acceptably in the marketplace because of a broad definition. Subsequent to release, the manager will spend precious launch energy modifying the product, repricing it, modifying the feature set, and retargeting the market. While this is taking place, the competition is gaining momentum with their product lines. As will be shown in the next section, placing these factors in a matrix is an easy way to keep track of the positioning of the product within the marketplace, and will aid in defining it.

5. TEST THE CUSTOMER ACCEPTANCE

Testing the customer acceptance as a means for positioning the product goes further than either a positive or negative response. It is desirable to establish a gradient of features, functionality, and customer cost. When launching a new program, it is important to refine the gradient because the product concept can serve a wide variety of people, however, not at the same product configuration and the same price. By establishing the gradient, one can find the proper breakpoints and versions as illustrated in Figure 3-1.

In Figure 3-1, the matrix gradient of features and pricing is the result of primary customer contact, conducted to refine a product definition. As shown, the overall gradient is organized to have increasing pricing along the *x*-axis and increasing features and functionality along the *y*-axis. The individual squares indicate two things: the number of responses that indicated features and pricing in that section of the matrix, and the percent of total that it represents. The actual number is in the upper left-hand corner of the square, and the percent is in the lower right-hand corner of the square. As one moves along the *x*-axis to the right and the down the *y*-axis, on a diagonal, the matrix gradient indicates increasing price and increasing functionality.

In this example a total of 200 responses were received as part of primary customer interviews. The respondents fell into one of six market segments. The bulk of the responses centered on a medium-priced, medium-featured solution. The customers wanting features

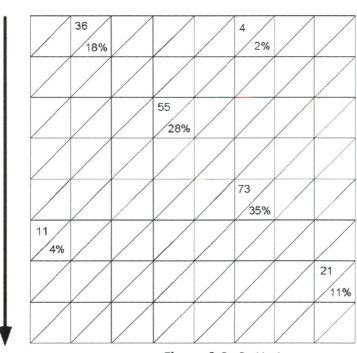

Figure 3-1. Positioning

desired in the lower end of this segment could be packaged with the features in the higher end in one product offering, perhaps. Together, they represent 128 (55+73) responses or 65% of the respondents. Conceivably a single product platform could serve both segments without incurring a cost penalty in the design.

Conversely, 36 respondents, or roughly 18% of the responses, were interested in the product concept having low functionality at a low price. Across the spectrum, at the high end, there was interest also. Roughly 21 respondents, or 11% of the respondents, were interested in a highly functional, highly featured, premium-priced solution.

If the platform would allow it, functionally, a separate product version could be created to serve this segment of the market. Since a premium price is acceptable, performance would be the issue, rather than platform cost.

At the low end, cost is the key issue because the market segment wants low features and low price. In this case a different, lower-cost platform may need to be used to serve that segment.

If this is the desired result from the customer contact, then the interviewing guide or questionnaire must be designed to produce this type of result. Consequently, the responses need to be directed into quantifiable criteria in order to be placed on the gradient with numerical weight. For example, the interview should elicit a response indicating, "This specific set of features are desirable at this price," and that can be entered into the matrix gradient.

To have these categories available to present to a respondent, it may be necessary initially to canvas several raw responses from the marketplace before designing the actual survey, to establish the categories first. Then the responses will fit the gradient with more accuracy because the survey will require a response to the query, "Would you buy the product under these circumstances with this feature set at this price level?"

COMPETITIVE ANALYSIS–STRUCTURING THE ADVANTAGE

1. EVALUATE THE COMPETITION PRIOR TO PRODUCT DEFINITION

Many companies make the same mistake when it comes to competition and new product development. The best time to do a comprehensive, competitive comparison chart is at the onset of a program. All too often this is done as part of the promotional package for the product launch, if at all. The comprehensive chart should be done at the project inception, and another one done at the time of product launch. The one done at inception will be required to define the product. The updated one at product launch will update your file and yield fresh information to the sales team, but it also will indicate the movement and nimbleness of the competitors. If they have made changes and introduced them in the time that your company has spent developing the product, it tells you something about their path, speed of development, incremental improvements, and market segmentation techniques.

A word about competitors in the marketplace: Competition is like a chess game, which is patterned after war, in effect. If the war analogy is used, then most competitors fall into one of four major groups forming the complexion of the marketplace. The individual elements are as follows:

A. **Defensive Orientation**
B. **Direct Pursuit Orientation**
C. **Oblique Pursuit Orientation**
D. **Opportunistic Pursuit Orientation**

The **Defensive Orientation** posture is based on the market leader's position whereby the major market share is already controlled and the objective is to prevent abdication of the share to anyone else. Their salient points of operation are that only the market leader can play a defensive game. They need to maintain market share by attacking themselves as others would attack them, only doing it first and establishing countermeasures to maintain share. Any strong competitive moves should be blocked immediately.

The **Direct Pursuit Orientation** posture is based on competitors that are generally number 2 or 3 in the marketplace. They are interested in growing market share by taking it from the leader's share. Their main consideration is the strength of the leader's position. They are most successful when they find the inherent weakness in the leader's strength and attack at that point. For example, a low-cost, volume-oriented producer of goods with a formidable, established sales channel would be difficult to take market share away from. However, their weakness is the inertia of the entire system, leaving a market opportunity for a mass customization offering to compete directly for the business. The offensive player needs to launch attacks of as narrow a front as possible or diffusion of effort will occur, causing failure of an initiative.

The **Oblique Pursuit Orientation** posture is an interesting one in that it is designed to seek out an uncontested area. Its major tactic is surprise. Since it is a surprise initiative, the pursuit of the business is as important as the attack itself. Many of these launches are manifested by developing functional alternatives to solving the problem, currently solved by the defensive player.

The fourth posture is the **Opportunistic Pursuit Orientation** posture. It is generally a small, company orientation and is manifested by ability to mobilize and serve market needs with originality and speed. The player needs to find a segment small enough to stake out and defend. They also must never forget that no matter how successful they are in a specific niche, they are to never act like a leader. If the playing field for that niche becomes hotly contested, they must be prepared to bug out at a moment's notice.

These represent the four major types of competitive players in the market place and are generally easily recognizable. It is important to categorize the individual competitors' posture on the competitive comparison chart, and understand their strategy. Know whom your competing against; it will clarify your actions.

2. LIST OF COMPETITORS AND ATTRIBUTES

The competitive comparison chart is an invaluable tool in positioning the new product in the marketplace. A properly prepared chart outlines the features, pricing, offering, business condition, and competitive standing.

With reference to its use as a sales tool; it has a tactical orientation, in which the product evolution flowchart is more strategic in nature. Its use as a sales tool is only for the benefit of the sales channel in training and communication of the current conditions. The document has its real value when used at the onset of a development program.

Figure 3-2 is a format that can be used to complete a competitive comparison chart. This is also included in the Tool Box to be used as a starting point to modify or embellish for your particular situation.

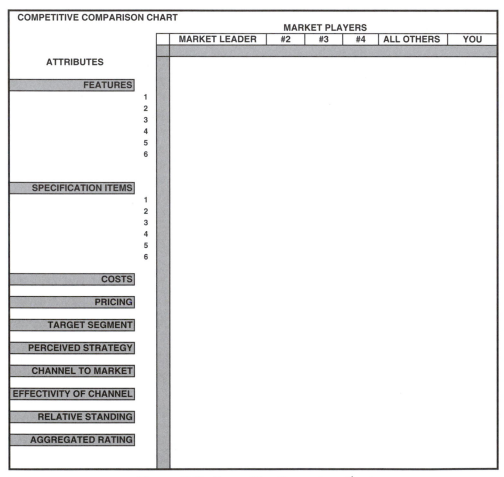

Figure 3-2. Competitive Comparison Chart

As shown in Figure 3-2, the vertical axis represents the product attributes. These are the tangible elements of the product line, which have value to the customer. Some are simple listings of features and pricing that can be entered in hard data. Some of the entries, such as target segment, strategies, effectiveness of the channel, relative standing, and aggregated rating, require additional research and tabulation of data. The target segment is a statement of where the product is aimed. Who will purchase it, and under what conditions? A simple sentence explanation here is sufficient. Perceived strategy is the market's assessment of the manufacturer's strategy and product positioning. Channel to market is fairly straightforward; simply chart the pathway from the manufacturer to the end user. Comment on any weakness that may be exploited, and comment on the overall effectiveness of the channel. Next, cite the relative standing of the manufacturer in terms of market share. As will be discussed next, these data will range from highest to lowest because of how we will organize the competitors.

Finally, based on the present assessment of market need, place an overall rating on each manufacturer with respect to their features, specification, pricing, and these other factors. A more-detailed analysis may require reducing these data to numerical values to be tabulated and posted, or an overall grade or rating may be used on observed objective data.

This essentially completes the vertical axis. The horizontal axis is arranged by manufacturer, whereby the market leader is in the first slot. Since they have the lion's share of the market, the assumption is that their package of values brought to the market is relatively complete. The second-, third-, and fourth-largest competitors are in the next slots.

Depending on the size of the market and the number of players, "all others" may be lumped together or considered separately. Next, place your company and the proposed product against all of the others. Remember, lead the target in new product development, which means your target specification for the product should exceed all others by a wide margin. This is because the collection of competitors will not remain motionless during your product development cycle.

Lead the target so that at product introduction, your firm will still be in a favorable position.

3. FUNCTIONAL ALTERNATIVES TO SOLVE THE PROBLEM

An entrenched competitor need not be invincible. The primary driver in the sales transaction is the satisfaction of the customer need. If there is an alternative way to solve the need, a new market niche may be created by those market constituents who accept the alternative solution mentioned previously in the discussion of flanking posture; this can allow participation in markets that otherwise would be inaccessible.

For example, the American station wagon lost market share and virtual existence by the development and introduction of the minivan. Designed from the ground up as an alternative with more functionality and less-design compromise as in the station wagon, the minivan dominates the landscape in highly functional, family vehicles.

4. ENGINEER THE COMPETITIVE ADVANTAGE

The competitive advantage cannot be added on at the end of a program. Added features, or other inducements to invoke market acceptance, generally adds cost or removes advantage in other ways.

To successfully beat the competition, a watchful eye must be kept on them during the development process, and a delicate balance must be struck in adding features to keep ahead and engaging in creeping functionalism, which will be discussed more in-depth later.

The competitive advantage must be implicit and engineered into the product. For example, many products are electronics based. As systems become more complex and needs vary, programming or setting of electronics has become more complex. Internal menu structures, which prompt the user through the choices, are a means in which the advantage is engineered. Generally, consuming memory space only, and adding no cost other than development time, these added features greatly enhance the usability of the device for the customer and aid acceptance.

Another example of competitive-advantage engineering is evident when product functionality begins to merge and formulate one new product. A good starting point could be hand-held calculators in the early 1970s. Starting off as four-function devices, each competitor built upon the other and added features. Memory functions, trigonometric functions, and programmability all were engineered in competitive advantages. The calculator has begun to merge with the day timer, text editors, and worksheets to form the palm-size personal computer. With each new introduction, additional features and operating systems to enhance usability were added. As the technology develops, these products are merging with the fax machine and cellular telephone technology to escort functionality and portability to an unprecedented level.

The key to market share is to satisfy the customer need better than the others in the arena. Engineering the competitive advantage has big payoffs at product introduction that cannot be achieved as well using other means.

5. ANTICIPATED COMPETITIVE ACTIONS

As discussed previously, the world we live in, and the marketplace we participate in, is not a fixed entity; it is dynamic in nature. Actions that your firm takes will be countered by actions that your competitor will take to counteract your salient. To win at the interchange, it is necessary to anticipate a reaction and plan for it. Then when your action occurs, and their reaction occurs, your position is in a net advantage because your action accounted for a possible future reaction.

To get a feel for the competitor's reaction to a specific situation, a reference to the strategic posture of each one is in order. This posture defines how the competitor conducts their business. If they are a guerrilla player, they will react a certain way versus if they are a leader playing a defensive position. By understanding the opponent and their posture, it becomes easier to predict their reactions.

The dimension that makes this more complex than ever is that reactions can be a result of one or more competitors working together to counteract your moves. Think of it as a chess game as you anticipate the market dynamics.

THE ROUTE TO MARKET

1. WHAT IS THE ROUTE?

Every company needs to have a sales channel. Are they company-owned stores? Are they distributors, or are they resellers, or agents? Each channel exists for a purpose, and each one has benefits and drawbacks. If you were beginning to think that new product success depended only on development, specifications, and features, then read on. The effectiveness of the channel will make or break your product introduction and sales effort.

Figure 3-3 diagrams the possible routes to market. It is by no means complete; however, it does show the links possible in the pathway of a product from manufacturer to customer. More important, each of these possible links can be a strength, or a weakness, in your marketing effort.

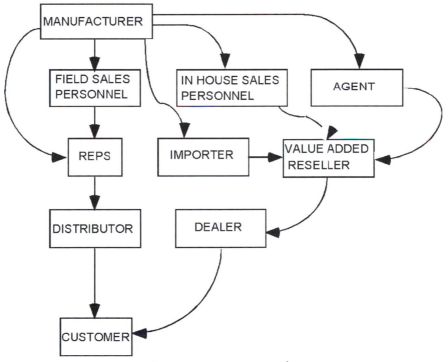

Figure 3-3. Route to Market

The ideal is to leverage off of their strengths, and add their strengths to your marketing effort. The important point is to partner with a formidable partner.

Beware of the fallacy of partnering with weak, but "hungry" partners. "Hungry cannot make up for presence, staying power, and a trained organization in place, and able to meet the demands of new product introduction. Weak will sap your corporate energy and drain resources, rather than accomplishing your objective of leveraging your resources.

Select your marketing partner to add strength and critical mass to the effort.

In many cases you will not have the luxury of selecting your route to market; however, go through the exercise of designing a route from scratch, and this will yield the requirements needed to improve the channel's effectiveness. Think of each of the links pictured as a question in the new product developer's eyes, asking the question, Is this the best route to the customer? What does this link contribute to the overall transaction? Is it a leverage point or is it a hindrance? It is important to understand the effect, as each link removes you further from the customer. It makes it harder to communicate information from the factory through the channel and it diffuses the feedback from the customer back to the manufacturer.

2. WHERE LIES ITS STRENGTH AND WEAKNESSES?

A sales route to market exists or has been established. Now determine where the strengths and the weaknesses are. As the product is being developed, the sales channel must be improved and trained to successfully launch the new product. This means the strengths must be built on and the weaknesses must be eliminated.

As a starting point, let's examine some of the elements of the channel and their responsibilities in the sales transaction.

The manufacturer has the responsibility to develop, manufacture, and ship product. They assist in creating the user demand through media advertising and market awareness. They train the channel in the product. They generally have sales personnel—either inside, outside, or both—to facilitate moving products through the channel.

The representative cultivates the demand among original equipment manufacturers and users. They may appoint distributors in cooperation with the manufacturer. In the case where there is a representative and a distributor, the product may be more engineered and need technical expertise to develop the market. The representative will report market intelligence gathered to the factory, as part of a feedback mechanism. They are independent business people who have other principles to generate income. They must balance effort among the lines. They provide local expertise and product knowledge. They serve an established group of industry-like or technology-like customer base. The representative does not have financial risk in the product, because they do not take title to it; rather, they are compensated by a commission.

The value-added reseller is similar to the original equipment manufacturer (OEM) in that they both add value to the product. However, the OEM incorporates the product in their product, whereas the value-added reseller completes the product offering in a way that the manufacturer cannot complete, or in some other specialized way the market accepts more readily. They take title to the goods and resell.

The distributor provides local stocking, availability, and collateral products as part of one-stop shopping. They have inside sales personnel and field sales personnel. They provide feedback on products to the representative or directly to the manufacturer.

3. HOW CAN YOU ACCOUNT FOR AND OFFSET THESE WEAKNESSES WITH THE NEW PRODUCT EMBODIMENT?

Assuming the sales channel is somewhat fixed and the product opportunity demands strengths in the channel not readily implementable, is there a way that the product can account for the shortfall in the channel? If the channel is technologically weak, can the product have a help screen, menu-driven diagnostics, or auto set up to make up for the shortfall in the channel? If the channel serves a wide variety of customer bases with different requirements, does the product have to be made modular and be configured at the point of sale or delivery? This is to satisfy the diversity in the marketplace, either by venue or type, which is not matched by corresponding diversity in the channel. An example would be personal care products in which the chemistry is formulated to be safe under mass-market sales campaigns where discount stores, not employing trained personnel in recommending application, are selling the goods.

4. DETERMINE MOTIVATION AT EACH STEP IN THE CHAIN; WHERE DOES THE MONEY FLOW?

To truly understand the motivation in each step of the route to market, it is important to understand the role each plays and where the pressure points in the transaction take place. Figure 3-4 outlines the relationship of each of the parties in the transaction, where the money and goods change hands, and where the value in the transaction takes place.

Although all of the players in the chart may not necessarily be required in a transaction, they are listed for reference. The chart may be interpreted by looking at the manufacturer first, then selecting the players for your particular route to market.

This chart is by no means complete, since one chart cannot cover all of the financial arrangements that may exist between a manufacturer and the members of the channel. To further illustrate the point a question mark was placed at the importer entry for payment trigger and the financial horizon. This is presented for effect primarily, since you may lack information as to their agenda in the transaction.

It is best to evaluate where the value is added, and if that value warrants the financial outlay for that part of the transaction.

As can be seen in Figure 3-4, the exchange of revenue happens at different points in the transaction, depending on where in the channel you are operating. It is also interesting

SUMMARY OF MARKET PARTICIPANTS							
	MFGR	**REP**	**DIST**	**AGENT**	**IMPORTER**	**DEALER**	**USER**
WHAT THEY SELL	Goods & svcs	Time,Knowledge	Availability	Connectability	Int'l Pathway	Local Support	N/A
PAYMENT TRIGGER	Shipment / Terms	Receipt of A/R	Shipment	Retainer/A/R	?	Delivery	Goods & Terms
WHAT THEY BUY	Supplies / Labor	Gasoline / Lunch	Goods	Goods	Goods	Goods	Complete Goods
MOTIVATION	Volume	Vol / Principle	Turnover	Mfgr's Shipment	Mfgr's Shipment	Cust. Delivery	Cust. Receipt
FINANCIAL HORIZON	Years	60 - 90 Days	Monthly	30 - 60 Days	?	30 - 60 Days	N/A
FINANCIAL STRENGTH	HIGH	LOW	MEDIUM	LOW	LOW	HIGH	N/A

Figure 3-4. Summary of Market Channel

to note the financial horizon for each and how they differ relative to each other. Along with the horizon can be seen the financial strength of each of the players.

These two factors contribute to the "commitment" factor in the overall relationship. This means, for example, that the relationship needs to satisfy the representative on a relatively short cycle. Their sales efforts must pay off in a cycle or two; otherwise, they will spend their earning power (time) with another principle.

The motivation for the rep is volume and principle. In other words, each day they ask themselves, "Which principle will contribute to my income 60 to 90 days in the future?" This is generally what they spend their time on. A distributor, by contrast, has inventory to sell and is interested in turns of that inventory. Their purpose is to ship product against local demand and assist customers in that regard. They are not motivated to "develop" a market for a manufacturer.

The overriding important difference is that all of the players from the manufacturer's perspective are independent businessmen. The only control your firm has over their time is to show them how they can make money with your product.

5. KEEP AS CLOSE TO THE CUSTOMER AS POSSIBLE WHILE MAINTAINING COVERAGE AND LEVERAGE

This is an important point. A repeated pattern, occurring as companies grow, historically can cause them to become further and further removed from the customer, and the all-important dialogue that must occur between the two. Figure 3-5 shows that as the channel becomes more complex and has more and more players in it, the company can lose touch with the customer needs and feedback on product performance.

In the illustration, the manufacturer gets to the customer via several players in the channel. Each participant (CH-1, CH-2, CH-3, etc.) diffuses the information flow. This diffusion affects the outward flow from the manufacturer to the customer and feedback to the manufacturer. This can isolate the manufacturer from the market and potentially place them in a dangerous position of losing business.

Furthermore, since members of the channel are often independent businessmen, the agendas and motivation of each factor into the overall communications integrity.

CUSTOMER FEEDBACK FLOW

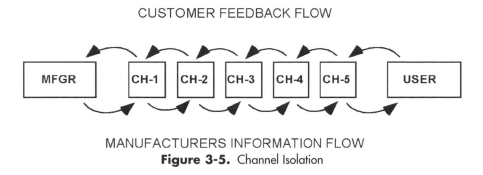

MANUFACTURERS INFORMATION FLOW
Figure 3-5. Channel Isolation

6. TEST THE ROUTE

Don't wait until the product is ready to launch to test the market. It's almost too late at that point. Start testing the market channel at the onset of the development program. Decide what information can be communicated to the field without compromising a competitive position. Then start testing the channel's response and enthusiasm to the program.

As part of the development, you need to test their commitment and resolve to execute tasks in promoting the product when it's released. There is no better measure than evaluating their behavior during the development. Select and assign tasks that you will need completed as part of the development. Draw them into the process and evaluate their performance. If they fail to perform during this stage, there is little chance of them performing to your expectations later.

STRATEGY AND TACTICS IN OPERATIONAL PLANNING

1. HOW THE NEW PRODUCT WILL BE PRESENTED TO THE MARKETPLACE

The product is identified, competition is analyzed, and the product positioning is accurate, but how will you prepare your organization to tell the new product story to the marketplace? What makes your firm think it is the firm of choice to develop and introduce this to the marketplace? The answers to these questions need to be woven into a strategic story that the marketplace will accept. If it cannot be told and sold as a story in the marketplace, then that marketplace will have difficulty in accepting your company as committed to the product and the industry segment. This makes it an uphill battle for recognition and leverage, as the same perception obstacles must be overcome with each new target customer.

2. WHAT MUST BE DONE WITHIN THE ORGANIZATION TO PREPARE FOR THE NEW VENTURE

If the market may need paradigm training in changing their perception of your firm, then perhaps your firm needs training in paradigm shifting to make way for the new product! Old organizations and complacency do not successfully generate new innovative products and businesses. They simply do not possess the skills to break the bonds of uncertainty to develop and platform new ideas into new products. To successfully generate new business through new product development, create an atmosphere in the organization to meet the demands of the new product. Don't be compromised by corporate complacency and preservation of a comfort zone.

3. WHAT PREREQUISITES ARE TO BE IMPLEMENTED

Consequently, what are the specific changes that the organization must undergo? At the stage the program is at, you are visualizing the future with the product, so it is easy to visualize the corporate shortfalls and how they can negatively impact the program. List the prerequisites for the program and institute the changes in the organization.

Not all of the issues are related to internal organizational changes. If the development program involves an external player in the form of a brand label partner or a joint venture, there are additional elements to consider. For example, if your corporate strategy has you collaborating with new and/or external players, review the section in Chapter 2 on partnering. Take the time to develop an understanding of the players, their personal and professional agenda, and their goals.

Also develop an understanding of the degree of vulnerability your organization has in embarking on this program.

An example of this comes from a personal experience. We disclosed a brochure for a product we were going to brand label to the supplier company as a courtesy. Once they saw the brochure was printed and 25,000 to 30,000 copies were in stock, they felt quite comfortable in raising the price slightly. We had little alternative, because the announcement was made to the sales organization, and the design was proprietary enough that it would have cost too much time to elevate our company's negotiation position by seeking an alternative. The lesson: Always understand the partner's motivation and the internal personnel agenda.

If it is beginning to sound to you as if the business of new product development effects change in all areas of the organization, congratulations. You are now undergoing the necessary paradigm shift to make the product a success.

4. WHAT TACTICS WILL BE USED TO INTRODUCE AND SHIFT LOYALTIES

There are built-in loyalties in the marketplace that have to be shifted to obtain market share with a new product. Human nature being what it is, there is a hesitation to change from the known to the unknown. As part of the new product development, the loyalties must be changed by your organization's actions. What will these tactics be—price, quality, service, knowledge, availability? During the development of the program, determine what these tactics are and attack at that point to shift purchasing patterns.

Bring something exciting, special, and of higher value to the customer base. Launch with this and keep the pace difficult for the competition to catch up.

5. HOW YOU WILL GET THE DUE ATTENTION TO THE MARKETPLACE TO BE EFFECTIVE

You have taken the necessary steps to ensure that the new product is not a 'me to' offering. It is unique and of high value. How do you get the marketplace's attention and focus on your firm's solution as opposed to the entrenched competition? Short of reciting a litany of media blitzes, promotional hype, and spin control, suffice it to say that the launch is your time in the sun. Make the most of it and use it to create a base to build on.

If a success story needs to be made, start planning it now at program initiation. If some other means for promotion needs to be made, make them now and start laying the groundwork early. Each step taken now will be valued later as it can be used or modified. This yields a great amount of flexibility, which you will need to offset the uncertainty encountered in the future.

BACKGROUND/FORMAT

1. TRACK RECORD OF THE BUSINESS

A. WHAT HAS THE COMPANY TRADITIONALY BEEN GOOD AT?

The issue of traditional success has been reviewed previously, in general terms, as they related to the overall strategic direction of the company; however, at this point you may want to take a more in-depth review of products that the company has been traditionally successful with. Is the new product you are ready to embark on developing consistent with these previous winners? How do the products differ? What are the significant differences that need to be understood and addressed?

B. WHAT TYPE OF PROGRAMS HAVE HISTORICALLY FAILED, AND WHY?

Conversely, what types of products have traditionally failed to meet expectations and how do they relate to the product under consideration? What were the root causes of failure for these incidences, and have they been corrected?

C. ARE THE SAME PRESUMPTIONS STILL TRUE TODAY?

Are there organizational attitudes that must be changed to allow success? Are there remnants of the root causes of failure that need to be eliminated? Is the organization ready to execute a program?

D. TRACE SUCCESSFUL PROGRAMS AND CITE THEIR CONTRIBUTIONS TO THE SUCCESS

Develop an in-depth understanding of the successes and the failures historically in the organization. More important, what were the contributions to success, and can they be built on? This series of questions is for your consideration, as a double check of your company direction, wants, needs, and tenacity. You are near the juncture where significant funds will be committed to develop a product and a business, so it is important to gain the understanding at this time.

2. PREPARING AN OPERATIONAL PLAN FOR THE NEW PRODUCT

Facilitating a new product into a company, and all the associated changes that come into play, is a formidable task. It in itself is a project that needs management. It is generally a good idea to formulate a plan to take care of all of the needed organizational items. These are not developmental items, but rather, are organizational changes and modifications that are needed to ensure success. It is best to take care of these changes as swiftly and as early as possible, because energy must be devoted to the development tasks at hand, not organizational tasks.

3. A DISCUSSION OF FORMATS

Set up a critical pathway with a list of critical changes and dates. This will ensure the organization changes to meet the needs of the new product. There are as many formats for program management as there are third-party software suppliers. Almost any format will do as long as the evaluation, measurement, and follow up are consistent. The difficult part is not keeping track of the changes; it is effecting the change through people and making it stick. An organization is people, and people have to make the organization work. Failure to decisively make a change and integrate it will cause problems all along the pathway to introduction.

4. INTEGRATION OF TACTICS TO EXECUTE A WORKABLE PLAN

What are the push buttons of the organization? How do the necessary changes get made and enforced? Do you work through executive decree, by influence, or by other means? Whatever the method, remember that you are sculpting the organization to accept the new product development program. Not everyone will share your vision and enthusiasm!

5. DETERMINING RISK

Each new product development carries with it a measure of risk to the corporation as an ongoing enterprise. There is the financial risk of the investment, the risk of lost opportunity, and the risk of any liability associated with the product and its performance. The measurement of risk falls into three areas that need to be evaluated. *A chart for Risk Determination is available in the Tool Box for your reference.*

A. The scope of the program (i.e., what portion of the total asset base of the company) is the investment. Some companies have placed all their hopes for the future in the development area and virtually lost control of the corporation, when the development dragged on with no return to offset the investment.

B. How achievable is the program, how likely is the chance of success?

C. What is the measure of acceptance of the program within the corporation? Who are the negative participants and what is their influence base?

6. MANAGING RISK

The other part of identifying the risk is to manage the risk to an acceptable level. This requires constant attention to keep the program on track. The mechanics of managing the risk of a program consist of the following three basic areas:

A. Be attentive to negative trends, issues, and problems that start out small and grow. Act early and decisively to contain them. You cannot be shy when containing risk.

B. Adopt a working philosophy that is central to the organization, such as, "Manage today and look ahead to tomorrow." Look for possible issues that have their origins in today's activities and could fester into tomorrow's problems.

C. Act quickly to mitigate any damage that may have already occurred.

AGREEMENT OF PRINCIPLES

1. AN OFTEN MISASSUMED AGREEMENT

This is more of a problem in smaller companies and closely held companies where strategy and direction are given by a select few people. An enthusiastic, new product champion may go through motions to solicit agreement among the principles; however, there may not be agreement. Companies are comprised of people, and people have human biases and fears of the unknown. They tend to protect their area of comfort in actions taken while verbalizing the need for dynamic change. It is therefore the responsibility of the manager playing the advocate role of champion to precipitate agreement among the principles and reinforce it throughout the program. All too often this agreement is tacit only and lacks the commitment behind it. Many times the agreement comes quick and shallow, only to be second-guessed and overruled later. Companies demonstrating this type of corporate behavior exhibit weak management and rarely complete a program through to success.

2. COMMITMENT VERSUS AGREEMENT: GETTING IT IS CRITICAL

There is a distinct difference between commitment and agreement. Saying the words to agree on a program is relatively easy and risk free. Committing corporate funds to back up the agreement is more difficult. It takes faith in the program, faith in the team, and faith in the new product manager. That is a lot of faith, whereas a more comfortable position is to accept the development and "see how it goes" approach. "If it progresses, we will fund more money to the program."

With development, however, it is a game of conquering uncertainty. The progress is measured by tasks (surrounded by uncertainty) being completed in a timely fashion. When uncertainty arises, that is the time to have the faith and the stamina to see it through. Unfortunately, the "see how it goes" approach generally has management viewing the program as going bad and seeking out other opportunities to fund. The result is completion of easy programs only, and loss of impact products to the organization.

3. TESTING THE COMMITMENT

The issue of commitment is so important; it must be tested for. Remember that you are leading the development group to accomplish a program. It is a corporate trust whereby you may not waste technical talent or corporate funds. If you are leading the charge, the organization must back you up to the extent they agreed to.

It is wise to test commitment with something that is noncritical, but important enough to be telling. Depending on the circumstances, you may want to test the organization as early as possible. The earlier you know the better.

The other dangerous game that can be played is that two or more projects may be initiated with funds for only one for completion. You have two basic choices, either fight a street fight for the balance of the funds or demand that in the interest of a successful development and fiscal responsibility, only one program be generated based on its own merit. Do not allow the organization to put off the new product decision or strategic choice by starting two developments and trimming back. Better to start one and prosecute it with the energy of two and complete it.

4. EFFECTING STAYING POWER

How do you effect staying power in the overall development? Constantly update management and financiers of the progress and problems encountered and tie into the strategic plan. Use resources of the organization to resolve problems, and make management and the organization part of the resolution. Bring them into the development to become part of it. If this is done effectively, management will have less of a desire to discontinue a program. It will also bring more power to bear on to the problem than if you go it alone.

REDUCING THE RISK OF NEW PRODUCT FAILURE

1. WE KNOW THE DON'TS, WHAT ARE THE DO'S?

In the Appendix 1, New Product Flops there was a discussion on the causes of new product failure. Although it may be advantageous to repeat the section here, it would be more helpful to list the positive steps needed to be taken at this juncture of the new product development process. It can serve as a recurring checklist that the manager can review periodically to ensure diligence in the program.

Figure 3-6 is the Diligence Checklist of each. *A separate checklist is also available in Tool Box.*

2. THE RECOMMENDATIONS

Start with the first listed item and assess the organization's and your performance in ensuring the new product opportunity fits the scope of the strategic plan and company capabilities. There is no honor in trying to develop something that is outside of the firm's capabilities. Discuss the idea with several people at the firm to get an idea on the ability to carry out the development.

Next, determine if the present criteria for evaluation are consistent with the overall plans of the company. If the strategic plan is forecasting one way and the product opportunity does not support it, the plan will not be met and the development will be undermined eventually. Make sure there are documented procedures, and that they are followed. Finally, ensure that operational plans have department-linking objectives so that individual departments are not at odds with each other.

Have you ensured that the company is capable of developing the product?

Market planning is somewhat of a misnomer, in that markets aren't planned generally. However, it is necessary to assess what the market direction is and what your firm's role, by way of the new product, will be in it. Has the product been positioned properly, such that

DILIGENCE CHECKLIST					
ITEM	DESCRIPTION	RESP	SUBSTANTIATION	DATE 1	DATE 2
1	PRODUCT OPPORTUNITY FITS SCOPE OF CORPORATE STRATEGY				
2	CONSISTENT ORGANIZATIONAL SYSTEM FOR EVALUATION				
3	CONSISTENT CRITERIA				
4	DOCUMENTED AND FOLLOWED PROCEDURES				
5	LINKING DEPARTMENT OBJECTIVES				
6	SOUND MARKET PLANNING				
7	POSITIONING				
8	SEGMENTATION				
9	CAPITALIZATION				
10	PRICING				
11	PACED DEVELOPMENT TO CAPTURE OPPORTUNITY				
12	PRODUCT DISTINCTIVENESS				
13	SUPPORTING EVIDENCE OF OPPORTUNITY				
14	SIZABLE MARKET WITH RESEARCH AND FORECAST				
15	PRECISE PRODUCT DEFINITION				
16	SOLID DESIGN CRITERIA AND REVIEW				
17	TECHNICAL FEASIBILITY IN THE COMPANY				
18	ACHIEVING TARGET FACTORY COST OBJECTIVES				
19	PROJECTED COMPETITIVE RESPONSE				
20	STABILITY OF THE TEAM/COMPANY				

Figure 3-6. Diligence Checklist

your firm's offering of it is plausible in the perception of the marketplace? Has the market been segmented such that the individual product embodiments represent the proper configurations?

Has your firm demonstrated serious commitment to the development by way of development monies and capitalization for manufacturing?

Is the pricing right? Are you positioned to charge a market level price and not suffer margin pressure?

Make sure that the pace of development is sufficient enough to capture the opportunity in the timetable that is actually being charted.

The product originally conceived was distinctive. As the development unfolds and compromises are made, reassess the distinctiveness. Has there been control of creeping functionalism during the development?

Have you diligently collected the required supporting evidence to justify the development? Has anything changed? Are the forecast numbers for volume still there?

What if any compromises did you make in the development? Did marginal results of tests go uncorrected? Are the proper tests, design qualifications, and production in place, and does the product pass these tests?

Has the technical feasibility initially assumed proved to be correct, or have compromises in the specification been made to overcome technical shortfalls within the company?

Manage the design team to meet the target factory cost. This is as important as any other specification item. The market price is given; the margin, profit, and retained earnings from the program are governed by the factory cost. Pay close attention to the cost issue and keep on track all the way through.

Try to anticipate the competitive response to your introduction and strategize on your next move. Competitors can be formidable opponents, so it is best to anticipate their reactions.

Finally, preserve stability in the team, the corporation, and the funds flow for the program. Interruptions sap energy and add time to the project, reducing your chances for success.

3. RELATE TO AN ORGANIZATION (INTERNAL STRUCTURAL PROBLEMS)

As can be seen, most if not all of these issues relate to the organization internally (i.e., they are under your control). This being the case, exercise control for the betterment of the program. Do not gloss over these issues because they will come back to haunt the program.

4. RELATE TO EXTERNAL PROBLEMS

Those issues that are external to the organization can only be addressed by market intelligence and knowledge of what is happening. Make it your mission to fully understand the market, the opportunity, and the areas of uncertainty.

THE ASSIGNMENT

1. REFINE THE FIT INTO THE ORGANIZATION FROM THE INITIAL MACRO LOOK

Initial plans are made in a macro outlook point of view. Now it is time to refine the prospect for new products into the organization with the detail. Where within the organization will the new product be placed? Decide which group is the best suited to champion the program and carry it to completion. This depends on their interest level, their training and understanding in the specific market segment, and degree of leverage achieved in the use of that particular group versus any other group.

Keep in mind that initial selection of a group within the organization is telling. If there is no sponsorship or acceptance of the program by the initially selected group, the program may become branded and avoided. This must be avoided at all cost. Do not undertake a program without the entire organization's support. You will be left out on a limb with no recourse, and more importantly, the organization has not bought into the program. Obstacles will take on another dimension of difficulty with this arrangement.

2. GET DETAILED ON GROUPS, RESPONSIBILITY CENTERS, ETC.

The concept of establishing responsibility centers is a good one. It presupposes agreement and commitment and names personnel and departments as responsible for certain critical outcomes. There is accountability and consequences for failure. It is a workable, practical means for assignment and follow up because it focuses on results, rather than behavior and tasks. A motivated group charged with responsibility and given the tools can accomplish the objective and conquer obstacles without excuses.

The responsibility center needs to identify the responsible parties, what they are to accomplish, when it is to be complete, a general direction of how it is to be accomplished, and where the activity is to take place. Be sure to include all of these elements, and be sure to amass a motivated group.

THE CONFIGURATION

1. COALESCING THE PRODUCT IDEA INTO A SET OF WORKABLE SPECIFICATIONS

This is one of the most important and difficult tasks in the new product development process. It is the essence of taking the loose collection of market data, wants, needs, preferences, and biases and evolving a specification that is achievable, cost competitive, and able to be developed by the organization. One can neither master this activity overnight, nor with one product development cycle. It takes several products, generated from concepts and developed and introduced to the marketplace and obtaining the feedback, that refines these skills. The mechanics may sound simple, however the "feel" of the market and the accuracy developed from that feel are what takes the time and numbers to cultivate.

The outputs from this exercise are answers to questions such as number of versions, packaging options, bundling with other products, configuration of platforms, and configuration mapping to the target audience.

Revisit this exercise often as the development progresses to verify positioning and targets

2. FOCUSING ON WANTS, NEEDS, AND MUSTS

"The customer is always right" is how the saying goes. This means that you must listen to the customer's conversation with you and glean out the wants and the needs. It is very important to segregate these out in today's marketing climate. In years past it was necessary to determine basic option structure of products. For example, an automobile had a need of basic transportation, with a want of climate comfort at a considerable extra expense. It was necessary to know which area of the customer base was interested in these options so as to design and structure an option system for the product line.

In today's computer age and flexible platforms it is important to know the segmentation from the perspective of how to market and bundle the options to make the product more appealing to the specific segments. For instance, it is virtually at zero cost that features can be added to a computer-based system by adding them in software. It is only the development cost. However, if all of the features researched and possible were to be included, it would make for a very cluttered function menu, or make the product too clumsy for the average user. Consequently, many products are segmented by these functions, and different versions are introduced targeted to these specific customer segments.

3. TIMING IS KING, COSTING IS QUEEN, ALL ELSE PALES IN COMPARISON

The issue of features specifications are only part of the new product equation, as mentioned earlier. As the heading states, timing is king in new product development. A product must be available to introduce in a specific time frame to capture the opportunity and to maintain advantage. And if timing is king, then factory cost is queen. It is the basis for financial success or failure.

4. PLATFORMING THE PRODUCT

When setting up the initial platform of the product, it is best to understand the technology that will be employed. In fact, it is best to be very conversant on a technological level to be able to make mental tradeoffs between cost adders and no cost impact items. If items do not add incremental cost and can be easily be absorbed in the user's tolerance overhead (i.e., the ability to tolerate the overhead that accompanies functionality), then add the feature. If, however, the cost is impacted or ease of use is impacted, reexamine the target customer base to be sure you are not overspecifying the product.

It is a natural consequence of development of a product to add functionality as you progress in development, so you will have to keep that in check without completely missing the target when initially setting up the product platform.

5. REFERRING TO THE PRODUCT EVOLUTION FLOWCHART, INTEGRATING INTO THE PLAN

At this point it is a good idea to revisit the product evolution flowchart to determine if the product concept is the same scope and basic configuration. Determine if the target is the one on the chart or if it has crept to the follow on version already. Next, integrate the identified product into the strategic plan and reexamine if the plan still makes sense. This will ensure that the plan is executed via the new product, rather than falling by the wayside as a result of the development.

MASS CUSTOMIZATION AND GENERIC PLATFORMS

1. THE KEY TO SUCCESSFUL EXPLOITATION OF MARKETS AND SEGMENTATION

As the number of participants in a given market increases and the market grows, it becomes more and more difficult for a company to be successful in that market. One way of competing in this arena is to adopt a strategy of targeting specific customer groups with specific customized offerings. There is a marketing caveat that you cannot be all things to all people and be effective. This strategy of mass customization allows your company to stretch the envelope a bit.

This strategy, however, cannot be executed in a vacuum of product development, since the concept of product platform must be the basis for design. By starting with platform design that is flexible to target different customer groups with customized feature sets, the company can meet increasing customer demands, expand market share, and increase revenue. This is by manufacturing a wide variety of products in small quantities while having a large overall unit volume.

2. DEFINING MASS CUSTOMIZATION PLATFORMS

Mass customization is a term based on the concept of product platform. A generic platform is designed and optimized for two, somewhat diverse requirements. They are the product range that will be offered and the preferential acceptability of features that combine to make up a unique product offering.

For this concept to work, there must be a thorough understanding of the marketplace requirements. This defines the scope of the offering and the scope of the basic platform design. Although the concept affords flexibility down the road, the scope must be exact or you will lock yourself out of certain market opportunities. If the scope is too wide, the cost of the basic platform will be too high and the configured offering will be noncompetitive.

Next, the individual features must be designed to be somewhat portable and combinable to create a unique offering. The overall scope should allow for adding features to capture individual market opportunities. Figure 3-7 is an illustration of how the new product progresses from a wide variety of market requirements to distinct product versions.

As shown in Figure 3-7, the market is made up of diverse requirements, represented by the different shaded boxes. Each represents a distinct need from the product by the customer. The product version #1 is conceptualized to establish the feature listing and configuration. From this, the basic platform is conceptualized. If the basic platform is too wide in scope, then the product will be too costly. If the manager undershoots by locking onto too narrow of a basic platform, the versions available for offering will be limited. Once the product platform is sized, all of the product versions can be generated, completing the product line.

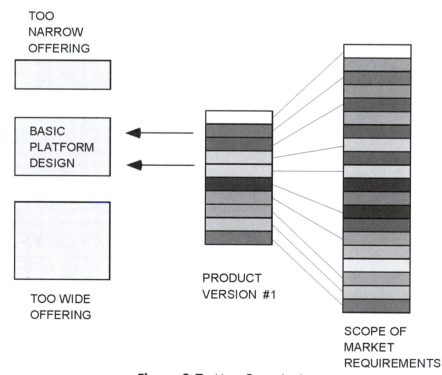

Figure 3-7. Mass Customization

3. HOW TO COMPETE WITH MASS CUSTOMIZATION

Competing with mass customization at your disposal is a very competitive means to obtain and retain market share. It allows you to target and secure new customer bases and to widen your company's offering beyond original expectations. It allows entry to markets you may traditionally be locked out of. It yields a great amount of product flexibility, in that if a new customer base is discovered, and an opportunity exists for the right configuration of features, the company can capture this customer expediently, rather than having to try to develop a product from scratch.

4. WORKING THE NICHES TO YOUR ADVANTAGE

Now that you have the flexibility it is necessary to work it to your advantage to grow the business. Select and target customers to secure and configure the products for that capture. It also affords you the opportunity to group feature sets to develop a tiered product line or an inclusive product line. The difference is in the way the features are grouped as shown in Figure 3-8. In the following there are three distinct product offerings, namely A, B, and C.

They can be configured in several ways as illustrated in Figure 3-8.

In the top part of Figure 3-8, products A, B, and C are generated from seven possible features. Product C has all of the features of product B, and likewise, product B has all the features of product A. Product C is an all-inclusive product configuration in this example.

In the bottom part of the illustration, product A is comprised of two features, configured to target a narrow audience. It consists of features #1 and #2. Product B consists of three features, namely features #1, #2, and #3. Product C has no more functionality than product B; however, it has a different target customer base, requiring that of features #1, #2, and #4. It is separate and distinct from product B, and can be launched separately and priced separately.

5. THE BENEFITS

The benefits of this type of arrangement are numerous. Mass customization offers flexibility, speed, and direct targeting of market segments. It offers an expanded customer base on which to build a business.

6. DISADVANTAGES

The benefits do come at a price, however. They require more effort on the part of product maintenance. This manifests itself in having to lay in updates to the basic product line in each of the versions. If a correction needs to be made to the basic platform, each version of the product generated from this platform needs to be updated. This requires strict revision

PRODUCT PLATFORM MANAGEMENT

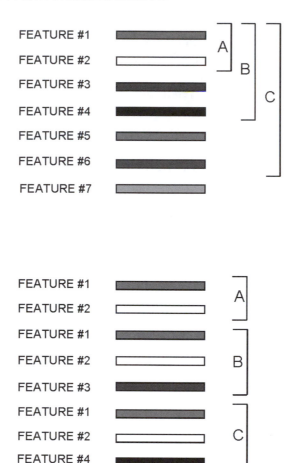

Figure 3-8. Product Platform Management

control and tracking through the product life cycle. Although suitable for some types of products, some companies cannot tolerate the product maintenance requirements.

7. TIMING

The mass customization impacts timing of the development program. Designing to target cost, with the flexibility of the mass customization platform, takes longer in product definition and in development, than with a single-dimensional offering. It offers the flexibility after the development; however, it will take more resources and maintenance to support. It should be used as a means for capturing diverse market share, not as an investigative means into a market.

8. COST IMPACT

The other impact item on the approach is the manufacture cost. The mass customization platforms are not as optimized as the single-dimensional offerings and therefore will have a higher factory cost to accommodate the requirements for flexibility. The company may need to lose a little margin to capture market share. It is the price to be paid for the flexibility. If the product line is a tiered offering with low- to high-price gradient, the overall performance of the product line will be good, whereas just the lower tier may generate a little less margin.

CREEPING FUNCTIONALISM

1. THERE IS A MONKEY ON MY BACK

Creeping functionalism is like a monkey on your back. It is added weight and distraction that can take a program off course from the original intended focus. The creep starts quite innocently and is generated by both the marketing and development personnel.

The marketing personnel, constantly watching developments in the marketplace, will have a tendency to add on to the specification to keep pace with the competitive offering. The other way this occurs is if the marketing people failed to accurately define the product initially. This is one of the most dangerous lacks of diligence that can occur, because it dispatches personnel and expense to chase a wandering dream.

Another way creeping functionalism occurs is when the development personnel discover interesting additions and features that add value to the product, and have a tendency to add them in. If the manager is not diligent in screening these ideas, creeping functionalism prevails and costs increase while the program drags on.

If the development team is talented and motivated properly, and the marketing team did their job properly, the manager's focus will be to screen and slot features, as a result of development creep, into the program where it makes sense, does not add cost, and doesn't result in changing the customer acceptability of the product. The objective is to encourage the creativity and harness the output, for the benefit of the overall program, without missing the window of opportunity.

2. CREEPING IS A COST ADDER, TIME WASTER/COMPLEXITY ADDER/RELIABILITY DEGRADER

Given the above, you may infer that creeping functionalism should be rejected at all cost. Contrary to this, the objective is to evaluate the various ideas that come up and decide if it makes sense to add these items into the requirement to be competitive in the marketplace at product introduction. There needs to be an objective means for evaluating these items for

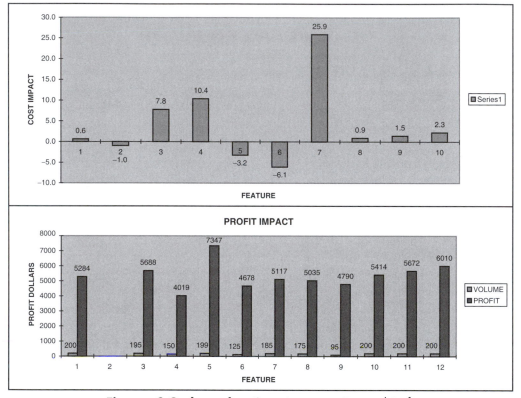

CREEPING FUNCTIONALISM AND PROGRAM IMPACT									
ITEM	COST		TOTAL COST	% IMPACT	SELL PRICE	GM	MARGIN $	VOLUME	PROFIT
BASIC PLATFORM COST	38.58	< +/−>			65	0.41	26.42	200	5284
FEATURE CREEP	COST IMPACT								
1 DESCRIPTION	0.25	+	38.83	0.6	68	0.43	29.17	195	5688
2 DESCRIPTION	0.37	−	38.21	−1.0	65	0.41	26.79	150	4019
3 DESCRIPTION	3	+	41.58	7.8	78.5	0.47	36.92	199	7347
4 DESCRIPTION	4	+	42.58	10.4	80	0.47	37.42	125	4678
5 DESCRIPTION	1.24	−	37.34	−3.2	65	0.43	27.66	185	5117
6 DESCRIPTION	2.35	−	36.23	−6.1	65	0.44	28.77	175	5035
7 DESCRIPTION	10	+	48.58	25.9	99	0.51	50.42	95	4790
8 DESCRIPTION	0.35	0	38.93	0.9	66	0.41	27.07	200	5414
9 DESCRIPTION	0.56	+	39.14	1.5	67.5	0.42	28.36	200	5672
10 DESCRIPTION	0.87	+	39.45	2.3	69.5	0.43	30.05	200	6010

Figures 3-9a b, and c. Creep Impact on Cost and Profit

inclusion or deferral to the next generation of device. Figure 3.9a illustrates the evaluation and planning method for feature inclusion into the product line.

The chart in Figure 3-9a lists the features, along with the basic cost for the product. Each feature is analyzed in terms of its factory cost impact and its acceptability to the customer base. This is indicated by a plus or minus in the column next to the feature. The factory cost impact and percent impact is noted also.

Next, the selling price is established. This was established previously by market research and product planning. Next to each feature is the assessment of the market price adder or detractor for each feature. The significance in the negative entries is that some features could be depleted and would require a price reduction in the marketplace.

The new projected margin, both in percent and dollars, would be calculated based on cost impact and market pricing evaluation. Next, evaluate and post to the analysis, the expected volume impact from the base assessment. In some cases for example, features #2, #5, and #6 resulted in less overall profit, even with a cost and price reduction, because of the volume reduction. Also, an added feature, such as in the cases of features #4 and #7, resulted in less-than expected profit because of increased factory cost, and less volume because of the higher price.

The one to seriously evaluate is feature #3. Although it adds cost, it has negligible impact on volume because the marketplace has recognized the added value and can support the price increase. It retains the volume and greatly enhances the profit even with the added cost.

As with any objective evaluation, the input data is most important to the output conclusion. The challenge is to be accurate in the assessment of the individual features that will be posted to the analysis.

Another way to look at the impact of the feature creep is the chart in Figure 3-9b, which shows the cost impact relative to each other.

Feature #7 shows a dramatic impact to the factory cost. As previously discussed, this represents only part of the overall decision. The other part is the profit impact, which is shown graphically in the chart in Figure 3-9c. In this analysis, feature #3 generates the most profit. This means of assessment can also be used in initial product feature selection.

For initial product planning purposes and feature creep valuation, a complete chart is available for your use in the Tool Box.

3. STRIKE BALANCE BETWEEN THE NEED FOR STRETCH AND THE NEED FOR EXPEDIENCY

There is a delicate balance that must be struck between the need for "freshness" in the design and product offering and the need for expediency in completing the development. The manager needs to keep creep in check and keep a watchful eye on the marketplace. This balance is best illustrated in Figure 3-10.

There is a baseline feature set that corresponds to the baseline market price tolerance and a baseline factory cost. The added features tend to creep in at a rate in excess of the market desire. It is a multidimensional issue because, as the development team adds features, factory cost is affected, and program schedule dictates how you may measure up to the ever-changing market demands.

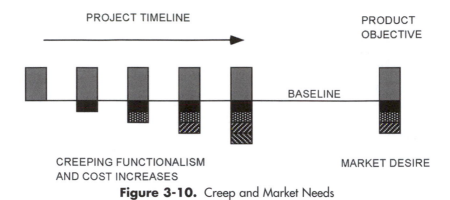

Figure 3-10. Creep and Market Needs

4. THE UNSEEN HARM: UNDUE COMPLEXITY TO THE USER

Even with features added to a product at virtually no cost, there may be still be impact to consider. There is the impact of time to develop the features, and the customer's acceptability of the more complex product. In our high-octane world, users are looking for immediate satisfaction and immediate solutions. They are not looking for research projects with each new product they use; consequently, manufacturers are driving more and more to simplify the complexity. This manifests itself in application-specific products, in which the feature set is customized to a specific application for the ease of the user.

5. THE CREEPING FUNCTIONALISM TRAP

There is a characteristic curve that describes creeping functionalism in a program. It can best be described as a chase in uncertainty. The curve is presented for review in Figure 3-11.

As shown in Figure 3-11, the product definition undergoes scrutiny and bears uncertainty a couple of times during the development. There is an original level of product understanding on which the program is launched. When it is in development, features get added to the product, and the team experiences uncertainty because of this creep of features. This is the first danger point.

As the project progresses, the manufacturing element needs to lock onto a design that can be tooled and set up for volume manufacturing. Whether the manufacturing-driven understanding bears close resemblance to the market requirements at this point is yet to be determined, however, the product is locked onto for manufacturing's sake.

At market introduction, the company obtains it first feedback on the product. Lackluster initial sales, or selective rejection in certain markets, may cause the team to second guess the product definition, and the second danger point occurs. Hopefully this introspection is unwarranted and the company can enjoy growth and volume and position themselves for the next development.

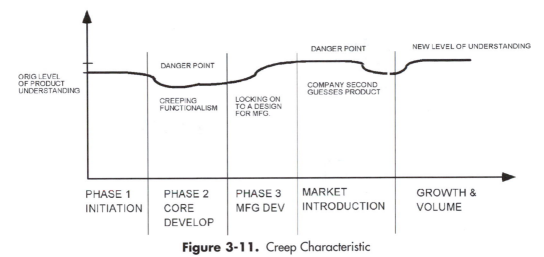

Figure 3-11. Creep Characteristic

DESIGNING TO COST

1. WHAT IS IT? IT'S A MUST IN ANY MARKET

Design to cost is a new product development philosophy that makes the factory cost a marketing requirement. It differs from traditional historical approaches, in which a product was conceived, designed, and priced based on cost plus basis, in which added value added cost. The design to cost philosophy is based on the principle that the customer has established a perceived value for goods and services, and the manufacturer must develop, manufacture, and market that product at a price not to exceed that perceived value. The difference in design philosophies is illustrated in Figure 3-12. As shown, and emphasized by the direction of the arrows to and from the marketplace, there is a vast difference in the two approaches.

The design to cost approach conducts accurate marketing research to determine customer needs and wants and attaches a value to these elements. It then works backward, from the customer, through the points of customer interface, back to the company, and eventually the development lab. By first determining the acceptable market price, the various added value elements are factored into the calculated maximum factory cost. The product produced by the organization must meet the market requirements.

In contrast, the cost plus approach differs in direction and starting point. It starts out with an idea of the customer wants, and accumulates rationalization along the pathway to product introduction. The development team starts with the concept and determines the approximate material cost to implement all the functionality. The labor is estimated based on the material content and burden is further estimated based on historical averages.

The estimated factory cost is then rolled up, and later in the development process, the actual factory cost is determined. I have seen instances in which the cost varies as much as 1.75 times as much as the original estimate.

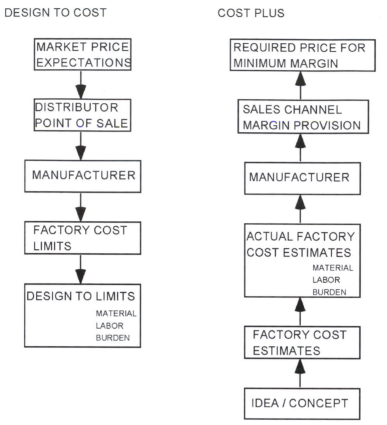

Figure 3-12. Design to Cost

The minimum acceptable gross margin is calculated and the minimum price the product will be shipped to the sales channel is determined. The pathway to market is then determined, and its associated costs are factored in to generate the minimum required price the product must sell for in the marketplace. Not withstanding the numbers involved in an example, it is the basic philosophy and direction of analysis that is flawed in the cost plus approach. It simply views the market as the end stage in the process, when it should be the driving force and focal point.

By viewing it as the end stage, it places the effect of all of the unbridled variability and uncertainty at the customer's door, when it should be factored and managed in the development lab.

The design to cost example in Figure 3-13 starts with a product that has an acceptable market price of $200. The various value-added activities are calculated in determining the maximum factory cost allowable at $88. The material labor and burden are then planned. The numbers at the left of the descriptions indicate the order in which the activities are executed. In design to cost, the spec is confirmed in step #10.

The cost plus example starts at the idea and is often a loosely defined specification. This prompts the conceptualization of the product to be wider than needed in an attempt to

DESIGN TO COST VS COST PLUS IMPLEMENTATION

	DESIGN TO COST	FINANCIAL		COST PLUS METHODOLOGY	FINANCIAL
1	MARKET PRICE EXPECTATIONS	200	9	MINIMUM SALEABLE PRICE	261.36
2	SALES CHANNEL PROFIT 20%	40	8	SALES CHANNEL PROFIT 20%	52.27
3	DISTRIBUTOR POINT OF SALE	160	7	DISTRIBUTOR POINT OF SALE	209.09
4	MANUFACTURERS PROFIT MARGIN 45%	72	6	MFGR'S MIN. REQ'D MARGIN 45%	94.09
5	MANUFACTURER'S FACTORY COST	88	5	MFGR'S ESTIMATED FACTORY COST	115
6	MFGR'S FACTORY COST LIMIT	88			
7	MATERIAL	61.6	2	MATERIAL	75
8	LABOR	13.2	3	LABOR	15
9	BURDEN	13.2	4	BURDEN	25
10	SPEC. AND PERFORMANCE REQMNTS	SATISFY	1	SPEC AND PERFORMANCE REQMNTS	IDEA

Figure 3-13. Design to Cost Vs Cost Plus

accommodate the uncertainty of the lack of accurate market information. An initial material estimate and factory cost estimate are made. The value-added activities are then factored in at traditional values and the minimum market price is determined. In this example, there is a significant difference between the two. Some customers can tolerate the difference; however, most will not. *A spreadsheet for your specific use is included in the Tool Box*

2. SOONER OR LATER EVERY PRODUCT MUST DEAL WITH THE COST ISSUE

Virtually no product is immune from the eventual market pressure associated with the design to cost issue. There is a short time in an emerging market fed by healthy demand in which a participating company may be shielded from the cost pressure issue temporarily; however, as these markets mature, and more competitors come into the market, the same characteristic forces come into play. Eventually the companies participating need to change over from cost plus to design to cost.

There is another proprietary reason to stay away from cost plus. This is the case in which the market price is established and the design and purchasing team effect cost reductions in material content. With a cost plus mentality, the savings in material will be immediately passed through to the customer without any benefit to the company, and if the company management only evaluates percentages, actual margin dollars can be lost.

3. LEARNING CURVE IS PROACTIVE DESIGN TO COST REDUCE CONTINUALLY

The learning curve cost reduction is an often misunderstood term, and for good reason. Its use lulls management into a false sense of security by sending a message to those responsible for a development program, that high costs today will be better tomorrow by virtue of the learning curve cost reduction. The fallacy of this assumption is that the same discipline required to hit the cost target initially is the same discipline required to effect these cost reductions. If the team cannot hit the cost target initially, there will be no "learning curve" thus exacerbating the eventual price pressure.

The learning curve requires consistent development work to effect a reduced factory cost. If there is no work, there will be no net reduction. In addition, the fallacy is that the purchasing function is the primary driver of the learning curve phenomenon, whereas in actuality it is a group effort of purchasing, manufacturing, and development that is needed to coordinate the reduction in cost.

4. SENSITIVITY OF DESIGN TO COST AND COST PLUS IN A COMPETITIVE SITUATION

The lack of design to cost control of a product line left unchallenged, either initially or through normal increases (i.e., vendor substitutions, increases in labor rates), can place pressure on the profit margin. Figure 3-14 illustrates the point that the manufacturer can be placed in a squeeze play in which they cannot respond.

With a fixed market price and increasing costs, the margin will deteriorate as shown in the chart in Figure 3-14. The margin problem is exacerbated when the market pricing levels are dynamic and also deteriorating. This is evidenced by the drastic change in slope of the margin with both the price and cost affected as shown in the second graph.

In any competitive situation, a company needs a certain amount of flexibility in price negotiations. If there were no controls on cost, or maintenance of costs, the amount of flexibility is reduced. In addition, the profit is impacted significantly depending on several factors as shown in Figure 3-15.

As shown in Figure 3-15, there is a range of products proposed in a bidding situation. The products have an initial baseline cost and gross margin based on a market price. For this example, the volume of units sold will be inelastic to price. The profit is generated from the difference between the original cost and the selling price. This is shown in the upper section of the spreadsheet. If the factory cost goes up the profit erodes, as shown.

If the pricing level deteriorates, the profit level will erode further. As shown in the example where the last three products in the list are outlined, the margin goes negative because of the increase in cost and the erosion of the pricing level. This puts the bid into a squeeze play whereby the manufacturer is locked out of participating.

Since a bidding situation is dynamic, and the market seems to follow the latest bidding price level, prices can deteriorate with time. If the cost creeps up, you may be in a situation where you cannot afford to take the business. Furthermore, if the customer is seeking a bundled solution of equipment, you may be prevented from even selling the profitable items. Consequently, the design to cost initiative is so important, because your company needs the latitude to navigate in the competitive arena. *A copy of a spreadsheet is available in the Tool Box for your use.*

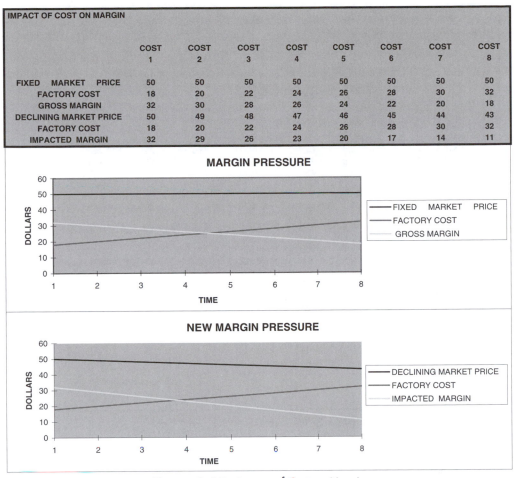

Figure 3-14. Impact of Cost on Margin

5. WHY THIS DOCTRINE DEMANDS ACCURACY AT PRODUCT DEFINITION; CHECKS AND BALANCES

As evidenced by the various examples, the issue of factory cost is essential to gain control of early in the development. Failure to do this is not recoverable later and prevents your ability to effectively compete. Since all subsequent development activity is generated from the initial definition of a product, it must be accurate so as not to send the development team in the wrong direction.

Design to cost is a good methodology for product development, but it cannot tolerate midcourse corrections or changes in the product concept. Generally, product development programs are not affected as much by wild inaccuracy in the initial stages as lacking sufficient definition. The problem occurs when companies ignore the uncertainty by not resolving

PROFIT IMPACT OF COST PRESSURE

ITEM	DESC	COST	BASE SELL PRICE	BASE GM	BASE MARGIN $	BASE VOLUME	PROFIT	$VOLUME
1	UNIT #1	40	68	0.4118	28.00	195	5460	13260
2	UNIT #2	38	65	0.4154	27.00	150	4050	9750
3	UNIT #3	26	78.5	0.6688	52.50	199	10448	15621.5
4	UNIT #4	50	80	0.3750	30.00	125	3750	10000
5	UNIT #5	23	65	0.6462	42.00	185	7770	12025
6	UNIT #6	55	65	0.1538	10.00	175	1750	11375
7	UNIT #7	78	99	0.2121	21.00	95	1995	9405
8	UNIT #8	67	68	0.0147	1.00	200	200	13600
9	UNIT #9	89	92	0.0326	3.00	200	600	18400
10	UNIT #10	76	78	0.0256	2.00	200	400	15600
							36423	129036.5

ITEM	DESC	% ADD	IMPACTED COST	IMPACTED SELL PRICE	IMPACTED GM	IMPACTED MARGIN $	CONSTANT VOLUME	IMPACTED PROFIT	
1	UNIT #1	5.0	42.0	63.92	0.343	21.9	195	4274.4	12464.4
2	UNIT #2	5.0	39.9	61.1	0.347	21.2	150	3180	9165
3	UNIT #3	5.0	27.3	73.79	0.630	46.5	199	9251.51	14684.21
4	UNIT #4	5.0	52.5	75.2	0.302	22.7	125	2837.5	9400
5	UNIT #5	5.0	24.2	61.1	0.605	37.0	185	6835.75	11303.5
6	UNIT #6	5.0	57.8	61.1	0.055	3.3	175	586.25	10692.5
7	UNIT #7	5.0	81.9	93.06	0.120	11.2	95	1060.2	8840.7
8	UNIT #8	5.0	70.4	63.92	−0.101	−6.4	200	−1286	12784
9	UNIT #9	5.0	93.5	86.48	−0.081	−7.0	200	−1394	17296
10	UNIT #10	5.0	79.8	73.32	−0.088	−6.5	200	−1296	14664
								24049.61	121294.31
							DIFFERENCE	12373	7742.19
							%DIFFERENCE	−33.970	

Figure 3-15. Profit Sensitivity with Cost

it early, proceed in the start of the development, and add the clarifications along the pathway of development. This changes direction, reduces momentum, and stalls the program.

6. DESIGNING FOR LOW MANUFACTURING COST

As stated in the previous section, the definition is critical to designing for a low manufacturing cost. The specification is the basis for which all the alternatives will be generated, and as such, if the specification is floating, the alternatives generated are not comparable. In addition, there is no pathway for the evolution of the product from an implementation standpoint. This makes it a one-time program with no subsequent leverage. The optimum means for product implementation is to start with a solid specification and determine the lowest-cost, most-flexible platform to implement the design, while factoring in future developments.

7. COSTING METHODOLOGIES

There are several different costing methodologies in use today. The issue of product costing and the various types will be discussed in more detail in Chapter 5. The important analysis to include is the relationship of the product's cost components from one product to another. If, for example, labor is accounted for, and is a larger percent of sales in the new product than in the existing product, then there may be cause for concern. At the same time, if the material is drastically reduced, the product cost may be able to tolerate the additional labor. More often then not, however, the trend in the industry is to absorb additional material cost to reduce labor.

8. THE TRIAD OF MODULARITY, COST, AND DESIGN SPEED

There is a triad used in product development that describes graphically the interrelationship of product factory cost, time to develop, and performance of the product. It is shown in Figure 3-16a. It is fundamentally based on the premise that once a product is established and a specification locked onto, the triad forces compromise into the process.

As shown in Figure 3-16a, the factory cost development time and product performance is in equilibrium. If any of these three parameters is more optimized, it will come at the expense of the others. The concept is that you can push any two of the parameters to optimization, but not all three. Figure 3-16b is an example of factory cost and product performance being pushed at the expense of time to develop.

Likewise, Figure 3-16c shows time to develop and product performance being pushed at the expense of product factory cost.

Finally, in Figure 3-16d, the factory cost and the time to develop are pushed to the expense of product performance, as shown.

DEVELOPMENT ENGINEER'S INFLUENCE ON FACTORY COST

There are several components that make up the factory cost of a product. This section discusses these components.

1. BILL OF MATERIAL COST

If design to cost is that critical to a new product success, then where does the majority of the cost occur and how is it levied on the product? The development engineer is capable of influencing the majority of the product factory cost, up to 70% in some cases. They select the materials, designate the platforms, and match components to features.

There is also a difference between prototype and initial volume and production costs that needs to be factored into the analysis. A Bill of Material (BOM) cost worksheet is available in the Tool Box.

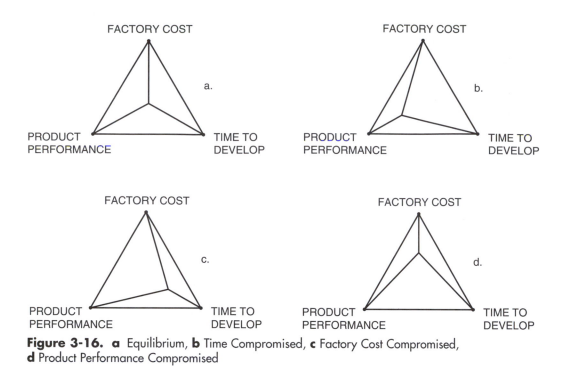

Figure 3-16. **a** Equilibrium, **b** Time Compromised, **c** Factory Cost Compromised, **d** Product Performance Compromised

2. PRODUCTION LABOR—DIRECT

The direct production labor can make or break a new product's cost structure. It is a measure of the development engineer's ability to apply materials and processes to embody an idea. The production labor is a measure of the effectiveness of these processes. If they are not accurate, effective, and deterministic, the labor will be excessive.

3. PRODUCTION LABOR—INDIRECT

The indirect production labor is a measure of the after-labor processes. It draws on the elements of the organization that are not specifically associated with the direct labor content; however, they are required for support of their labor. If a high involvement is required here, then it may be a sign that the processes are poorly defined and need remediation to be effective.

4. TEST

Test, being a part of the production direct, is a good example of this. If test is required in interim checking of the product during manufacture, and production labor is not performing this function, there may be a problem with the process being poorly defined or not in control. If the test function is indiscriminately applied to solve production problems and anomalies, you are not in production; you're still in the development stage. This is a serious problem for quality and consistency.

5. NEW PRODUCT DEVELOPMENT TOUCHES EACH

New product development touches each of the areas of production labor (either direct or indirect labor), the processes, and the manufacturing issues. If there are structural problems in the manufacturing organization, this will point them out. Often this is why new product development requires so much time and energy. The organization must be redesigned on several fronts to accommodate the product, the assumptions, and the planning that was done initially.

The management statement, "That's ok, we will just have to handle it when we get to that point," tells the story. If that was the prevailing attitude initially, and no changes were made as part of the program, management can expect longer than expected delays while the organization changes to accommodate the new product.

6. COMPARATIVE COST STRUCTURES

One of the best ways to highlight the issues being discussed is to do a comparative cost analysis. Figure 3-17 will clearly indicate the difference between actual design and production capability and assumed capability for the new product and the need for improvement.

As illustrated in Figure 3-17, the various components of the cost structure can indicate organizational differences or required changes. For the analysis to be truly comparative, the Y-axis must be normalized, and be presented at a percentage of net sales revenue. To facilitate this example, the existing product is on the right-hand side and the new product is on the left. Refer then, from right to left.

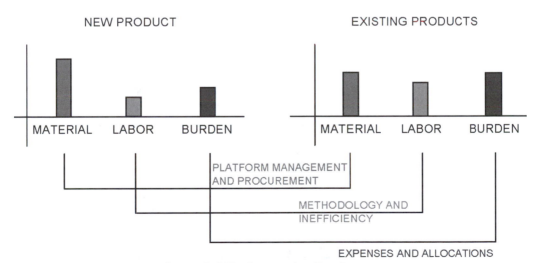

Figure 3-17. Comparative Cost Structure

The change in material content from the existing product to the proposed is an indication of platform management by the development engineers. In this case greater material cost for the new product is offset by the decrease in labor to assemble it. This can also be an indication of investment in tooling.

The change in labor goes to the issue of improved methodology and efficiency in the labor force. Will the new product demand personnel being directed in a more efficient manner with processes and procedures that remove variability?

The third area is the burden and manufacturing expenses. This is often a result of allocations and extra expenses associated with the manufacturing organization.

7. VALUE ENGINEERING AT THE PRODUCT'S INCEPTION

The value engineering exercise needs to be completed at the beginning of a program. It can be the means by which value is attached to the various features of the product. It is a structured way to differentiate the needs from the wants and to determine the degree to which the customer will pay money for them. The earlier section in this chapter on creeping functionalism, when modified, outlines the means by which value can be placed on the individual features.

As shown in Figure 3-18, adding priced features to the basic product will affect volume and profit. Improved gross margin, based solely on cost and price without the volume, will not drive profit, looking at feature #4. However, an incremental price adder with incremental cost can drive overexpected volume to generate more profit.

This type of analysis and planning is useful when developing the requirement's specification for the development team. *A separate spreadsheet is available in the Tool Box for your use.*

MANUFACTURING

1. INTEGRAL PART OF THE NEW PRODUCT DEVELOPMENT SYSTEM

Manufacturing is an integral part of the new product development system. The term *system* is wholly descriptive, because new product development is not an isolated series of tasks performed

VALUE ENGINEERING

ITEM		SELL PRICE	ADDED COST	TOTAL COST	GM	MARGIN $	VOLUME	PROFIT
BASIC PLATFORM PRICE		87.5	0	45	0.49	42.5	200	8500
1	ADD FEATURE 1	91	0.25	45.25	0.51	45.75	200	9150
2	ADD FEATURE 2	93	0.37	45.37	0.52	47.63	200	9526
3	ADD FEATURE 3	94	3	48	0.52	46	189	8694
4	ADD FEATURE 4	104	4	49	0.57	55	125	6875
5	ADD FEATURE 5	89	0.56	45.56	0.49	43.44	225	9774
6	ADD FEATURE 6	95	0.35	45.35	0.53	49.65	175	8689

Figure 3-18. Value Engineering

in a vacuum. It is an integrated interdependent program requiring the cooperation and commitment of the organization. This is why it is critical to involve the various functional areas at the onset of a program. This subject will be covered in more detail in Chapter 8.

One of the most critical and so-often–overlooked functional areas is the manufacturing element. They are overlooked until the project reaches a certain point where they are contacted. Usually by then, it is too late to benefit from their input. The manufacturing element can affect the factory cost, the production quantity, and the quality of the goods produced.

The manufacturing element, whether it is internal to the organization or external to the organization, is a partner in the new product. There must be give and take in the interplay between development and the manufacturing people.

2. DON'T UNDERESTMATE ITS IMPACT OR VALUE

There is a term used in the new product development arena referred to as *over–the-wall syndrome*. It is best characterized as a fractured development team in which the development engineer designs a product in the vacuum of manufacturing. When the bills of material and drawings are complete, the engineers turn them over to manufacturing for their first look. Manufacturing then must develop manufacturing strategies to build the product. There are no tradeoffs at this point, so what you get is what you get.

3. THE KEY TO MOMENTUM

The key to momentum is to get very proficient at moving a product concept through development and manufacturing. Rather than this being a sequential process, this must be more integral, in which both disciplines contribute to facilitate the program through to completion.

As shown in Figure 3-19, the integrated approach will execute a program in a more complete manner than the sequential manner. In addition, the sequential method always results in a timeline longer than originally anticipated because it is fraught with delays, restarts, cost overruns, lack of optimized designs for manufacture, and tooling compromises. This is because of surprises at the development to manufacturing and the manufacturing interfaces.

4. PURCHASING, PARTS SPECS

Development engineers design products based on forecast volumes. The purchasing and procurement function aligns all of their activities based on these numbers. The manufacturing discipline, via this procurement function, affects the factory cost of the products produced. In addition, the quality of the product is affected by purchasing, by virtue of configuration control of purchased parts and substitutions. For a long-term look, they affect quality by managing the conversion of obsolete parts in continuing to build product.

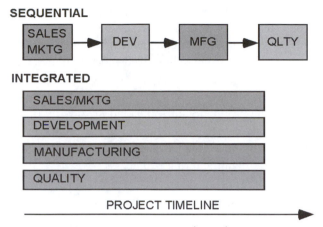

Figure 3-19. Integrated Development

5. CONTROL

Engineering change control processes are also an integral means for affecting the quality and cost of a product. They need to be managed to retain cost structure and performance of the product.

6. PRODUCT CONFIGURATION

A large portion of the problems encountered in manufacturing can be designed out at the product definition stage. There are methods for product configuration that can facilitate the product line. These fall into following three basic areas:

A. Raw parts count
B. Portability of assemblies over the product line
C. Forward compatibility

At this point in configuring the product, always try to minimize the raw parts count. Undue complexity involving human interaction through assembly generates a host of training, retraining, and variability issues. This causes difficulty when trying to trace down root causes of problems.

Try to leverage subassemblies so they can be used over the product range or line. This will increase volume in custom assemblies thus lowering the cost, as well as assist in balancing the production line.

Design the product line such that actions taken today do not lock you out of supporting the product in the future. This means that original decisions need to be forward-compatible to subsequent product support decisions. The three factors to consider here are duration of the life of the product line, degree of supportability, and cost.

7. REPORTING, FEEDBACK, ANALYSIS, CORRECTIVE ACTIONS

Once a product is designed, manufactured, produced, and sent out to the field, there needs to be a feedback mechanism to allow customer comments (clear of interpretation from outside sources) to be channeled back to the development people. The field feedback serves the following two major purposes:

First, it provides the necessary information for product weaknesses, performance, and ease of use back to the source. The second, and in a longer-term sense, more important purpose is it plugs the development group into the customer. A specification can articulate needs and wants, and internal people can try to influence development; however, nothing has as much effect as a customer visit or direct feedback to the development personnel. For improvement, nothing can influence the urgency of a corrective action faster or more effective than direct customer feedback to development.

8. FORECASTING AND SCHEDULING

These are two basic requirements for the manufacturing sector that directly contribute to the success of a product line. The manufacturing sector depends on an accurate forecast to drive materials and manpower allocation. This is one of the most difficult areas to get resolution on in some companies. The difficulty is encountered when the initial market development work is not done properly. At the time of product introduction, there is no accurate information on the volume or mix of the line. If the proper work was completed, it is a simple matter to update the forecast, and issue to manufacturing.

The forecast also directly affects the company's financial performance through the balance sheet item known as inventory. If the forecast is inaccurate, the incorrect materials will be driven and there will be no consumption to offset the procurement.

Additionally, if the forecast is incorrect, the purchasing function does not purchase in patterns originally planned on which the factory cost was based. Consequently, material variances show up due to expedited purchases, wrong parts, changing quantities, and missed shipments.

GLOBAL MARKETING

1. WILL THE NEW PRODUCT BE MARKETED GLOBALLY?

It is difficult to conceive of a product that is developed and won't be marketed globally. The world is becoming more homogeneous, and products developed for the world markets are becoming more prevalent. With counter-cyclical behavior in several markets, most manufacturers are making export an increasing part of their product portfolio. Dependence on export relies on products defined and developed for the global market.

2. HOW IS THE SPEC AFFECTED?

If a product is to be globally focused, the specification must include that requirement at the onset of the development. Few products can be accepted globally without being developed as such. The requirements simply cannot be laid on top of an existing design. If you're going global, go early and go prepared for it.

3. HOW IS THE COST AFFECTED?

A product line capable of domestic adherence, and also foreign adherence to specifications, may require a cost structure that may be higher than the separately focused products. This needs to be reevaluated against breaking out the product line to devote to each and keep track of all of the support and maintenance issues of product management.

4. MANUFACTURING VENUE

Another thought to keep in mind is the venue of manufacturing. Are all versions manufactured in one location for distribution throughout the world, or will each market be served by a separate manufacturing venue? Where will the parts come from? These issues need to be addressed initially, which will negate the need for product rationalization later.

5. SENSITIVITY TO CURRENCY EXCHANGE RATES

The issue of exchange rates was discussed in a previous section, and the issue of import duty can also affect the success of a product. Unfortunately, strides in cost reduction through engineering effort can be overridden by changes in tariff, import duty, and changes in relative standing of currency. A sensitivity analysis at this point may be in order as part of the overall plan.

6. CONTROL OF TECHNOLOGICAL ADVANTAGE

There is also another factor to consider when marketing products in foreign countries. This is the issue of proprietary technology. How do you maintain control of competitive advantage through technology when engineering and manufacturing may be overseas. The control of this advantage is critical to the long-term prospects of growing the business. There are several examples that could be cited here; however, it is sufficient to just be aware of the issue at this point.

7. MOTIVATION FOR PROFIT ABROAD AND FAR EAST PARTNERS

If there is a partner involved, again, evaluate their motivation for profit and their tax structure. These elements will govern their primary actions. If taxes are high, the partner may be motivated to operate at no profit, content to only fund their growth, making repatriation of funds difficult in a timely and worthwhile manner.

8. EXPORT COMPLIANCE REGULATION

This issue is not one to take lightly. Export compliance is one of the strictest, most unforgiving arms of the federal government. Violations go to the security and the foreign agenda of the federal government. If your company deals in goods that are prohibited from being shipped to certain foreign countries, you must adhere to the law regardless of knowledge of the law, changes in controlled products, and destination countries. For exporting, your company must set up a real-time evaluation system to maintain compliance.

REQUIREMENTS SPECIFICATION

1. THE PATHWAY TO UNDERSTANDING THE CUSTOMERS' EXPECTATIONS

The requirement's specification is a vehicle to organize and document a list of various customers' wants and needs into a specification defining the new product to be designed and built. It is not just the sum total of all of the suggestions, ideas, and new product nuances; it is, rather, the compilation and selection of those features and requirements that are most appropriate to the success of the product.

The pathway to the marketing requirement's specification is best illustrated by Figure 3-20.

As shown, the new product is birthed as an idea that may originate anywhere, from the most internal parts of the organization, to the customer. This idea must then be coalesced in a new product opportunity. This opportunity can then be qualified in terms of secondary market research and then primary market research. The combination of these two studies defines the company's perceived value of the product to the marketplace.

Next is the definition of the value to the individual customers. This occurs in the iterative process of product positioning, cost evaluation, and determination of customer value. This is one of the most important steps as the product definition must be accurate and a measure of what the customer will purchase.

This generates and finalizes the marketing requirement's specification, which is now ready to issue to development.

2. BASIC ELEMENTS OF A REQUIREMENT'S SPECIFICATION

There are several basic elements of a requirement's specification that can be reviewed. The basic outline of these elements is as follows:

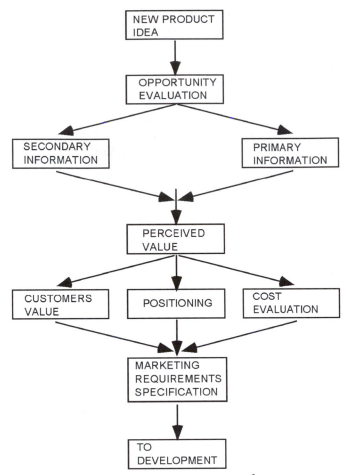

Figure 3-20. Requirements Specification

Background

The background section of the specification should identify the general market and customer base that the new product will be involved in. It serves to direct the audience into the specific area for consideration. A historical perspective is in order here, as well as a general projection of what will be in the future. This will help the reader focus on the specific product under investigation and serve to better evaluate the rest of the specification.

Industry Trends

The industry trends section allows the manager to discuss the general trends in the industry under investigation. Specific references to products and technologies should be included here. This area should be specific enough so the reader can understand the basic industry, learn the trends, and picture the opportunity. There are several items to be included here, namely: the general application of technology, the speed of transition, the industry's acceptance of change and improvement, and tolerance for learning.

Market Opportunity

This section outlines the selected opportunity from all of the background information. It discusses how the development of this new product will capture market share. Make sure to be specific in the identified opportunity, rather than dabbling in generalizations. If the specification is to have meaning and be useful, the identification of the opportunity should be accurate and specific enough to be able to be referred back to and redirect any company or development efforts. If the team should get "lost" they should be able to find their way back via this area of the specification.

Tie to Strategic Plan

The tie to the strategic plan is also critical to the specification's value. It lends legitimacy to the development effort. It also will be invaluable to reiterate the program's importance and positioning after the program starts and second thoughts may be occurring. Again, the information needs to be brief, tangible, and specific.

Scope of Product Line

The opportunity also needs to be framed in scope. Identify the product's positioning, the markets served, the breadth of the product line, and the number of versions. Outline size and rating performance requirements in terms of market segments, and identify the scope of how the product will satisfy the market demand. Take care to be specific in identifying how selected versions address specific segments.

Component Parts and Product Configuration

This section discusses how the product line needs to be configured to address the market. It outlines the component parts and how they will fit together to comprise the line. It needs to be as accurate as possible because this part of the document will be used to generate the manufacturing scope and plans and any other downstream activities.

Functional Sequence of Operation

This is a detailed description of the operational characteristics of the product. It should be written from the user's perspective, and be as complete as possible. In addition to describing the function of the product from a positive aspect, it should describe the function of the product under misuse and wrongful use (i.e., What happens when the wrong key sequence is entered?). An often-overlooked aspect, this is useful to the development engineers to perform their function in a more complete manner.

Performance Requirements

The performance requirements are the complement to the functional specification. It specifies the operating envelope for the product. It is the responsibility of the specifying engineer to take care in this part of the specification, since over-conservative specifications and superlatives here can cause undue complexity at the development level. In addition, the degree of over-design is specified here. The degree of over-design is that amount of additional capacity beyond normal advertised use. It has correlation to several items, such as technology employed, clarity of channel, relationship between the manufacturer and the customer, standards, and manufacturing variability.

The Operating Envelope

1. ELECTRICAL

   A. PERFORMANCE

   Describe how the product should perform electrically

   B. ENVIRONMENTAL

   Describe how the product should endure in an electrical environment (i.e., spikes, noise, etc.) and perform to specifications

2. MECHANICAL

   A. PERFORMANCE

   Describe how the product should perform mechanically

   B. ENVIRONMENTAL

   Describe how the product should endure in a mechanical environment and perform to specification for the product's expected life.

3. TEMPERATURE

   A. PERFORMANCE

   Describe how the product should perform over a temperature range and still meet specifications

B. ENVIRONMENTAL

Describe how the product should perform when subjected to temperature extremes

4. HUMIDITY

A. PERFORMANCE

Describe how the product should perform over a humidity range and still meet specifications

B. ENVIRONMENTAL

Describe how the product should perform when subjected to humidity extremes

Standards

What are the relevant specifications that the product must adhere to in the markets to be served? This is a critical element on which to be accurate, because designing to the wrong specifications can waste development time and create a situation in which compliance is not possible or practical after the design is complete.

Unfortunately compliance to standards cannot be overlaid on to a product line to achieve compliance. It must be integrated as part of the original design. If not, the tradeoffs made during a design process, by an individual engineer or engineering team, will now have to be executed through engineering change control. In addition, compliance to several standards after the design fact may not be possible given an existing design, parts availability, and cost constraints. It is critical to get this accurate at the onset.

Cost Target

Product opportunities are only financial opportunities if the factory cost is consistent with all of the other assumptions. To that end, the factory cost target is as important a specification item as any other. Failure to achieve this can ruin a program in the short and long term. Be sure to factor in the trends of cost items and forecast the factory cost target in the time frame when the product will be produced. This is especially true of electronics-based products in which the cost decreases as rapidly as it does.

Timing to Introduction

If cost is critical, so is meeting the window of opportunity. A product introduced late cannot be offset by improvement of the other elements. You simply must deliver a design on time and at cost to be effective in the marketplace. In fact, time is usually the most critical and least recoverable element in the new product development arena.

Human Factors Engineering

If humans use the product, then factor in their needs and wants for ease of use, accuracy in use, and overall acceptance. For example, complex electronics-based products with complex readouts and menu structures are simply not good choices for elderly people. Consequently, ignoring these human factors in engineering the product will result in nonacceptance. Conversely, designing a product for ease of use by the elderly will prove to be a success in the marketplace.

Safety

Aside from the altruistic reasons for engineering a product for safety, there are financial reasons, and in some countries, personal liability reasons for engineering safety into the product line. The marketing requirement's specification should outline the safety reviews and their expected measure of effectiveness. In addition, when documenting a program during development, these safety issues and resolutions and safety reviews should be recorded for possible future use in a defense.

Longevity and Product Life

Every product has a useful life that is finite. It performs its function for some time and then it fails. If it is a repairable product, it can then be repaired and have continued life. That is the meaning of the following terms:

1. MTBF: The MTBF assumes a repairable product, and is the mean time between failures.
2. MTTR: The MTTR also assumes a repairable product, and is the mean time to repair.

The requirement's specification should outline these issues and prescribe the expected life of the product.

Service Plan

If the product is repairable, then there needs to be a plan to integrate the service of the product. Is it designed so a third-party repair facility can repair the unit or does it have to be repaired at the factory?. Where do repair parts come from? How is the design controlled? What about parts substitutions? Is there a warranty on the repair? All of these issues need to be defined, and the design needs to reflect the plan.

Field Replacement Parts

Following on with the service plan, are there going to be field replacement parts or sub-assemblies? If so, these products need to be designed as part of the product, and be factored into the manufacturing strategy. What is the control plan for these parts and subassemblies?

Corporate Standards

Finally, identify what specific corporate standards must be followed in the development of the product. Resolve any conflicts between the market requirements for standards and internal requirements. For example, if the market standards allow for lower quality, and price pressure causes little or no margin or de-rating, and your company's standards are higher, ask yourself if this is the market segment that your company can be successful in. It represents a divergence from traditional positioning of products by your company, which goes to the overall strategy and product planning.

THE INTERRELATIONSHIP OF THE SPECS

During the course of development of a product line, there are several specifications that the development team needs to concern themselves with. Each basically traces and documents the pathway of the product though the development and manufacturing process.

1. MARKETING REQUIREMENTS SPEC

As discussed earlier in the section, the marketing requirement's specification identifies the customer's needs and wants and coalesces these requirements with internal corporate and external market requirements and conditions. It is the basic foundation of the program and must be accurate to be effective and achievable. It is issued to development for analysis and planning purposes. Most important, it initiates a discussion between the marketing, development, and manufacturing elements of the company.

2. DESIGN SPEC

The design specification is the development team's response to the marketing requirement's specification. If the marketing requirement's specification is the summary of what is wanted, the design specification is a summary of what can be done. The challenge is to lead the target by time-phasing the two to take advantage of changes in fast-moving technologies.

3. SPAF: DESIGN VERIFICATION, QUALIFICATION, AND PRODUCTION TESTING

SPAF stands for the *s*ignificant *p*erformance and *a*pplication *f*eatures document. This is the performance document produced after the development is complete and the product is released to manufacturing. It outlines the specifications, from an operating perspective, that the product actually does. For example, if market requirements specified, "needs to tolerate a temperature of 65° centigrade," the design specification outlines tolerance of 70° centigrade and the SPAF shows test data that the actual unit can tolerate 68° centigrade. In this example the manufacturer is reputable in the advertised ratings, but the actual performance fell short of the design spec; however, the unit was still ahead of the marketing requirement's specification.

4. TYPICAL SPECIFICATION

The typical specification is a sales tool that describes the operating envelope of the product. It is generally used as a means for third-party project engineers to specify your product in projects. It is generally written feature specific without specific reference to names of manufacturers. This means that it is written such that, the degree to which the typical specification is accepted by the specifying project engineer, is the degree to which your company can bid the project and not have competition without that competition having to take exception to the specification. A cleverly prepared specification can lock out competition to a great extent.

SUMMARY

This chapter had several objectives spanning those activities in taking a concept from identification of an opportunity, through all of the aspects, to position it as a revenue- and profit-generating opportunity for the company. It started off with how to generate the new product idea and coalesce it into a business prospect for the company. Competition was discussed in terms of its effect on development.

The all-important route to market for the product was reviewed and outlined. Also discussed was how the new product will be integrated into the company, and all of the caveats involved.

The product configuration was next, along with mass customization, the discussion leading now into the hard features and functions of the product. The issues of creeping functionalism and its impact were reviewed, along with tools to mitigate its effect.

The concept of design to cost was reviewed in detail in the effort to enhance chances of the new product's success on the marketplace.

Finally, the basic elements in producing a specification were presented, as well as a discussion of the interrelationship of the various specifications encountered in a development program.

The reader should now have a good understanding of how to determine the market opportunity and coalesce it into a workable specification from which the company can develop, introduce, and generate planned revenue.

This provides the basis for the next section, Creating the Business Plan: The Road Map to the Company's Future!

CHAPTER 4

THE PRODUCT AND BUSINESS PLAN

ABSTRACT: This chapter deals with the planning of the new product opportunity within the framework of the business. You now have generated the new product concept and refined it into a product opportunity capable of generating revenue and capturing market share. Now the activity focuses on preparing the business and product plan, which will be the road map to success. The objective of this chapter is to act as a guide in structuring an approach to the business plan, and ensuring all requisite steps are addressed so that the company can realize a return for their investment.

THE PLAN

1. ELEMENTS OF A BUSINESS PLAN

The business plan for new product development can take several forms, depending on the circumstances and the type of product and businesses. They do, however, share many common traits and elements, and it is these elements that are mandatory for its value. The business plan is a means of obtaining funds, in addition to serving as a road map for the development of the business. Within this context, the business plan needs to be written such that it can initiate the appropriation of sufficient funding to execute a program. This is true for a group within a larger company, as well as for a start-up company seeking venture capital. The story told in the business plan should be effective and complete enough to secure financing from whatever means.

2. FORMULA FOR SUCCESS

No one format for a business plan will generate instant or guaranteed success. Each product opportunity will require different treatment, analysis, and planning. If there is a formula for success in preparing a business plan, it is in factoring in the objectives, the opportunities, the caveats, and the contingencies into a cohesive, executable plan and sticking to it through completion.

3. INCLUDE ALL OF THE ELEMENTS

There are temptations to gloss over certain parts of a plan. Depending on the circumstances, however, this is never advisable. It is these loose assumptions that will haunt the program later. For example, in a larger company with an established sales channel, it may be tempting to gloss over the route to market. This may prove to be fatal for the program, if the product does not fit into the existing channel or the existing channel is too expensive to secure the market. The lesson here is to include all of the basic elements of the plan so as to double-check all of the assumptions.

4. TOUCH ALL AREAS OF BUSINESS INITIALLY SO AS TO INTEGRATE; OTHERWISE, NEW PRODUCT DEVELOPMENT WILL NOT BE ABSORBED BY THE BUSINESS

If there is such a thing as a trial ballooning a business plan, then it should effectively trigger the various organizational immune systems in the corporation. The desire here is to see where the organizational boundaries lie and what corrective actions need to be taken to secure commitment to the endeavor. Make sure that all areas of the plan address all functional areas of the business. It is an effective way to garner support for the program. Failure to address this early will cause problems later because the business will not take ownership for the product, as they should. The program will be best served by determining the organizational issues. The business plan is an easy way to surface these issues and gather support.

5. REVIEW OF THE PLAN ELEMENTS

The following represents a typical example of the basic elements of a business plan. The purpose is to include all of the information required to assuage the concerns of the investor. This is important for acceptance whether it is in seeking funds from a venture capitalist, mezzanine financing, or from the next layer of management in a corporation.

A. EXECUTIVE SUMMARY

The executive summary is the quick overview of the entire plan. It is designed to allow quick review of several plans in a short period in order for the manager or investor to narrow down a list of opportunities to more closely scrutinize. It consists of the following four elements:

1. Business mission statement

The business mission statement provides the reader several critical pieces of information. It positions the business in terms of its products and services and how they relate to the customers and marketplace. It also states the general objectives of the concern and how it perceives itself in the cadre of competition. It also defines the scope of what the company's operations is about and will serve as a check against the proposed opportunity. Finally, it provides specific focus for the company to discourage wild deviations to tempting, but off strategy opportunities.

2. General market information

The next part of the summary is a legitimization of the market conditions that the plan is predicated on. General market data supporting the case for the new product development should be documented here.

3. Brief statement about competition

Since no marketing action is without reaction, and competition is always at the doorstep, a brief statement about the competitive climate is in order. The objective here is to convince the reader that the idea, business, and plan will overpower any competitive threat.

4. Critical factors for success

Every program has critical factors for success. These are usually company specific and must be reconciled as part of the initiative. Cite these factors so the reader can weigh the financial risk.

B. STATE OF THE COMPANY

This section of the business plan establishes the state of the company; its health (financially), the people, and the direction.

1. Status of the business

The first element to include is the current status of the business. It must be reviewed from two perspectives, internal and external. The format should consist of two basic elements:

a. External review perspective

The external review is a summary of how the market and competitors view the business and the company's participation in it.

b. Internal review perspective

The internal review is an introspective assessment of how the company and its managers view themselves, with respect to the marketplace and its customer base.

2. Corporate objectives

The plan also needs to have the company objectives in it. Be sure to include both the near-term and the long-term objectives. This will indicate a certain amount of resolve on the management team's part, as well as formulate an opinion in the investor's mind. This will be important if the program should founder a bit. In addition, state the long-term ownership objectives of the company. Is it growth for near-term sale or long-term measured growth and stability?

3. Management team description

This should not be a listing of résumés, since that will be included as detail information in an appendix. Rather, this should be a description of how the management team has accomplished something together. Again, this will give reassurance to the investor that the people in place can actually carry off the program. As part of this section, be sure to include the management team's objectives and how they perceive themselves as part of the enterprise.

C. THE MARKET AND ITS NEED

The plan now needs to focus on that segment that will eventually generate the revenue though the new product development program. Define the market need in terms of specifics. What set of customers will buy what, when, under what conditions, and at what price level? The following are the parts to this section, which characterize the market:

1. Introduction

The introduction is simply a means for the reader to become acquainted with the general area of the market under scrutiny.

2. Industry trends

The industry trends should give the reader an idea that the industry is a vital market to participate in, as opposed to a declining market, which may be overcrowded with players.

3. Market description

This section describes the general market in terms of wants and needs. It should outline what the customers buy and why they buy those products. It is designed to create familiarity on behalf of the reader with the patterns of market behavior.

4. Market trends

The market trends section needs to accomplish two basic things. The first is to cite and document specific trends in the marketplace from a historical perspective. The second is to project what the future trends will be based on past behavior and new technology available.

5. Market numbers/segments

This is the area to be specific and accurate. It is, in fact, the basis for the investor's due diligence in making the investment. The number presented here will be corroborated by a third-party source. Failure to be accurate and specific here will lead to rejection of the plan.

6. Market growth and stability

The investor needs assurances that the program they invest in will have a certain amount of growth stability. The market numbers and segments analysis can substantiate the conclusions in this section; however, it will be helpful to prepare a narrative explaining the company's flexibility in creating revenue in this market with the new product.

7. Competition analysis

Review what is happening with the competition in this area. Describe the competitive trends and how they may affect the new product plans. Include an explanation on the following:

a. Domestic versus global

Are there global players who may displace domestic development activities? Are there any threats to consider?

b. Functional displacement

Can the new product be functionally displaced by a flanking market move?

c. Sensitivity to external partnerships

Are you at risk by two or more competitors teaming up against your new product?

D. THE PRODUCTS

In this section of the business plan you need to clearly articulate the scope and configuration of the products. In addition, the plan must contain a description of how you will develop these products on time and on budget, and how you go from an opportunity to profitable products. What are the product descriptions? Define the scope, boundary, and use of the products in this description. What is the cost structure of the products and will they generate profit? Explain why the products are a better long-term solution than what is currently available.

Reemphasize the difference between research and development. Reassure the investor that the team understands the difference. Also, show that the product needs development, not research. This will go a long way to assuage their fears about funding endless activity without return.

Finally, develop and present a detailed product development schedule and timeline showing expected progress and measurable milestones. The development plan should include the following:

1. Schedule with dates, and timeline

2. Funds expenditures with timeline

3. Critical factors for success

4. Technology basis for the products

E. MANUFACTURING

In this section the business plan needs to describe the manufacturing system that will be used to manufacture the product. Hopefully it will be consistent with the company's present manufacturing system; otherwise it could be perceived as an additional development in these systems. If the technology needs to be competitive on the world market, then the manufacturing system also needs to be competitive.

1. Describe who will manufacturer the product.

2. Describe what manufacturing systems will be used.

3. Describe how the labor force will be trained.

4. Describe how the certification of the labor force will be maintained.

F. MARKETING AND SALES

In this section describe how the products will be taken to market. Cite the route to market and its cost, as well as the strategy used in that route. Show how the customers will be targeted and include a typical expected timeline for a sale, from the introduction to the acceptance of the product. Describe who will market and sell the products. Factor in how you will capture your company's share of the channel participant's time and have them devote it to your product.

G. PRODUCT LAUNCH

Describe how and when the product launch will take place. Cite expected initial sales and forecasted demand. Project how the competition will react to the product launch. Develop and present countermeasure plans to sustain momentum. Describe the expectations of the channel participants and their expected performance for the company. Review the competitor's route to market and cite any differential advantages you may have. If there is an international element to the program, factor this into your presentation.

H. PROGRAM FINANCIAL SPECIFICS

The format for the business plan has largely been comprised of a narrative at this point. In this section specific numbers are expected to be presented for the investor's evaluation. The following are some of the essential elements:

1. Development budget requirements

2. Manufacturing budget requirements

3. Marketing budget requirements

4. Sales forecast (unit sales, not dollars, growth patterns in units, and timing)

5. Revenue projections

6. Profit projections

7. Balance sheet impact

8. Returns on investment and other measurables

9. Risk analysis

I. CONTINGENCIES

Develop and cite contingency plans for each critical aspect of the plan. This will demonstrate that the plan was well considered, that the management team who authors the plan understands that things can and do go awry, and that recovery needs to be in the new product vocabulary. Make sure to cover the contingencies that are affected by people, finances, and technical issues.

J. AFTERMARKET PLANS

If the nature of the business has aftermarket requirements, these also need to be included in the plan. If there are incremental additions of manpower required for the product line, they need to be factored in at this time. Include elements such as repair, service, and customer support

APPENDICES

1. PRODUCT SPECIFICATIONS

Include a detailed product specification for the investor's due diligence.

2. RESULTS OF MARKET ASSESSMENT

Include the summarized results of any questionnaires, market assessments, or general feedback to the product concept. Use primary and secondary results.

PROGRAM TIMING

1. TIMING IS EVERYTHING

If timing is everything for the development and introduction of new products, then it is also everything in obtaining funds to execute a program. Despite the commonly used phrases, funding is not always readily available for good opportunities. Many good ideas go unnoticed and unfunded because the timing was inappropriate or the presentation was ineffective.

The concept of timing needs to be viewed as a window of opportunity to align with, rather than a blind sense of urgency. The manager needs to seek funding from the sources, at a time where and when they are available, respectively. They also must be consistent with the investor's parameters for investing. If the investment medium is a short-term player who is used to producing immediate results, it is foolhardy to approach him or her with a long-term program that will require massive funding.

A common misconception, and a source of the misunderstanding about funding, is that small funding is easier and more readily available than large funding. One needs to remember that the parameters for the investors funding are based on timing, amount of returns, and risk, not the absolute amount of funds. If the parameters are in line, the specific amount of cash invested may not mean that much to the speed of securing the funds.

2. CONTINUITY OF PROGRAMS

Funding of a business plan may require several phases, or at least, the funds will be released in discreet steps. It is wise to align with an investor who can fund an entire program and allow you to finish what you start. One of the worst things that a manager can do is secure funding for only a part of a program and fail to provide the team with the necessary continuity to complete it.

The problem manifests itself where different investors will have different requirements and agendas. They may be at odds with the initial fund provider, or worse yet, the program will have to be compromised to secure the funding.

3. HIT THE OPPORTUNITY WINDOW

The best way to secure funding is to hit the opportunity window of the investor with a program requirement that is consistent with their investment parameters. Keep in mind that as odd as it may sound, the investor is in a situation in which they are under pressure to invest funds in viable programs. Your job in preparing and presenting the business plan is to help them in selecting and funding your plan by demonstrating that it is the best choice. From a product-planning perspective, the plan must sell the investor into a full program, not a halfhearted attempt at one. If it is not funded fully, it may compromises the development, as depicted in Figure 4-1.

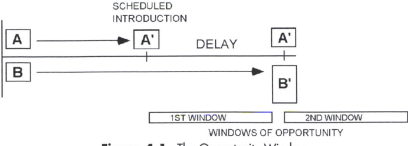

Figure 4-1. The Opportunity Window

In this example (Figure 4-1), the first window of opportunity is available for product A'. This is the result of a development in iterating from A to A'. If the program is delayed, as shown in the bottom section of the illustration, the available window of opportunity for A' ceases to exist. Unfortunately, many companies see the second window of opportunity as a continuation of the first and assume the market will be there for a long time.

What is really happening is that a second window of opportunity has been made available for product B'. This product can address more of the market than A' can, so A' lost the window of opportunity, and more important, will not generate the revenue it needs. From a pure business perspective, the business plan failed to secure adequate financing, which slowed development and execution; this in turn caused delaying product introduction, thus losing the window, resulting in the reduced revenue.

4. THE STATISTICS OF TRYING TO CATCH UP

The importance of timeliness to the marketplace cannot be overstated, especially when the funding stage is involved. The entire enterprise needs to create financial and development progress, a momentum that cannot be substituted by other means. Doubling efforts, reorganizing the program, or throwing additional personnel at the problems cannot be a substitute for steady progress. It simply takes 9 months for the gestational period of a human being. It cannot be foreshortened by impregnating nine women to have a child in 1 month!

5. THE FINANCIAL IMPACT

To get a feel for the financial impact of slipped schedules caused by underfunding, refer to Figure 4-2.

Figure 4-2 shows two equivalent programs. The top line in the graph shows the on-time program bringing in the revenue as planned. It initiates at t_1 and contributes to t6 time period. The bottom line in the graph shows the same program, but delayed. It doesn't start until t3 and stops contributions at t6. The time base is fixed because this represents the "window of opportunity"; the area under the curves is the totalized revenue. It shows a signifi-

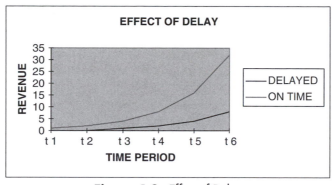

Figure 4-2. Effect of Delay

cant loss because of the delay, which was a result of poor funding or program management. Keep this graph in mind when securing funds to make a case for adequate funding, and also as a reminder to keep programs on track.

STRUCTURING THE BUSINESS PLAN

1. MARKET MATURITY

One of the most valuable exercises is to determine the level of market maturity that exists in the specific market under consideration. Depending on the degree of maturity, the market can be flat growth, low growth and high growth. A mature market will be flat or no growth. Conversely, an emerging market will be more growth oriented. Figure 4-3a illustrates the difference, and will be used as a basis for illustrating the market share and profit potential of each.

As shown in Figure 4-3a, the differences in emerging and mature markets manifest themselves differently. This affects the business plan and the program. The high-growth markets have more opportunity associated with them. They generally effect lifestyle changes in the customers. An example of this is the automobile industry in the early 1900s. It was a growth market that caused sweeping changes in the lifestyles of the customers; it was very lucrative and fast-changing, which challenged the manufacturers to keep up with the changing market demand.

The low growth market has less opportunity, is more price sensitive, and can be thought of as a zero sum game. What market share you increase is obtained from the competitor only, not a growing number of customers using the new product or technology. Company growth is determined by swings in the economy, more than other factors, and there is stability in the marketplace.

Figure 4-3b shows the impact on the companies participating in these markets.

The benefit shown in the emerging market is that the company can gain absolute shipment dollar revenue through increasing its market share in an increasing market easier than in the low-growth markets.

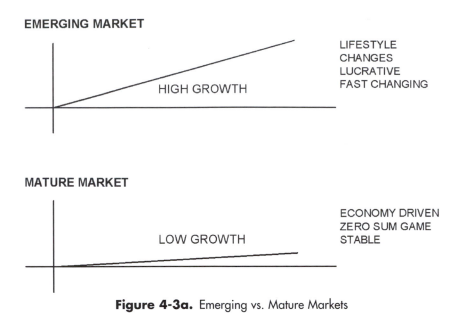

Figure 4-3a. Emerging vs. Mature Markets

In either case, as part of the business plan, there should be a clear understanding of what the market dynamics are and how you intend to participate in them.

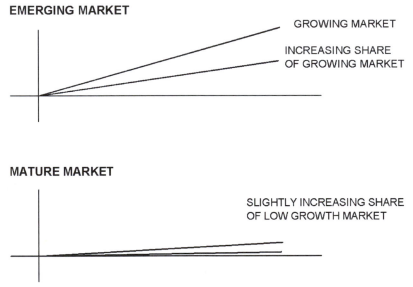

Figure 4-3b. Company Participation

2. LIFE CYCLE AND MARKET CYCLE IMPLICATIONS OF PROFIT

Figure 4-4a illustrates the specific impact on a company operating in these diverse markets. In the emerging market the economic benefit to the firm in dollars of margin far outweigh what can be generated in a mature market.

These examples are for illustrative purposes only, and the absolute values or percentages are somewhat exaggerated for effect in illustrating the difference. Also, the scale of market size and revenues has been scaled for the example. However, the analysis shows the vast difference in actual gross margin between the two markets. In the emerging market there is a growing percentage of market share to work with. In the mature market the behavior is relatively fixed. The graphs show the difference.

Another critical point to remember is that markets are not exclusively either mature or emerging. They start out as emerging and matriculate into mature markets. Therefore the business plan developed for new products and business has to factor this into the development and account for the transformation.

Figure 4-5 indicates how this may look in terms of market growth per time. The market growth starts out small and increases dramatically, and then remains somewhat constant to a peak growth rate. The growth rate then diminishes to essentially zero, where the market size remains constant. This assumes, of course, a normal distribution and behavior without overt manipulation by any one dominant competitor, who might be large enough to unilaterally effect a change in the marketplace.

3. LOW-COST PRODUCER

There are certain strategies that a company can use to navigate the changes in markets and use them to their advantage. In the following graphs the various behaviors will be presented in the framework of the various markets, to show why it is desirable—from a survival point of view—to be a low-cost producer by the time the market has matured.

Consider the set of graphs in Figure 4-6, where a contrast is drawn between a mature market and an emerging market implementing older, less cost-effective technology and newer, cost-effective state-of-the-art technology. The behavior of prices and costs tells the story of company profits.

4. LEADING-EDGE TECHNOLOGY

In the first scenario, a mature market being addressed with older technology, the cost remains relatively flat while the prices are driven down. In the second example, a mature market with state-of-the-art technology, the decreasing prices are offset by decreasing costs. The third example shows, in a growth market, the prices remain stable while the demand for functionality can quickly outstrip the older technology's ability to supply. In this example

GROWTH PATTERNS

	YEAR 1	YEAR 2	YEAR 3	YEAR 4	YEAR 5
EMERGING MARKET					
AVAILABLE MKT	1000	2500	4500	8500	12500
MARKET SHARE	1	3	7	12	18
REVENUE	10	75	315	1020	2250
COST OF SALES	0.6	0.65	0.7	0.725	0.75
MARGIN	4	26.25	94.5	280.5	562.5
MATURE MARKET					
AVAILABLE MKT	1000	1500	2000	2500	3000
MARKET SHARE	1	2	4	6	8
REVENUE	10	30	80	150	240
COST OF SALES	0.7	0.725	0.75	0.775	0.8
MARGIN	3	8.25	20	33.75	48

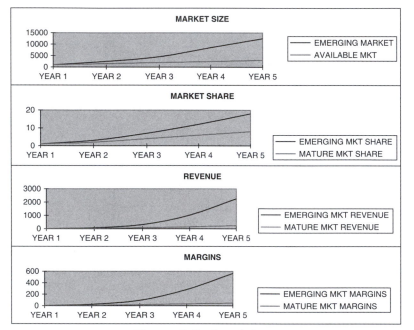

Figure 4-4. Growth Patterns

sales cease after a time because the older technology is limited and cannot satisfy the market needs.

In the fourth and final example, a growth market with state-of-the-art technology, shows prices slightly declining with costs declining at a greater rate, thus preserving the margin.

As shown in Figure 4-6, the best overall strategy a company can take is to be at the leading edge of technology in order to keep current with the market requirements for functionality and price, as well as cost containment, to maximize profits. It simply puts your company ahead of the competition initially, so that you can establish and maintain a market position.

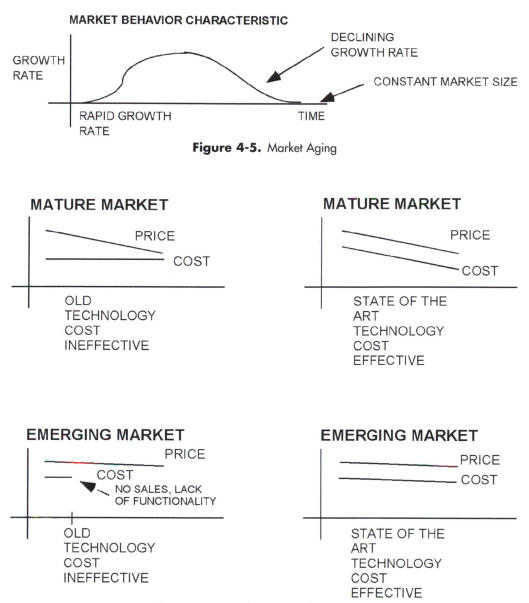

Figure 4-5. Market Aging

Figure 4-6. Market Aging Characteristics

The danger is in allowing the company to be placed in a "squeeze" situation whereby the market is demanding a low-cost, highly functional design and you have a high-cost, barely functional design to offer. There will be lackluster sales and minimal profits, resulting in thin profit margins and eventual inability to reinvest properly.

PRODUCT MIX/OFFERING

1. HOW WILL THE PRODUCTS BE BROKEN OUT (i.e., COSTS, STRUCTURE, MODULES, LINE COMPLETERS, SIZES)?

The determination of how the product will be configured, and what will comprise the product line, will help determine the financial risk. A product line consists of several parts, although it has a basic product as its core. These line extensions and peripherals add to the risk of development and the cost to develop. How the product will evolve and how the pieces will be added are decisions that must be made at some point in the development..

Figure 4-7 illustrates the risk and management of uncertainty in configuring the product line. As shown, there are four line completion products numbered #2, #3, #4, #5 generated from the #1 product. The increasing line indicates the investment increasing with each product segment. The top (increasing) line indicates the ability to manage the uncertainty. The more invested and the further through the development of the product line, the greater the ability to manage the uncertainty. The sawtooth (decreasing) line indicates the uncertainty itself. It decreases through the development cycle, and rises to a new level with each start of a successive development. Overall the uncertainty reduces through the development period, but spikes with new sub-projects introduced.

From the development standpoint, it is desirable to design all the pieces concurrently with all of the requisite resources. In this way the lowest cost structure can be selected

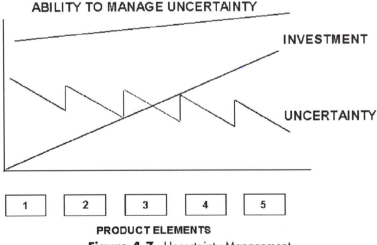

Figure 4-7. Uncertainty Management

and implemented. Unfortunately, it is rare that all of the pieces can be developed concurrently. Rather, they are developed sequentially with planning for future models and forward compatibility. In other words, forecast the added models and the "hooks" it will need to operate with the flagship unit and standardize on those elements in the flagship unit. Care must be taken to ensure the standardized platform will be a cost-effective and competitive one throughout the term of the subsequent developments and the life of the product run.

2. MARGIN PLANNING AND ENHANCEMENT

When configuring a product line, there should be some effort to plan the margin initially and a plan to enhance the margin after initial launch. This margin enhancement can generally be effected through a cost reduction rather than a price increase. Depending on the commercial aspects of the development and the product marketing strategy, there may be price pressure immediately after launch, so the cost reduction is generally in the best long-term interests of the long term of the company.

3. CONFIGURING THE MIX FOR FUTURE ADVANTAGE-MANUFACTURING ISSUES

One of the chief contributors to low cost and manufacturing success of the product comes from orchestrating the design to leverage off of commonality between versions of the product. By reducing costs of the product and uncertainty of future inventory, future manufacturing success can be gained. If the design team fails to implement this effectively, inventory will swell and lead times for the product will extend as product versions and forecasts will rarely match. This contributes to always having shortages or the wrong subassembly in inventory to satisfy an order requirement.

4. DISCUSS THE VULNERABILITY OF THE MIX

The danger of missed implementation is positioning the company in a vulnerable position because of the product mix. Consider the example in Figure 4-8.

The market price for the product is linear with the size or rating of the product. The cost structure is as shown with decreasing cost for decreasing size or rating, down to a floor cost. The quantity is high at the low end of the rating with the fixed cost (at a low market price level). Profits here are low to possibly negative. Don't get caught being a supplier to the product 1 scenario, in which the higher margin, lower quantity segments like product 2 and product 3 may not offset the losses generated by the lower end of the market (product 1).

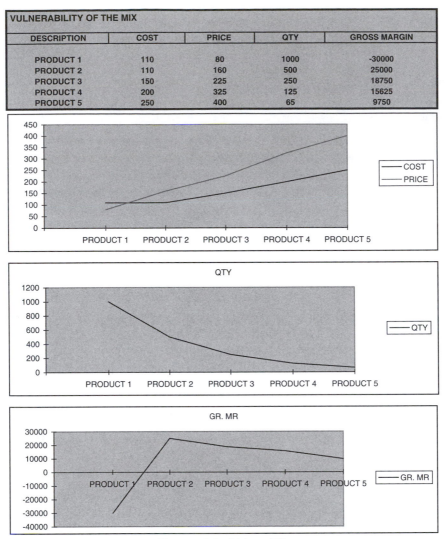

VULNERABILITY OF THE MIX				
DESCRIPTION	COST	PRICE	QTY	GROSS MARGIN
PRODUCT 1	110	80	1000	-30000
PRODUCT 2	110	160	500	25000
PRODUCT 3	150	225	250	18750
PRODUCT 4	200	325	125	15625
PRODUCT 5	250	400	65	9750

Figure 4-8. Vulnerability

PRICING POLICY

1. BACKGROUND OF PRICING

"So what's it worth to you?" This old adage, which has almost folklore status, clearly articulates the basic concept behind pricing of a product and service. As will be presented subsequently, there are numerous strategies and approaches to pricing, depending on the circumstances of the new product. One tenet should ring through: Pricing should reflect the perceived value in the eyes of the consumer, or what it is worth to the consumer! The pricing is a fundamental component to the successful execution of the business plan

All to often, product-marketing personnel make the mistake of setting pricing without clearly understanding the dynamics of the buying decision, the buyer's alternatives, and the value the buyer places on the goods and services at the time of purchase. In addition, the conditions that may lend validity to that value can change or shift, causing a price to be less reflective of the consumer's value placed on the product.

The key element in setting pricing is to understand the customer's value system and structure your package of values (the product) accordingly.

2. (EXTERNAL VERSUS INTERNAL) STRUCTURE/LONG-TERM GOALS

There are two facets to consider when setting pricing—the internal requirements for the organization and the external market price level. These two are opposing, because the internal corporate need for profit is placed in check by the external pricing limits the market has tolerance for (as shown in Figure 4-9).

A good starting point for a pricing exercise is to take the market price and determine the costs and overhead, then see how the company can make money with the product at the particular market price level. For example, the pricing breakdown (see Figure 4-10) for a manufactured product shows a $100 price. The profit before provision for tax is at 9% of the sales price.

Can the ownership structure's requirements and profit motive be satisfied with 9% of sales? This is the internal orientation of pricing when the market sets the level.

There are rare circumstances in which the manufacturer can dictate the price level in the vacuum of market pricing. This occurs where the manufacturer or service provider is a market leader; there are no tangible customer alternatives, and time, space, and venue circumstances work to the goods and services provider's advantage. In this case, albeit rare, the world is in fact beating a pathway to your door to buy that better mousetrap!

There is also the issue of the morality of pricing. Pricing should be an equitable and even exchange between the buyer and the seller. Profits realized in the transaction are generated from the firm's efficiency, know-how, skills, and ability to market. Price gouging and taking advantage of a buyer's no-alternative buying decision will generate bad will and eventually work against you.

There are three basic, but differing, objectives in setting pricing. They are as follows:

Figure 4-9. Internal Versus External

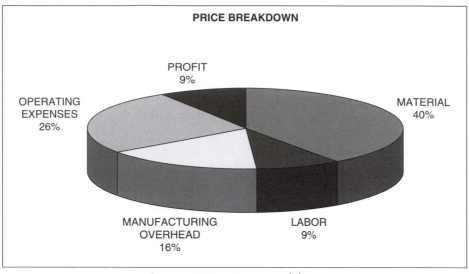

Figure 4-10. Pricing Breakdown

A. Sales objectives
B. Profit objectives
C. Competitive objectives

The sales objective focuses on top-line bookings and shipment figures. It may factor in profit requirements or competitive pressures; however, its main focus is to generate volume through attractive pricing.

The profit objective's main focus is the profit for the corporation. It may factor in other elements; however, maximizing profits is the main goal here. The profit objective, if left unchecked, can wreak havoc on the reputation of the organization.

The competitive objective's main focus is beating the competition in the market-place by offering the most attractive price for the goods and services. This as well as the sales objective, when taken solely, can have devastating effects on the organization.

The pricing objectives are supported by a cadre of price strategies. As a starting point, it is helpful to think of pricing in a triad. The three vertices of the triad are costs, demand, and competition, as illustrated in Figure 4-11. This triad formulates the basic operating arena for pricing strategies and market pressures. From this, each of the strategies will be defined and reviewed.

3. RELATING PRICING TO VALUE

The process of relating pricing to the value the customer places on the product is similar to the value-engineering process when configuring the product. It is based on the ability of

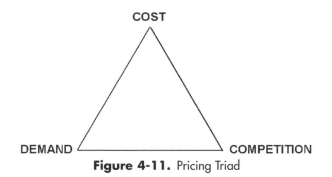

Figure 4-11. Pricing Triad

the company to understand the mindset of the customer, and to attach value pricing to each feature and to the entire package of values embodied in the product. As a starting point, let's examine the process in simplistic terms.

Figure 4-12 is a chart to assist in evaluating the features of the product and relating them to a price. *An additional value-pricing chart is available in the Tool Box for your specific use.* Each feature is listed on the left. The estimated price relating to each feature is established. The product is then considered in total and a total price is established. Then each competitor is evaluated in a similar manner, this time factoring in the features, which may or may not be included (as denoted by the inserted X's).

Next, the positioning of the product, with respect to the competition, must be decided and established.

The overall complexion of the marketplace is now in place so that the pricing may be set.

4. PRICE VERSUS QUANTITY

The concept of a price/volume relationship, in which the volume increases if the price is lowered, has often been misapplied. In certain markets and under certain circumstances, this

RELATING VALUE TO PRICING

FEATURE	PRICE EST	COMPETITOR A	COMPETITOR B	COMPETITOR C	POSITIONING	PRICE LEVEL SET
1		x		x		
2		x	x			
3			x	x		
4		x	x			
5				x		
6		x				
7				x		
8		x	x	x		
AS A PACKAGE						

Figure 4-12. Value Pricing

methodology does play out. However, it is not a universal concept, and indiscriminate use of it will only sacrifice gross margin on each unit sold, without generating additional sales.

If you are attempting to use this concept, it is imperative to understand the sales transaction thoroughly, first. Understand where and why the motivation for the exchange of funds for product takes place; otherwise you may draw the incorrect conclusions from someone else's history.

The other caveat to remember, when being tempted to drive volume by cutting price, is that price is only one component of the transaction. It cannot be the sole driver of a buying decision in a vacuum of other factors. Since the buying decision is not a single-dimensional issue, the single dimension of price reduction cannot necessarily drive huge volume increases.

Cost-Based Strategies

The cost-based strategies consist of several approaches, as listed. They all ignore the demand side of the marketing equation and focus solely on internal needs.

A. Formula pricing

   This is a strategy in which the cost elements or subcomponents are manipulated in a mathematical arrangement to generate a price large enough to absorb all other costs, transaction uncertainty, and provide for profit. It has pure internal focus and neglects market acceptance criteria. For example, if the product cost is $50 and the formula is a 3x multiple, the price is $150. Fifty dollars is the cost, and the balance of $100 is for other costs and profit.

B. Cost plus

   The cost plus strategy is like a snowplow, in that it simply adds the required expense numbers and the required profit to the base cost to generate the price. This absorbs the uncertainty and provides for profit by decree. The premise is that the general level of consumption is stable, known, and understood. It also assumes that the market for the product is less sensitive to price.

C. Cost plus by intermediaries

   This can be characterized by accumulating price as the product progresses through the sales channel. Each intermediary will get his or her "cut" on the way to the customer. The dynamics are channel dependent and involve "customary" markups. They can be different for different classes of product within the same sales channel.

D. Target return

   The target return approach seeks a specific return on the transaction, either in percentage or raw dollars. It is based on an assumed volume and quantity. The actual versus planned volume and quantity is a critical issue in securing the return. Overhead and other costs are absorbed across the volume (qty) of products sold and are a key factor in the calculation of the return.

E. Break even

Break even is a strategy, usually short term for healthy programs, in which the total revenue minus the total costs is zero, leaving no profit for the corporation. The caveat here is to thoroughly understand your entire costs so as not to have one product selling at the expense of another. There are numerous methods to calculate the break even base on absorbed costs; however, the basic philosophy is the same: move product, generate volume, and secure zero profit.

F. Experience curve

The experience curve strategy is based on two cost-related phenomena: economies of scale and the experience curve of the organization. Simply stated, the greater the volume of product manufactured by an organization, the greater the economies of scale. This is the application of the fixed costs over a larger volume base. In addition, other economies of scale can also be gained in procurement of goods, manufacturing setup, and cost reductions. This experience cost basis can then be applied to pricing in several ways, depending on the competitive circumstances.

The other component is the experience curve the organization takes with the increased volume. As the organization's rate of learning increases, the pathway to lower cost gets shorter. It is important to note that pure quantity of produced product does not in itself reduce costs, rather, it is the ability of the organization to absorb and learn from the volume that reduces cost.

G. Marginal cost price

The marginal-cost pricing strategy is based on a price that allows an organization to recover primary or direct costs as a minimum. It is not the total cost recovery and therefore should be used temporarily, and with care.

As an example, the manufacturer may want to lead the experience curve with this type of pricing; however, they need to ensure the company does in fact progress down the learning curve.

Demand-Based Strategies

The demand-based strategies consist of several approaches. These strategies focus on a different perspective, that being of demand-side pricing.

A. Prestige pricing strategy

The prestige strategy is based on the premise that some product categories that can command higher prices can bring higher sales volumes. Particularly popular with safety items and purchases based on quality (e.g., food), the prestige strategy indicates to the buyer that increased price indicates increased value; in those categories, increased volumes are the result.

B. Price elasticity strategy

This strategy applies to products in which the price directly affects the volume of the product. There are three types of price elasticity strategies: elastic demand, inelastic demand, and unitary demand (see Figure 4-13).
   The elastic demand behaves in a manner in which a small decrease in price generates a large increase in volume.
   The inelastic demand behaves in a manner in which changes in price, either up or down, will not materially affect the volume.
   The unitary demand is a more idealistic behavior in which a percentage change in price generates an equal percentage change in volume.

C. Price range strategy

The price range strategy can also be thought of as a class of product. For example, the Mercedes-Benz automobile is in the high-end, high-price category and is in demand by a few select consumers. It, however, has a price range for the consumer. The lower limit of the price range still sets the perceived quality level as high.

Competition

The competition-based focus is an active pricing strategy that must be nimble and react to the rapidly changing marketplace. As competition changes their price, your company must alter your price or change the package of values. Sometimes the competition does not initiate a pricing action; they initiate a maneuver to gain market share. Your company needs to respond to maintain a presence in the game.

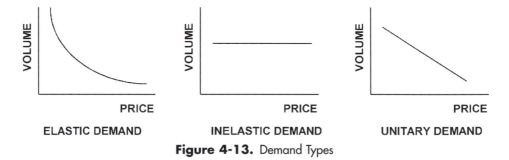

Figure 4-13. Demand Types

Figure 4-14 illustrates some of the possible dynamics of a typical situation. There are three competitors other than your company in the marketplace. Competitor #1 initiates an offering with X as the price. Competitor #2 responds with a revised price (X'), and you respond to the marketplace with (X"). Competitor #3 is nonreactive in keeping price (X) during the interchange.

In this way the market is dynamic and competitors are participating in pricing exercises based on competitive price levels.

Other Strategies: Situational and Value Pricing

There are other pricing strategies that are situation specific and are based on value, which are summarized as follows:

A. New product introductions
B. Pricing when intangibles are important
C. Pricing in oligopolies
D. Pricing when buying is habitual
E. Pricing to reflect buyer-behavior attitudes

These are basically offshoots or temporary applications of the previously discussed examples in order to gain advantage in certain situations. They, along with pricing over the product life cycle, will be covered in more detail in Chapter 11.

Whatever the strategy or for whatever the goal, pricing must reflect value to the customer and allow for profit to the corporation. It is therefore a key component to the business plan and must be well considered.

5. BUNDLED VERSUS UNBUNDLED PRICING STRATEGIES

The other consideration in pricing is the use of bundled and unbundled pricing. In this case the package of values can be adjusted to suit the market, by including collateral goods

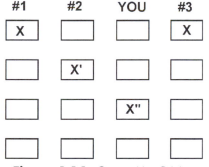

Figure 4-14. Competition Pricing

and services with the product to make the entire package more appealing to the customer. These can be very popular and very successful as consumers begin to examine total installed costs in which bundling can lower their outlay for the entire package of goods and services.

6. SCRUTINIZE PACKAGE OF VALUES CAREFULLY; KNOW WHAT IS WORKING AND WHAT IS NOT BY MARKET FEEDBACK

One final note about pricing: Each assumption and each plan may appear flawless on paper; however, it must be verified in the field with actual transactions under the planned or assumed scenarios. This adds credibility to the assumptions and prevents marching down a path that will not yield results. Get early feedback as a point of reference and monitor periodically for any required changes.

FACILITATING CHANGE IN THE BUSINESS TO EXECUTE THE PLAN

1. NOT ALL COMPANIES ARE STRUCTURED TO EXECUTE THE PLAN WRITTEN FOR THE NEW BUSINESS

As was stated in previous chapters, the marketplace is a fast-changing place with interwoven dynamics. A company positioned for a specific type of business today may not be positioned well tomorrow. If the business and product plan and the company's present direction diverge, then the company must be brought in line to execute the plan. The key is to recognize this at the time of the preparation of the business and product plan, and to integrate the necessary changes into the company as the development progresses.

2. THERE GENERALLY ARE REQUIREMENTS FOR SLIGHT MODS TO MAJOR OVERHAULS

The type of changes can vary in scope and by functional area. They do, however, fall into one or more of the following categories:
  A. People
  B. Systems
  C. Equipment

Is the current workforce equipped from a tools-and-training standpoint to prosecute the new development and the new business? Are the manufacturing, management information systems, and quality systems that are in place functioning and sophisticated enough to support the new business? And is the company's equipment state-of-the-art enough to support operations? These questions must be addressed and any concerns resolved.

3. EACH AREA MAY NEED TO BE IMPROVED

The business plan should outline the scope and extent of the changes required. There should be a cautionary note, however. If there are widespread changes required in each and every area of the business across all of the functional areas, then the business plan may be too much of a stretch for the organization. Initiating change in all areas of the company at the same time and making it stick is a very difficult proposition that requires a level of management commitment far beyond the basic commitment for a new development. If the business plan calls for sweeping changes, rethink your objectives; the company may not be able to cope with them.

4. SALES/MANUFACTURING/SERVICE/ACCOUNTING ALL REQUIRED TO BE IN LOCK STEP

The key point and recurring theme in new product development is that it cannot be initiated and successfully prosecuted without all parties being in lock step in executing the plan. A resource-consuming new product development is so important to the company; you must have all areas working toward the same goal. Do not let non-cooperation and non-contribution go uncorrected. It demoralizes the contributing groups and draws energy to combat.

MANAGEMENT FOCUS

1. SENIOR MANAGEMENT FOCUS IS NOT A GIVEN IN NEW PRODUCT DEVELOPMENT

There are two areas of management involved with new product development, namely senior management commitment and focus. We have previously discussed this management commitment in earlier sections, and would now like to focus on management focus. The senior management focus is that single-minded prioritization of program initiatives that supports the goal. Unfortunately, not every senior manager can have this single-minded focus. Their day is filled with new opportunities for products and businesses. Each may appear initially better than the current program. However, it is just as important to identify a program and see it through to completion as it is to select the best program. Focus is important because you cannot have restarts over and over again and complete a project. The issue of focus is a management problem, not an engineering problem.

2. THE DIFFERENCE BETWEEN MANAGERS AND VOCATIONAL ENGINEERS

The role of the manager differs from that of the engineer. The manager generally has a wider scope of interests for the business and does not have nearly the depth in any one of the interests. Management traditionally focuses on multiple priorities, juggling of resources, and managing the day-to-day affairs of the business. The engineer, or development personnel, needs to focus on the development tasks at hand and have a tremendous amount of depth in these areas. They need to have a single-minded focus to complete the program on time and at cost.

3. SENIOR MANAGEMENT COMMITMENT IS ONE OF THE HARDEST TO GARNER AND MAINTAIN

The subject of management commitment and focus is one of the most fleeting concepts in new product development. It sounds easy, straightforward, and permanent; however, in actuality, it is difficult to garner and establish a consistent effort toward. Multiple agenda, changing priority, and the nature of short-term profit goals make it difficult for the senior manager to support programs.

4. PRODUCT CHAMPIONS

The fact that senior management has a hard time supporting programs has prompted the emergence of product champions within the business to provide the necessary focus and sense of urgency to the product development team. They are the "spark plug" to help keep momentum during the project. The product champion has an almost-sacred trust with management to deliver the program on time, at budget, and under any prevailing conditions. The terms *zealot* and *crusader* apply to these personality types. Care should be taken in identifying the product champion in your organization to ensure the individual's personality type is consistent with the program. There are numerous stories about skunk works programs, which are successful programs in the face of adversity and the like. The important point to remember is that product champions have the focus to complete programs. They are not distracted by other elements of the business and can devote 100% of their energies to the program execution.

5. LISTEN TO THE WORDS SPOKEN, WATCH THE ACTIONS, AND CONSIDER THE EXPERIENCE LEVEL OF COGNIZANT MANAGER WHEN CLAIMS AND PROJECTIONS ARE MADE

For the project to go through the various obstacles of program justification, product development, launch, and product management, consider the manager that is selected to lead the program. An experienced manager who has completed several programs is a good choice. To get a sense of how this manager will execute the program, evaluate how the initial pitfalls are handled. This will give a sense of the approach to the handling of major problems. If you are the manager, keep this in mind because your management is evaluating your actions.

6. THE HOUSE-OF-FIRE MANAGER AND THE WELL-CONSIDERED APPROACH

Although it sounds cliché, a competent, new product development manager should anticipate the problems and have generated alternative approaches ready to resolve them. The energy expended personally and collectively, by the product development team in anticipating problems and accounting for them, will be less than the energy expended by a totally reactive manager. These people burn corporate energy at every turn and do not provide the cohesive momentum required for a program. Do not write them into your plans.

THE IMPORTANCE OF THE ACCOUNTING FUNCTION

1. ENGINEERS ARE TRADITIONALLY CONCERNED WITH PRODUCT FOCUS

Development engineering is a highly specialized field of endeavor. As such, few engineers broaden their scope of corporate understanding to include a solid background in accounting. It is, however, a good idea to round out the engineer with this discipline. The accounting function is the score-keeper and the checks and balances within the corporation. A development team should never underestimate the power of the accounting function in an organization. All the plans and great product ideas can vanish in the wake of an accountant's fiscal recommendation. They are the keepers of the funds used for projects. They can turn off the money faucet as quickly as it is turned on. This is not meant to be alarmist in any way. The accounting function is not to be feared in an organization; rather, the accounting team is to be respected for the fiscal job they have to do.

2. TEAM MEMBERS NEED TO UNDERSTAND THE ACCOUNTING FUNCTION

To help make the people on the development team more productive and holistic in terms of their understanding of the entire scope of the program, it is key to integrate the accounting function into the process. Let team members log hours and post costs to the project. Have them chart progress and keep track of lost opportunity costs. This will go further to imbue a sense of urgency within the team, rather than using the program's manager to preach the urgency.

A thorough knowledge of accounting is also required to design a successful business plan. It has been made abundantly clear to most development people that the yardstick in the measurement of success in the business field is money! It is what initiates the program, what keeps it going through the gestational period, and what is used to measure the return. In many cases it is what is used to reward the contributing members of the team.

Like it or not, accounting is one of the few objective, dispassionate, and universally accepted methods for evaluation. This being the case the project team members should become conversant with the dynamics and the language of accounting.

3. COST TRACKING

The accounting function is basically concerned with three things: the expenses, the return, and the timing. Consequently, you can expect that when funds are released for development, the accounting function will be watching the amount of funding, the speed at which funds are consumed, and when the return will start coming in. They simply will be reviewing the business plan you prepared and will expect performance to it. The operative point here is to design the business plan as close to what you believe the actual pathway for the program will be. Most planners do not overstate the revenue or understate the expenses; however, they do misread the timing, and as such, risk getting caught in a situation of not meeting the business plan.

4. USING THE ACCOUNTING FUNCTION AS A PARTNER, NOT AN ADVERSARY

All to often, development engineers see the accounting function as an adversary and not as the partner they should be. The accounting function can be your ally, both in good corporate times and in bad times. They can sway the cutbacks away from a program or toward it. In preparing the business plan, bring the accounting function into the process of the plan development. They can steer you toward positive input and away from negative input. In the presentation they will be your advocates, if they were part of the plan development. This can assist you in selling the plan.

5. GET CLOSE TO THE MONEY; IT OFTEN SPEAKS LOUDER THAN YOU CAN

If there is one conclusion or idea that you need to draw on from this section, it is that the manager/development team leader needs to understand the money flow in the organization. Watch where the money comes in and where it flows out. This will round out the manager to be more effective in their role. The best idea for a new product cannot sell itself to management without the accounting function as part of the justification, and the accounting function cannot solely generate cash and profit for the company. Consequently, it is the combination of both, that can generate and initiate a program with management support.

WHEN TO SAVE A PROGRAM AND WHEN TO KILL IT

1. YOUR FIDUCIARY RESPONSIBILITY

As a new product manager, your first priority is your fiduciary responsibility to the company. Management is entrusting you with the funds to prosecute a program, and as such, you need to monitor performance of the team and effect appropriate action. As discussed earlier, the development expense burns the resource candle at both ends. It consumes financial resources and has lost-opportunity costs associated with it. Therefore expending it requires diligence.

2. COST PATTERNS AND PROFILES OF A FOUNDERING PROGRAM

The business of project management has two basic items that must be controlled: expenditures and time. The expenses, as related to time, can be outlined for each stage of a program and should be, as part of the business and product plan. The profile generated can therefore be placed on a graph and actual expenses per time can be compared to planned expenses. The comparison therefore can give an indication of program progress.

Each stage of a project has an appropriation request. This is the amount of funds appropriated to complete that stage, and is the spending limit for that stage. There are certain milestones that must be completed during the stage, and each requires a certain amount of funds. The progress should track the milestones and the expenditures.

Some programs are front-end loaded. In this case a large amount of cash outlay occurs initially for equipment, and the balance is labor to experiment with the equipment, for example. Others are back-end loaded, where the expenses are at the end of the stage. Yet others are linear in which the expenses basically track the milestones. The linear is generally the case in which pure engineering labor talent is used throughout the stage. (See Figure 4-15 for the three types of loading programs and their characteristics.)

3. SOME PROGRAMS SHOULDN'T BE SAVED

It is not a popular position to take; however, the statement is true. Some programs may be ill conceived, or go sour and need to be redefined in a major way or terminated. There are some signs that the program may be in trouble based on the expense profile. Figure 4-16 is an example profile, indicating trouble spots.

The 3 horizontal lines indicate the different appropriations limits for the stages of the project. The lower profile is the planned profile and the upper profile is the actual profile. As shown, the first sign of trouble comes in stage one where expenses are clearly out of line from the front-end loaded projection. Corrective action is taken to mitigate the damage, as indicated by the change in slope, but the program is already off to a poor start. Worse yet, is if the project milestones slip also. This would be indicated by a normalized x-axis for both the actual and planned

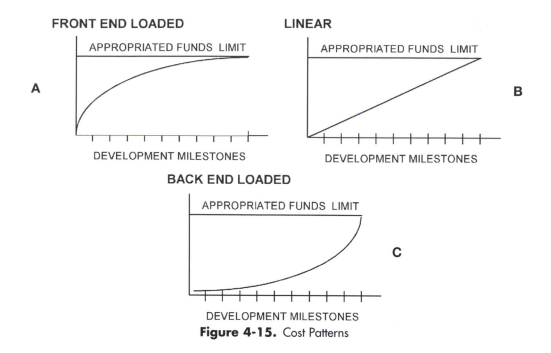

Figure 4-15. Cost Patterns

expenses/milestones curves. In the second and third appropriation, the rapid increase in cash outlay would indicate a possible premature capital expenditure.

Since this is a valuable tool for planning a sample project cost profile is included in the Tool Box for your use.

4. DON'T GET SADDLED WITH A DUD; INITIATE A FIX OR KILL IT

From a pure career-management perspective, and also a corporate employee perspective, do not get saddled with a dud. It is bad for your career; it wastes corporate assets and resources and demoralizes an otherwise-productive development team. Initiate a fix, relay out the program, or kill it. Do not let it founder. It can tear down your career by undermining any future decisions or recommendations you may have to make to management. Remember: Management is looking to you to watch out for their interests. Do not disappoint them.

5. EARNING MANAGEMENT RESPECT BY TERMINATING A PROGRAM, SETTING STAGE FOR FUTURE FUNDING

It then follows that if you demonstrate diligence for the company in the decisiveness to terminate a program when it is warranted, you will earn management's respect for your judgment. This can set the stage for future requests and grants of funds for programs. One

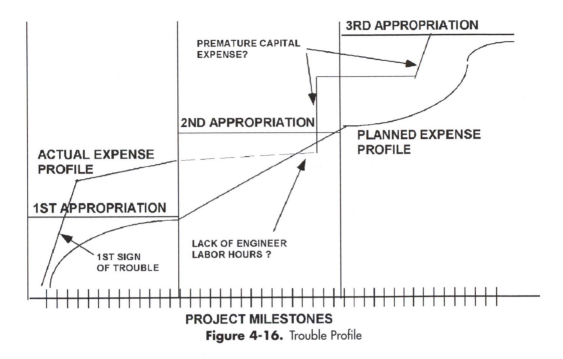

Figure 4-16. Trouble Profile

of the most difficult aspects of management of new product development is the need to request and secure additional funds beyond the budget. If you demonstrate decisive action in protecting company assets and resources, management will more readily trust your assessment of a situation and respect your wishes.

TESTING THE MARKET

1. CONSTANTLY ASSESS YOUR NEW PRODUCT POSITION

The process of planning and prosecuting a new product development effort is rarely initiated and completed using the same information, assumptions, and data. More realistically the team needs to reassess the product positioning, the competition, and the cost and pricing. As we stated earlier, the business of new product development is a fast-changing, dynamic environment where fixed plans give way to navigation. The new product must be navigated through the competitive maze and the dynamics of the market, to introduction.

Failure to adequately assess the market during the development phase may cause a mis-introduction of the product, overlooked or missed opportunities, or incorrect positioning depending on what the competition may be doing.

2. TEST THE MARKETABILITY, PRICING, AND TIMING

Specific items to be reevaluated during the process are the general marketability of the product, the demand for it; the all-important quantity of produced units, the pricing of the product, and the timing of the program and introduction. As a matter of course, it is a good idea to keep a running list of these items, since it is easier to establish trends and potential pressure points.

3. UPDATE COMPETITIVE ANALYSIS AND EXTRAPOLATE CHANGES, PROJECT COMPETITION AT LAUNCH

A properly executed program shouldn't involve changing the product during the development phase. The development should be fast enough to capitalize on an opportunity. Alternatively, the program's objective should lead the market enough, such that a unique enough product satisfying the demand still exists at the end of the longer development cycle.

The product stays stable; however, the business plan may change to suit changing market scenarios. If price, availability, or usage pattern shift, the plan, the financial profile, or the distribution arrangement may be modified to suit conditions. Therefore it is crucial to understand what is happening and to project the effects into the future when you will introduce the new product.

Monitor the market and accommodate material changes, in order to position yourself.

CONFIRMING THE TECHNOLOGICAL FIT

1. INTERNAL FIT: ARE ALL ELEMENTS OF THE BUSINESS EQUIPPED TO HANDLE THE TECHNOLOGY?

Previously we discussed the ability of the organization to absorb the new product, its technology, and its needs and requirements. This is especially a factor when new technology is brought into the organization from outside, rather than grown within the organization.

This goes beyond the human resource issue and preference issue to include capital equipment procurement required to support the new product. If capital, talent to operate the new equipment, or training for the personnel are required, they need to be factored into the business plan.

How will the technology base be supported? Think in terms of pathways. This question needs to be thought out in detail. Not every company can afford the expense of leaping into a new state-of-the-art technology. Many times there needs to be a pathway to the new technology. This pathway allows the affordability, as well as the time, to assimilate it into the organization and create the learning required to implement it effectively.

2. EXTERNAL FIT: IS THE TECHNOLOGY RIGHT FOR THE MARKETPLACE? CAN IT HANDLE IT?

At the same time it is critical to assess if the technology embodied in the new product is a good fit with the marketplace. For example, it is a mistake to use a technology in a product serving a market that may have an aversion to it. It simply becomes an uphill battle for market acceptance, and energy will be spent on selling the embodied technology, rather than the solution the product brings to the customer.

3. ARE YOU ON THE RIGHT GENERATION OF TECHNOLOGY?

The optimum time for evaluating this question is at the beginning of the project. As discussed earlier, the technological base cannot be changed mid-project and still meet timing requirements. The planning of the technology must be well considered, as you want to be operating on a platform that has longevity and will be in demand at the time of introduction.

An extreme example of failure to do this is in the case of instant-developing photographic film, and extending it to movie film. The extension was a natural conclusion for the chemical-oriented product planners. What they didn't count on was the development of the magnetic tape medium for movies and its embodiment of the video camera for the mass market. Coupled with the VCR's popularity gains to play the movies, the video camera was the end for wet chemistry in movie film, instant developing or not.

The lesson: Be sure the technology used in the product is the one that has market acceptance at the time of introduction and thereafter.

4. TECHNOLOGY EVOLUTION FLOWCHART

In preparing the product and business plan, be sure to forecast the technology evolution for the product line. There needs to be a balance struck between time to develop and capture the market, degree of difficulty in implementing a technology, market dynamics, acceptance of new technology, and longevity of the technology employed. The forecasting of technology is good to include in a business plan, even to the point of breaking it out separately, since the overall technology needs of the corporation can be evaluated and like requirements can be funded, as part of a corporate research initiative to support the businesses.

5. IS THE COST OF THIS TECHNOLOGY ON THE APPROPRIATE LEARNING AND COST-REDUCTION CURVE?

As a final check, make sure the technology employed will be on a cost-reduction or learning curve that will be consistent with the market expectations for pricing. As an example, many products have changed from pure mechanical form to a combination of mechanical and electronics.

This is driven by the market demand for long life, added functionality, and lower cost. As the transition to electronics hardware was taking place, there was also a trend to embed microprocessor technology and implement functionality in the use of software. This can be shown in Figure 4-17.

Taking it a step further, the hardware electronics will progress down an integration pathway whereby more and more complex circuitry will be integrated onto decreasing areas of silicon substrate. The software will have its own pathway whereby structured techniques for software development will eventually give way to adaptive learning in systems, in which the operational characteristics will readjust automatically.

Each of these leaps into the newer technology eventually drive product costs down while increasing functionality. In many cases this increase in functionality opens up new market and customer bases for increased sales.

The market will communicate the need for these leaps into the new technology, sometimes in an indirect way. The challenge is to be able to sense the need and react, without leading the technology too quickly. Doing so can lead to product cost problems, reliability problems with immaturely developed technology, and general market acceptance issues.

EVOLUTION OF TECHNOLOGIES

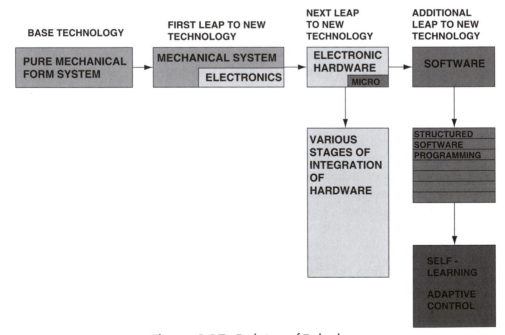

Figure 4-17. Evolutions of Technology

TRADING TIME SAVED FOR TECHNOLOGY

1. REPEATING: TIME IS OF THE ESSENCE

As discussed earlier, time, in new product development, is a precious commodity. It is an increasingly shorter supply, and the speed at which competitors can assimilate information to capture new opportunities is decreasing. Therefore the implementation of technology needs to be on a predictable schedule. There is little tolerance for the development of raw technology and the application of it during product development. The technology should ideally be well understood and mastered before the application of it in the product.

2. TECHNOLOGY DOESN'T SELL; FUNCTION, PRICE, AND PERFORMANCE SELLS; BE PRACTICAL

Many times the lure of new technology can take over the focus of a product development. When this happens, the development of the technology, not the product, becomes the driving force. This is a prescription for failure because customers do not buy technology per se, they buy products with features that create value and benefit as a result of their use.

This is an important point, because it is a vocational hazard for a product development engineer to become enamored with technology and lose sight of the program's objective. If the technology is not mastered, it is manifested by development engineer's trying to "work around" issues by placing constraints and qualifiers on the application and use of the product. In its worst form, the marketing and sales part of the corporation end up "holding the product's hand through life" while engineering tries to understand the technology. This is a certain prescription for product and new-business failure.

Figure 4-18 illustrates the concept of targeting a technology for a product that is (or soon will be) mastered by the corporation and implementable in the new product. As time

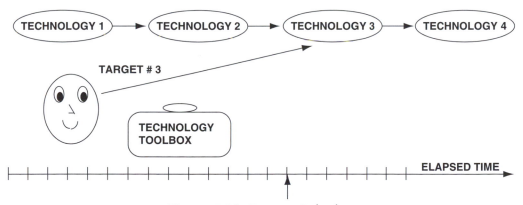

Figure 4-18. Targeting Technology

progresses, the technologies evolve from the technology base 1 through to the technology base 4. The product development manager then targets technology 3. This ensures that, at the point in the future when the product is introduced, the corporation will have mastered the technology, and that the product is still in demand by the market. In this example Technology 4 is too much of a stretch because it may be too new and not understood enough for product plat-forming. Technologies 1 and 2 will be less than state-of-the-art or minimally acceptable by the customer base.

3. DONT CHASE THE HOLY GRAIL OF TECHNOLOGY

This goes without saying, but will be repeated here for emphasis: "Do not chase new raw technology as part of product development. Rather, you need to implement new technologies that are understood by the corporation, as part of the product development, with the primary focus being product development.

Consequently, you need to be practical in its selection and application. The new product development manager and the team should strike a balance between implementing the latest state-of-the-art in technology and project timing.

Many companies have made the fatal error of betting on the implementation of new technology in product development before the technology is internally mastered within the corporation. The degree to which companies have bet their future in this manner is the degree to which many companies have had to endure financial losses, changes in ownership structure, and product recalls.

4. SOME MARKETS DON'T CARE

The primary function of development is to service the customer in the best way possible. The team needs to assess the markets' needs and desires. For some markets, the customer simply doesn't care about the technology. They are interested only in the functionality and price. Take this into account.

THE CUSTOMER IS PART OF THE PLAN

1. DEPENDING ON THE PRODUCT, INVOLVE CUSTOMERS EARLY IN THE PROCESS

There is no substitute for direct market feedback in understanding the market need for the product being developed. To the extent that you can disclose information without affecting security, the direct customer disclosure and feedback method is better than any secondary measurement. It serves several purposes: it keeps good, loyal customers informed, it fosters trust, and it allows an honest engagement with the customer to solicit unbiased feedback on the product in its current state.

Feedback also permits the customer to "buy in" to the product and the development program. It also allows the company to "test drive" their marketing story on a real prospect. Finally, it allows the manufacturer to double-check their strategy and begin to weave the manufacturer's strategy with their own.

There are several caveats, in this engagement that must be noted: You need to guard against information leaks, protect your intellectual property by confidentiality agreements, and double-check the potential customer's competitive affiliation.

2. ADDING VALIDITY TO THE PLAN

The use of the direct engagement technique is an excellent way to add credibility to the business plan. It provides incontrovertible evidence of a market for the new product, which can be used to nullify any doubts present in the audience.

3. A GOOD CUSTOMER WILL TELL YOU THINGS YOU CAN'T LEARN ELSEWHERE

One of the most desirable aspects of the direct engagement technique is that you can obtain information from the customer that is not available elsewhere.

For example, let's say you are a manufacturer of cement mixers. Your secondary marketing information has shown that there is a worldwide market for 1000 cement mixers each year. By direct interview, you may find out that only 350 cement mixers are sold each year in the free world. The balance is sold by third-party payoffs and other methods different from the traditional methods you use. Consequently, these 650 mixers are unavailable to your company. If your business plan depended on a break even of 350 or 400 cement mixers, your plan would be in trouble. You would have to obtain 100% market share of the "available market." This is not apparent in the secondary market survey. It was only knowledge passed directly from customer to vendor. By obtaining this critical information, you can redesign your business plan to suit the business condition.

4. PUT YOUR TEAM ON THE FIRING LINE

Finally, one of the most important reasons to use the direct engagement technique is that it puts your team directly into the firing line of a customer. All the requesting and specsmanship you may bring to the marketing requirement's specification cannot outweigh the benefit of exposing the development team to the customer firsthand. Once they hear it from the customer, you will no longer need to repeat yourself.

SELLING THE PLAN

1. GETTING THE AUDIENCE

A fair amount of preparation is needed to prepare a fundable business plan. On completion of the plan, it now becomes time to present it to senior management for review and approval. The process is less about employee/management relationship than it is about investor and entrepreneur. At this point, it's strictly business, and your task is to secure funds for development that will generate future returns for the investor.

However, it can be the "fun" part of the process. The first step is to identify and secure the audience.

Who is the audience? How is the funding controlled? Under what conditions are the funds released? What are the expected returns? Gather all this information as you are preparing the presentation. You need to control the meeting and presentation and its outcome, so the more information the better.

Find out the "hot buttons" and biases of the members of the decision-making group. Prepare the presentation in such a way as to appeal to the members. In addition, you also need to assess the competition.

There are other groups usually seeking funding for their programs at the same time you are seeking funds for yours. What are their strategies for obtaining funds, and how does your program stack up against theirs? Position yourself to prevail against your competition.

2. HAVING THE REPUTATION PRECEDE YOU

An individual has an individual's energy. The best that one person can do is to transfer positive energy or enthusiasm about a program on a one-on-one basis or in front of a group. On a one-on-one basis the energy transfer is arithmetic (i.e., one for one). To get three people enthused, however, you need to expend three times the energy of getting one enthused. In front of a group, you can get more people enthused; however, the energy transfer is only for the length of time of the presentation and is subject to any scrutiny after the presentation.

Therefore to generate enthusiasm or positive energy about a program, you have to do it geometrically. This means influencing several people who can influence others. In this manner consensus can be built for your program. You convince one person, who convinces two people, who convinces four people, and so on.

Generating enthusiasm geometrically multiplies the effect for you. This needs to be done before the pivotal presentation so at that presentation, the audience is ready and eager to "hear" your message. The presentation then becomes a confirmation of what everyone else agrees with.

Having your reputation precede you is even better as you have a demonstrated track record of success.

There is also the concept of consistency of inputs. Most senior managers gain knowledge indirectly from a variety of sources in making their decision. Make sure that the sources feeding the information about your program are consistent and positive to the senior manager. In this way the presentation becomes a confirmation of known data, with no surprises.

3. TRIAL BALLOONING

Trial ballooning is a technique whereby a concept can be tested in the audience to gain understanding or trigger an initial reaction. It is useful when the presenter may not have any prior knowledge of how the audience will react to an idea and wants to "trial balloon" it to see the results. It also can be an effective tool to build consensus by releasing information a little at a time, and allow for modification at the subsequent releases. These techniques are used when the presenter has uncertainty in the idea.

The caveat in the trial ballooning technique is that audience reaction may not always be accurate. The conditions under which the information is transferred may not support the desired conclusion, or the lack of a complete, cohesive story may cause an adverse reaction. In addition, it is more difficult to recover from a negative reaction once the information is out and conclusions are formulated.

4. THE PRESENTATION

The presentation needs to follow a basic format, regardless of how much embellishment you may want to include. The following is the basic format to be followed for the presentation. It should also be noted that each of the basic elements needs to have detail to back it up.

A. Overview

This is the first section of the presentation. It needs to provide focus to the audience to allow their scope of examination to go from a macro view of many opportunities and markets to the business view of which the product and business plan are based. Bring them into focus fast and clearly. Do not confuse the audience by outlining many choices or options for the same opportunity. Pick your story and articulate it!

B. Opportunity

In this section, clearly articulate the business opportunity and the business condition. Now that you have focused the audience to a specific area, show them how you intend to make money for them with your business opportunity. Define it in terms of narrative and numbers. Use just enough detail to communicate the opportunity and to substantiate it. The audience is interested in your vision for how you intend to evolve this new business for them. They are not interested in grading you on all of your details.

C. Product Description

If it is your intention to create the new business by embodying the opportunity in a product, here is the opportunity to articulate what the product is and how it will serve the business condition. Be as complete as possible in outlining the product, and relate the features to the market needs and requirements. Explain the cost structure and what gross margins are expected.

Explain how the company will implement the technology and what the technology for the product is. Create a perception in the mind of the decision-maker that this program is doable and should be done without delay.

D. Strategy and Implementation

You must have a clear and concise plan of how to go about prosecuting the new product opportunity. In this section of the presentation, present the plan you have developed to make the new business a reality, and show how you intend to do it. Times, dates, places, responsibility centers, linking objectives, funding, management support, and commitment are all the items that must be integrally linked in this part of the presentation. The issue of linking objectives is a very important one and not to be underestimated. The product needs to be presented in what can be referred to as a 5-for-4 format. Cite 5 benefits the organization can obtain in pursuing this program that will only cost 4 cost units. In other words, show how this program will help others in the company. It will create the desire to fund this program because the management can get more for their money in this one as opposed to others.

E. Call to Action

All the preparation work is done and presented in a positive light to the management. Now is the time to wrap up the presentation and ask for funding. This now becomes a bi-directional interface between management and the project team seeking funds. As the leader, do not leave this discussion without getting a firm answer.

You have done all the homework required and deserve an answer one way or another. Management can request, and you can provide clarity of detail, additional detail or supporting facts; however, it is decision time for management, and they should give you an answer.

Call for action and drive it to a conclusion—yes, no, or follow up review date.

5. SECURING THE COMMITMENT

Nothing speaks louder than money! The objective of the presentation of the business plan is to obtain funding. This being the case, there is no statement, reassurance, or conversation that can occur between management and the product team that can substitute for money. If management cannot fund the program, they won't be able to support it either. Also, do not be lulled into a false sense of security or buy in by management with the suggestion that the program be partially funded. Statements and compromises like these are an indication they are unsure and concerned. As discussed earlier, management may choose to partially fund several programs and select the best one at a later date. This only starves good programs and inappropriately funds poor ones. Funding is what you came for; don't leave without it!

6. ASSIGNMENT OF APPROPRIATED FUNDS

Generally, an entire program will not be funded from start through to completion. You will, however, want to obtain funds to initiate and complete a stage of development. This serves several purposes. First, it allows for the accurate conclusions to be drawn from work done. Second, it allows for an appropriate termination point for work completed. This may be required for later start up or transfer of technology to another group. Therefore it is important to obtain the funding and to post it to a specific project with specific deliverables in a specific time frame. Otherwise, it can have a tendency to become a catch-all project to gather expenses for non-approved programs.

7. FOLLOW ON FUNDING

On successful funding of an initial stage of a program, immediately start campaigning for subsequent stages. In this way a momentum can be established for the entire program. Do this for each stage of the program.

SUMMARY

In this chapter we discussed the organization of the new product concept into a business plan. A format for the business plan was presented and discussed. The importance of timeliness was stressed, and the relative standing of the company, in terms of where the market is headed, was also presented.

The relationship between market maturity and the company's expectations for profits was reviewed, in light of product costs and platforms. A section on pricing and the philosophy and justification for pricing was presented, all for the purpose of constructing a workable business plan.

Internal issues, such as management focus and organizational changes, were included to be part of the business plan.

A section on the importance of the accounting function was included to serve as a background in preparing the business plan.

Market testing and technological fit were also included to provide continuity to the business and product plan.

Finally, we discussed in detail, how to sell the plan to management and obtain the funding.

In the next chapter we will now look at the program from the accountant's viewpoint, and give specific tools to ensure the accounting function embraces the project.

JUSTIFYING A PROGRAM: THE ACCOUNTING VIEWPOINT

ABSTRACT: The objective of this chapter is to present an accountant's viewpoint of new product development. The accountant's viewpoint is quite different from the new product developer's viewpoint. When adopting the stance of the accountant, you must address all of the nuts and bolts of financial analysis and understand the financial perspective in decision making. Conservative estimates are key when dealing with financial people. As the discussion in this chapter will show, the pressure points, the timing, and the tolerance for adversity are different from those of the product development team. Consequently, it is a good idea for the new product manager to understand the accountant's focus so that both can cooperate to transform a new product opportunity into a reality.

BACKGROUND

1. THE FINANCIAL MODEL

The financial model for a new product is different from that of a development engineer or team member. The finance model is characterized by funds flow, direction, exposure, and timing. Finance generally has limited understanding of the technical, marketing, or sales aspects of the transaction. Accountants' view of a new product development program can be represented in terms best accepted by accounting, as shown in Figure 5-1.

The basic model for financial analysis is that there is an initial investment in development that will return future dollars at an increasing rate in subsequent time periods. In addition—between time period three and four, for example—an additional investment is required to update the product or to respond to some competitive action. The funds initially

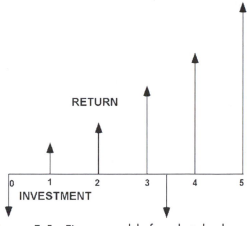

Figure 5-1. Finance model of product development

allocated and subsequently allocated are offset by the returned revenue stream. The value of the revenue, the time to revenue generation, and the initial investment all factor into the quality of the investment from the financial perspective.

2. THE NEW PRODUCT STORY IN TERMS OF BOTH NARRATIVE AND NUMBERS

When presenting the program to the financial team, be sure to use both numbers and narrative to ensure understanding in the accounting functional area. The numerical data will serve the analytical requirements of accounting, while the narrative will assist the team in understanding strategy, product planning, and the company's overall marketing direction. The combination of data and narrative will best serve the information needs of all concerned.

ACCOUNTING AND FINANCE AS PARTNERS, NOT ADVERSARIES

1. ACCOUNTING'S PERCEPTION OF NEW PRODUCT DEVELOPMENT

The accountants' perception of new product development is strictly a financial one. Consequently, they have empathy neither for the problems encountered nor for the means by which they are surmounted. The lack of concern for these issues can cause tension between the finance functional area and the engineering functional area.

It must be stressed that the cooperation of engineering and accounting is what sells management on a program. Therefore, make the financial team partners, not adversaries, in the development program.

2. FORMULAS, RESULTS, EXPECTATIONS

The financial orientation evaluates things in terms of equations and checksums. This means that for every two pounds of invested effort, there is an equated result of some measure of progress. Unfortunately, actual development doesn't always occur in this manner. There are breakthroughs followed by little or no progress at all, followed by further progress. The degree to which this process occurs is related to the uncertainty in the development program. Equations and checksums are not necessarily compatible with uncertainty analysis.

3. ZERO TOLERANCE FOR FAILURE OR DELAY

Because this equation orientation exists in the minds of the accounting team, there is limited understanding or tolerance for delays, failures, or restarts on product design. The perception is that the funds and effort were expended: Where are the results?

To a large extent, this orientation is valid; there are times, however, when progress is neither linear nor forward. It is during these times that you need the partnership of the finance people the most. Because they have one eye on the rest of the business and receive input from marketing on general market conditions, they make fiscal recommendations to senior management. They have the ability to influence funding or project termination. The operative lesson is to make finance and accounting your partners; they have the power to intervene to save a program during lean times.

4. TECHNOLOGICALLY ADVANTAGED VERSUS DISADVANTAGED PERSONNEL

Because the development engineers are technologically advantaged relative to all others in the enterprise, things that are obvious to the development team might not be readily understandable to others. For example, at a certain point in a development when a road block can be removed with the procurement of outside services or capital equipment, a large expenditure required to make the breakthrough might be obvious to the technically oriented, but not to anyone else. This is where the trust relationship with the financial and other non-technical people needs to transcend functional boundaries.

5. NARRATIVE OF SIMILAR PROGRAMS

To help the cause of understanding along, it might be helpful to cite similar programs that have experienced similar circumstances. In this way, a connection can be made between what you are requesting and a previous success story. Parallels can then be drawn, differences discussed, and resolution can be sought.

6. PRODUCT DEVELOPMENT'S ROLE: FACILITATE UNDERSTANDING

The role that the new product development manager has, then, is to facilitate understanding between the financial community and the development community and to forge a partnership that will last through product introduction and beyond.

FINANCIAL AND ECONOMIC ANALYSIS

1. BACKGROUND AND GENERAL MODEL FOR PRODUCT DEVELOPMENT AND SALE

The mechanics of financial analysis for a new product are straightforward. They involve investment, revenue, and a specified time frame. The most difficult task is to determine the company's expectations and minimum acceptable performance for a program. Referring back to the section on strategic planning and assessment of operations in Chapter 1, the company has an established track record for projects and an expectation of the minimum acceptable return. These should serve as gauges for what would be acceptable for the new product opportunity under consideration.

There might need to be a rationalization or normalization of historical data with the present to compare the two effectively. Figure 5-2 represents the general model for a product sale that will serve as a basis for discussion in this chapter.

This model deviates from a traditional income statement and balance sheet format for illustrative purposes only. The data presented is a designed example to show the investment and return and some of the expenses along the way toward growth. This model recognizes a booking concurrently with a shipment, and collection is immediate. Materials are also procured concurrently with bookings. In actuality, these dates can differ based on several internal and external factors as well as on requested ship dates and component lead times.

As the figure shows, the model begins with a marketing product planning expense. This is the expenditure to determine the new product opportunity. The next step is to expend the funds in development to develop the new product to be sold. This occurs during the first time period with no revenue to offset.

The sales expenses are based on a fixed amount of $4000 and 11 percent of booking value. This means that as the booking value increases, the sales expenses increase also. Orders or bookings are a hoped-for result of the sales expenses. The company then responds by ordering material and adding labor to assemble it into the product. Burden is a calculated value based on 1.75 times the labor expenses. The product is shipped and a collection is made. Administrative expenses are deducted and a profit is realized.

As shown in the graph, the allure of large sales numbers in the forecast quickly descend to modest profit figures. These profits are the foundation on which the enterprise must operate and with which it must grow. Another factor to consider is that the company is spending money on an ongoing basis—purchasing materials, buying labor, and paying other expenses. In addition, all of the profit added up throughout the "run" of the product barely offsets the development expenses and the marketing expenses. This calculated profit does not even factor in the temporal value of funds that we will discuss later.

FINANCIAL MODEL FOR A PRODUCT SALE

ITEM	DESCRIPTION		QTY 2	QTY 8	PRICE EACH 5000 QTY 16	QTY 35	QTY 65	QTY 105	QTY 135	
1	MKTG / PLANNING EXPENSE	−75000	0	0	0	0	0	0	0	
2	DEVELOPMENT INVESTMENT	−200000	0	0	0	0	0	0	0	
3	SALES EXPENSES	0	−5100	−8400	−12800	−23250	−39750	−61750	−78250	
4	BOOKINGS	0	10000	40000	80000	175000	325000	525000	675000	
5	MATERIAL	0	−4250	−17000	−34000	−74375	−138125	−223125	−286875	
6	LABOR	0	−900	−3600	−7200	−15750	−29250	−47250	−60750	
7	BURDEN	0	−1575	−6300	−12600	−27562.5	−51187.5	−82687.5	−106312.5	
8	SALES	0	10000	40000	80000	175000	325000	525000	675000	
9	ADMINISTRATIVE EXPENSES	0	−500	−2000	−4000	−8750	−16250	−26250	−33750	
10	PROFIT	0	−2325	2700	9400	25312.5	50437.5	83937.5	109062.5	278525

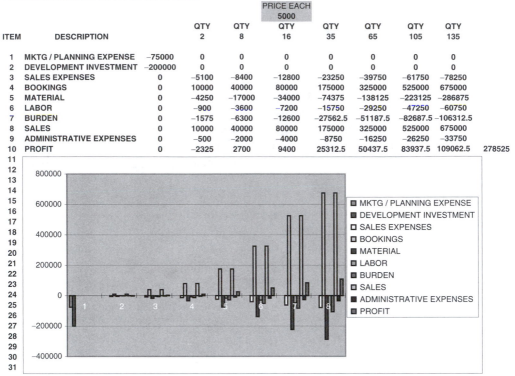

Figure 5-2. Financial model for a product sale

2. FINANCE DEFINITIONS: ROI, NPV, IRR, RONA, ROE, ETC.

The accounting and finance function have a specific terminology for evaluating performance of a company. Here, we present for review and in definition format some of these financial terms as they relate to new product development and operations.

A. RETURN ON INVESTMENT (ROI)

The return on investment is a measure of the business' performance based on the entire investment. In general, it is expressed as the ratio of the return divided by the investment. It places value on the amount of money invested and compares that amount with the annual funds returned as a gauge for each year.

B. NET PRESENT VALUE (NPV)

The net present value is a measure of the discounted time value of money. It is based on the premise that money earned in the future is worth less than money today due to inflationary factors. The NPV is used to evaluate the present value of future funds by reflecting them back to today's values at the discounted rate caused by inflation. The value is discounted each year by the estimated inflation rate of that year.

C. COST OF CAPITAL

The cost of capital is a measure of the corporation's cost to employ capital. This capital is used for equipment and can generally be the cost of the longer-term debt of a company. If the company needs capital to operate and grow, this value is the measure of what it will cost the company to employ it.

D. INTERNAL RATE OF RETURN

The internal rate of return is also referred to as the time-adjusted rate of return. It is simply the actual interest yield of an investment program over the life of the project. It is computed by determining the discount rate that will equate the present value of the cash inflows generated by the program with the cash outflows required by the program. It is the discount rate that causes the net present value of an investment to be zero. The internal rate of return is used as a measure of the present rate of return on an investment, analogous to a time deposit account in a bank.

E. NET INCOME

The net income is the amount of income that is left after all expenses in the business. It is the sum of material, labor, manufacturing overhead, operating expenses, and taxes and can be expressed as a percentage or as a dollar amount.

F. OWNERS' EQUITY

The owners' equity is the amount of money the owner actually has after any liabilities have been subtracted. As the enterprise grows, the owners' equity should grow, as this value represents the owners' value stake in the business.

G. TOTAL ASSETS

The total assets are the sum of cash, investments, accounts receivable (monies owed to the corporation), inventories, any prepaid expenses, and property plant and equipment.

H. RETURN ON EQUITY

Return on equity is a measure of the amount of revenue returned compared with the equity or ownership of the company. It relates the profits to the owners' stake in the business. The return on equity is a measure of how the annual return compares with the owners' equity in the business at that time. In general, it is expressed as the ratio of the annual dollar return to the shareholders' actual dollar equity in the business.

I. RETURN ON TOTAL ASSETS EMPLOYED

The return on total assets employed is a measure of the return of the operations to the average total company assets within the specified period. It is calculated by dividing the net income and interest expense by the average total assets, factoring in the interest expense, which is adjusted for taxation. In this way, the measurement is an evaluation of the total return compared with the total assets, with no measure of how those assets are financed.

These are only a few of the many measurements that are used to evaluate an organization's performance. There are additional ones to measure specific items that circumstances might dictate.

3. RELATING FINANCIAL TO ACTUAL EVENTS

From the development perspective, one of the most unacceptable accounting practices is to manage exclusively "by the numbers." There are instances in countless companies in which accounting renders judgments based solely on numerics, without relating these to the operations of the business. This can be a mistake, as can evaluating a program solely for its contribution to technology with little regard to the numerics involved. The best way to manage the business is to evaluate both aspects.

To do this, the accounting function needs to become knowledgeable about products, the operation of the business, and new product development. In addition, development personnel need to become conversant with accounting principles, driving forces, and corporate concerns. Both groups need to accommodate one another's vocational, parochial interests. For example, a lack of sales of a newly introduced product needs to be understood to effect corrective action properly. Both sides must understand whether the cause is a temporary delay in driving the introduction through the channel sales or a permanent trend.

Figure 5-3 illustrates the financial analysis of sales progress and relates the numbers being reported to actual events. Understanding within the corporation takes place based on both of these entities.

4. PRODUCT COST SYSTEMS

The product cost system used by the organization should reflect accurate information about the product cost and provide management information to effect appropriate action. It is selected for use from an operations standpoint. There are three basic types of cost systems in use today: job order cost systems, standard cost systems, and activity-based cost systems.

A. Job Order Cost Systems

The job order cost system is a cost-gathering mechanism that trails the product through manufacture and posts accumulated costs in terms of material labor and burden. As costs increase, the job order cost system trails the increase without reference to a maximum allowable cost or market pricing. As the product moves through the production processes, it simply gathers cost.

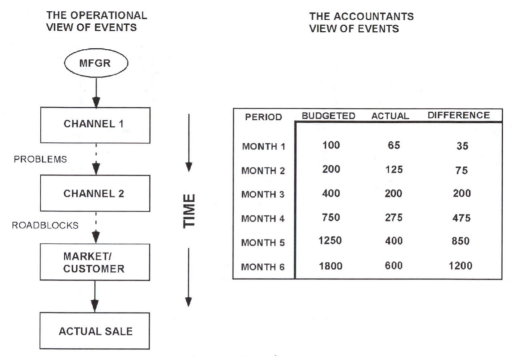

Figure 5-3. Sale progress

Job order cost systems are generally used with a cost-plus pricing strategy. The two approaches can get a company into trouble quickly from a profit and volume perspective. Such systems are, however, adequate in terms of posting the product's cost to record. A job order cost system example summary is shown in Figure 5-4a.

B. Standard Cost Systems

The standard cost system differs philosophically from the job order cost system in that it is designed to set a specific "rolled-up" cost. The cost roll-up is the planned cost anticipated as a result of the product design process. Periodically during the production cycle, the actual incurred costs are gathered with the appropriate overhead allocations and absorptions included. These are then compared with the standard costs, and any difference is referred to as a variance. The variance is used as a prompt for management action to contain costs and align them with the standard. A standard cost system is illustrated in Figure 5-4b.

C. Activity-Based Cost Systems

The activity-based cost system generates a product's cost by focusing on the underlying activities that are necessary to produce the product. These underlying activities consume the resources. The activity-based system is appropriate when manufacturing

A

JOB ORDER COST SUMMARY

	DESCRIPTION	VENDOR		COST EA	QTY	EXT. COST
MATERIAL #1	PART 1	ABC		23	2	46
MATERIAL #2	PART 2	ABC		34	5	170
MATERIAL #3	PART 3	DEF		45	7	315
MATERIAL #4	PART 4	DEF		56	10	560
MATERIAL #5	PART 5	FGH		67	12	804
	HOURS	BURDEN 150%	COST/HR			
LABOR #1	10	15	50		LABOR	1250
LABOR #2	12	18	37.5		LABOR	1125
LABOR #3	2	3	24.5		LABOR	122.5
LABOR #4	3	4.5	12.6		LABOR	94.5
LABOR #5	6	9	78.45		LABOR	1176.75
ENG. TIME	24		55		ENGINEERING	1320
MISC. TIME	2		30		TOTAL MISC.	60
					TOTAL COST	7043.75

B

STANDARD COST SUMMARY

	STANDARD COST 19XX	19XY	VARIANCE	19XZ	VARIANCE
MATERIAL	50	45	5	θ	−10
LABOR	35	37	−2	32	3
BURDEN	110	120	−10	90	20
TOTAL	195	202	−7	182	13
			UNFAVORABLE		FAVORABLE

Figure 5-4. Cost systems

direct costs shift away from labor dominance to material-dominated or other elements. With labor becoming less and less of a factor in the cost makeup, the manufacturing expenses normally burdened to labor become such a large percentage that it becomes difficult to generate a meaningful cost and understand the absorption of expenses.

The cost system employed will not make or break a program materially by itself. The accuracy relative to its association to the operations is negligible, in most cases, compared with the other uncertainty aspects in a new development. Sample cost roll-ups for the standard cost system and the job order cost system are available for your use in the Toolbox.

5. COSTING A PRODUCT AS PART OF A PROJECT (QUICK TEST)

As part of the market investigation and the product presentation, a short-form test could be required to prequalify a program. This qualification is a quick test of the project's development cost, the product cost, and the revenue expected. It also incorporates means for evaluating prototype costs. This quick test can be used easily with a few data to determine project viability. Figure 5-5 shows a sample of the spreadsheet for the evaluation.

In the example costing of a project shown in Figure 5-5, separate cost roll-ups are used for the prototype and the production units. Labor costs are determined by the number of minutes required to assemble the product multiplied by the cost per minute. Burden is calculated by the burden rate entered at the beginning of the chart multiplied by the direct labor content. The indirect cost is factored at base value. The material, labor, and burden costs are totaled for both the prototype and the production units.

The cost to develop the product is calculated by evaluating each of the engineering elements multiplied by its cost per hour. This value is entered at the beginning of the chart for engineering costs. Additional development expense is recognized in the next section and later totaled in the sales, general, and administrative expense entries. The capital equipment is also factored into the payback calculation in the next section.

The next section takes the data on costs and allows a forecast for the prototype and the production units to be entered. The revenue per unit at the beginning of the chart is used to determine the revenue forecast. The expenses, product cost, and capital equipment costs are deducted to determine net profit. Although this sample is neither an elegant return on investment calculation nor an in-depth cost analysis appraisal, it is a quick calculator for determining the viability of programs under consideration.

A separate spreadsheet is available for your use in the Toolbox to evaluate your own specific programs.

6. THE FALLACY OF THE HOCKEY STICK

Whenever a program justification is prepared, there is always a tendency by the product planners to skew the forecast of product sales to the later years of the program. This is referred to as a back-end loaded forecast or a hockey-stick forecast. The name "hockey stick" derives from the characteristic graphic appearance of the forecast.

For managers of new product development, it is critical to avoid justifying a program using the back-end loaded forecast. If the characteristic sales volume is truly expected to follow this path, then temper the forecast. If, however, a hockey-stick forecast is being used (as happens in some cases) to justify expenditure and stall for time after development, be aware that it can have a devastating effect on the product development team, management, and careers. In many cases, a linear forecast can yield the same level of success as a back-end loaded forecast from a net present value standpoint.

In the following example, the product price is valued at 500 dollars each. The gross profit is 50 percent or half of the revenue. Forecast revenue values are generated for both the linear case of volume buildup and the back-end loaded case of volume buildup. For both cases, the net present value is calculated based on the investment of 1.5 million dollars and the gross profit stream generated throughout the 5-year product run. Figure 5-6 illustrates this scenario.

PROJECT DEVELOPMENT WORKSHEET

LABOR COST/MIN.	0.25
ENG. COST/HOUR	55
BURDEN	2.33
REVENUE EACH	10000

PRODUCT COSTS

MATERIAL

ITEM	DESCRIPTION	PROTOTYPE	PRODUCTION
1	DRIVE 50HP	1870	1870
2	ENCLOSURE	1600	600
3			0
4	AMPLIFIER	1250	1360
5			
6			
7			
8			
9			
10			
TOT MATERIAL		4720	3830

LABOR

	PROTOTYPE		PRODUCTION	
DIRECT MIN.	2500	625.00	1750	437.50
INDIRECT MIN	850	212.50	450	112.50
TOT LABOR	3350	837.50	2200	550.00
BURDEN		1456.25		1019.375

TOT FAC COST	7013.75	5399.38

ENGINEERING

	HOURS	DOLLARS
APPLICATION	0	0
HARDWARE	2	110
ALGORITHM	24	1320
SOFTWARE	4	220
DESIGN SPEC	1	55
SPAF	1	55
MANUAL	3	165
BOM	0.5	27.5
COST ROLL UP	0.25	13.75
TEST DATA	4	220
ROUTINGS	1	55
TEST PROCED	1	55
TEST SW	24	1320
TOTAL HRS	65.75	3616.25

EXPENSES   CAP EQUIP

TRAVEL	4000		TEST FIX	0
SERVICES	0		TOOLING	5000
MISC. EQUIP	500		EQUIPMENT	0
TOTAL EXP	4500		TOTAL CAP	5000

SUMMARY

	PROTOTYPE	PRODUCTION
FAC COST	7013.75	5399.38
S,G&A	8116.25	8116.25
TOOLING	5000	5000
NET/UNIT	10000	10000
FORECASTED UNITS	1	2350
REVENUE	10000	23500000
COS	7013.75	12688531.25
GROSS PROFIT	2986.25	10811468.75
REQ'D SALES	4.39	2.85

NET PROFIT	($10,130)	$10,798,353

Figure 5-5. Project evaluation worksheet

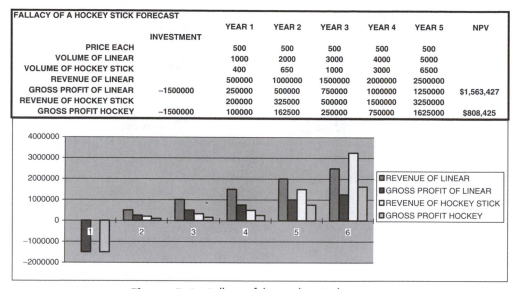

FALLACY OF A HOCKEY STICK FORECAST		YEAR 1	YEAR 2	YEAR 3	YEAR 4	YEAR 5	NPV
	INVESTMENT						
PRICE EACH		500	500	500	500	500	
VOLUME OF LINEAR		1000	2000	3000	4000	5000	
VOLUME OF HOCKEY STICK		400	650	1000	3000	6500	
REVENUE OF LINEAR		500000	1000000	1500000	2000000	2500000	
GROSS PROFIT OF LINEAR	−1500000	250000	500000	750000	1000000	1250000	$1,563,427
REVENUE OF HOCKEY STICK		200000	325000	500000	1500000	3250000	
GROSS PROFIT HOCKEY	−1500000	100000	162500	250000	750000	1625000	$808,425

Figure 5-6. Fallacy of the Hockey-Stick Forecast

In our example, the net present value generated from the investment and gross profit circumstances shows a higher value for the linear case than for the back-end loaded case. There is, however, a more important point here. The forecast should be an accurate representation of the expected sales volume; it will be used in the program as a measure of success and as a justification for further funding. It is also important to show early progress and success to management. A forecast that is purposely back-end loaded is a warning sign to management that the program must absorb much more uncertainty in the marketplace before the corporation achieves payoff.

Understandably, management finds it difficult to support back-end loaded scenarios, as time and uncertainty are the enemies of new product success. Management wants to know what magic will happen to generate such strong sales in years 3, 4, 5 at a greater growth rate than in years 1 and 2. Management is justified in its position on this issue.

7. IMPACT OF OVERBUDGET ON DEVELOPMENT COST VS. FACTORY COST AND SALES VOLUME

In any development program that must result in volume sales to justify the investment, there are concerns about budgets, expenses, factory costs, volume of sales, and timing. The key to successful planning and product management is to focus on those items that will have a significant affect on the success of the program. As we will see in the next example, these parameters are not weighted equally in relation to their respective impacts. The example will show that volume of sales, factory cost, and timing are the key pressure points in a program. The development cost is less important if the market can support the projected sales volume. This is why it is critical to understand the market, the company's capabilities, and the company's effectiveness in securing orders. No amount of cost cutting or other compensation can make up for lack of sales volume.

Let's review the following examples (Figure 5-7). The examples are based on several assumptions:

- The factory cost is $250 dollars.
- The sell price is $500 dollars, yielding a gross margin of 50 percent.
- The net present value is the time-adjusted value of money reflected to the beginning of the project.
- Cases 1,2,3,4, and 5 are decreasing in volume of sales.

BASELINE EXAMPLE	INVESTMENT	YEAR 1	YEAR 2	YEAR 3	YEAR 4	YEAR 5	NPV	IRR
PRICE EACH		500	500	500	500	500	BASED ON	
COST EACH		250	250	250	250	250	5% INFLATION	
CASE 1 VOLUME	−1500000	1000	2000	3000	4000	5000	$1,563,427	30%
CASE 2 VOLUME		800	1600	2400	3200	4000	$965,027	22%
CASE 3 VOLUME		600	1200	1800	2400	3000	$366,628	12%
CASE 4 VOLUME		400	800	1200	1600	2000	($231,772)	0%
CASE 5 VOLUME		200	400	600	800	1000	($830,172)	−17%
GROSS PROFIT CASE 1	−1500000	250000	500000	750000	1000000	1250000	$1,563,427	30%
GROSS PROFIT CASE 2	−1500000	200000	400000	600000	800000	1000000	$965,027	22%
GROSS PROFIT CASE 3	−1500000	150000	300000	450000	600000	750000	$366,628	12%
GROSS PROFIT CASE 4	−1500000	100000	200000	300000	400000	500000	($231,772)	0%
GROSS PROFIT CASE 5	−1500000	50000	100000	150000	200000	250000	($830,172)	−17%

FACTORY COST GROWS	INVESTMENT	YEAR 1	YEAR 2	YEAR 3	YEAR 4	YEAR 5	NPV	IRR
PRICE EACH		500	500	500	500	500	BASED ON	
COST EACH		250	290	330	370	410	5% INFLATION	
CASE 1 VOLUME	−1500000	1000	2000	3000	4000	5000	$323,807	12%
CASE 2 VOLUME		800	1600	2400	3200	4000	($26,669)	4%
CASE 3 VOLUME		600	1200	1800	2400	3000	($377,145)	−5%
CASE 4 VOLUME		400	800	1200	1600	2000	($727,620)	−15%
CASE 5 VOLUME		200	400	600	800	1000	($1,078,096)	−30%
GROSS PROFIT CASE 1	−1500000	250000	420000	510000	520000	450000	$323,807	12%
GROSS PROFIT CASE 2	−1500000	200000	336000	408000	416000	360000	($26,669)	4%
GROSS PROFIT CASE 3	−1500000	150000	252000	306000	312000	270000	($377,145)	−5%
GROSS PROFIT CASE 4	−1500000	100000	168000	204000	208000	180000	($727,620)	−15%
GROSS PROFIT CASE 5	−1500000	50000	84000	102000	104000	90000	($1,078,096)	−30%

DEV EXPENSE HIGH	INVESTMENT	YEAR 1	YEAR 2	YEAR 3	YEAR 4	YEAR 5	NPV	IRR
PRICE EACH		500	500	500	500	500	BASED ON	
COST EACH		250	250	250	250	250	5% INFLATION	
CASE 1 VOLUME	−1750000	1000	2000	3000	4000	5000	$1,325,332	24%
CASE 2 VOLUME		800	1600	2400	3200	4000	$726,932	16%
CASE 3 VOLUME		600	1200	1800	2400	3000	$128,532	7%
CASE 4 VOLUME		400	800	1200	1600	2000	($469,867)	−4%
CASE 5 VOLUME		200	400	600	800	1000	($1,068,267)	−20%
GROSS PROFIT CASE 1	−1750000	250000	500000	750000	1000000	1250000	$1,325,332	24%
GROSS PROFIT CASE 2	−1750000	200000	400000	600000	800000	1000000	$726,932	16%
GROSS PROFIT CASE 3	−1750000	150000	300000	450000	600000	750000	$128,532	7%
GROSS PROFIT CASE 4	−1750000	100000	200000	300000	400000	500000	($469,867)	−4%
GROSS PROFIT CASE 5	−1750000	50000	100000	150000	200000	250000	($1,068,267)	−20%

Figure 5-7. Return on investment calculator

As shown in the baseline example in Figure 5-7, the investment of 1.5 million dollars is amortized with good return in cases 1 and 2 and only marginally in case 3. The decreasing sales volume affects the program significantly in cases 4 and 5. This set of cases will serve as the baseline example showing the impact of sales volume reduction.

The "Factory cost grows" example illustrates the impact on factory cost increases as time goes on. This example shows immediate and deep impact on the net present value due to the increased manufacturing cost and fixed market price.

Case 3 shows the impact of increased development expense on the overall program. As shown in Figure 5-7, this has much less of an effect than the manufacturing cost increase example in case 2. The third example, "Development expense high," supports the conclusion that if the market opportunity is real and the company can achieve market penetration within the anticipated time frame, the cost of development is a less important factor in the overall equation. Keep in mind that this conclusion applies only to development cost, not to the time required to develop the product, which is still critical to market acceptance and success.

For your specific program, you might want to try a few scenarios to experiment with the potential impact that each parameter might have. A sample return on investment spreadsheet is available for your use in the Toolbox.

8. COLLATERAL COSTS: DIRECT AND INDIRECT

The cases presented in this chapter are simplified for illustrative purposes. In actual situations, there are additional costs to take into account. Some of these costs are obvious and can be posted directly to the product line. Others are indirect and are considered elsewhere. Be sure to factor these costs into the calculation of sales return. Some of these costs are hidden.

Cost assignment is determined by the culture of the organization. The manner in which costs are absorbed and posted to individual product lines can be misleading. The operative lesson is to be consistent. If comparable products do not include all of the costs, then do not post them to your disadvantage. You might want to keep track of them separately, however, for your own use.

TIMING AND LOST OPPORTUNITY COSTS

1. TIMING IS INDEED EVERYTHING

In some strategic market situations, timing is the key element to success. It is more important to get into the game and sacrifice profits than to miss being in the game. If this is the case, then timing is key to the extent that sacrificing profits is acceptable to gain immediate participation. Long-term sacrifice of profits will have devastating effects on the organization and limit any future market response. To repeat an earlier statement, you must be able to deliver the product at specification at the market's cost and allow profits to amortize the development in a market-oriented time frame.

2. HIERARCHY OF SUCCESS FACTORS PERTAINING TO NEW PRODUCT DEVELOPMENT

In terms of importance to product development, there exists a hierarchy for critical success factors. This hierarchy starts with the health and the size of the business, which determine the relative affordability for particular development programs.

Next, it is important to examine the profitability of the business to determine investment endurance. Can the organization carry off a long-term investment and fund it sufficiently throughout periods of uncertainty?

Equally important are management's spending priorities and habits. Is management consistent in what it spends funds on, and is there an established pattern of investment for the business? The answers to these questions determine the actual disbursable cash that management can support for a given program.

The time frame for investment is also important to determine. Does the organization have a track record for investing and ensuring that investment through to completion? Is it averse to navigating uncertainty? Does management change direction in midstream? Knowing the historical pattern of corporate response sets the sense of urgency for the program. Finally, what is the organization's tolerance for error or failure, and what are the company's patterns of recovery?

Determine the enterprise's approach to cost management. How does the organization approach manufacturing cost roll-ups? How are costs contained?

These issues need to be explored and understood, as at least some of them are likely to affect the program at some point during its execution. Among critical success factors, timing is a key issue.

3. SLIDING WINDOWS OF OPPORTUNITY REVIEWED

Previously, we reviewed the sliding windows of opportunity for new products and saw that a marketplace opportunity does not remain in existence for a long period of time. Rather, it is available to capitalize on for a short time only and generally triggers activity by many competitors.

To be a long-term player, it is critical to be selective in choosing opportunities and to amass the horsepower needed to execute projects effectively. In years past, these opportunities evolved slowly, and companies had more time to evaluate whether they wanted to go in a specific direction. Today, opportunities are intense but shorter lived, with the spoils going to the competitor who can mobilize its forces most effectively.

4. THE MOVING TARGET AND THE SPEED OF THE BULLET

Previously, we discussed "leading the target" in product planning. The development effort must be coordinated with business cycles and accounting's financial plan for the business. If we use the moving target analogy, then the "bullet" fired is the new product. The speed of the bullet's movement from the gun to the target is analogous to the speed of development. Each year, the market moves a little faster, so product development must improve in speed to lead the target and secure a market share.

5. COMPARATIVE ANALYSIS FOR PROGRAMS

One of the tools for deciding which program to fund is a comparative analysis of several programs. Comparative analysis allows dispassionate evaluation of the programs on a strictly financial basis of projection. It consists of estimations of investment and revenue and computes a return in a consistent, measurable way. Figure 5-8 shows three product programs with various investment scenarios and calculates returns for each.

As the first figure shows, an investment of $2.0 million dollars results in three possible scenarios for return. In this analysis, as in the other analyses we will review, the volume of product sold in the best, most likely, and worst cases is held consistent for each of the three

PRODUCT OPPORTUNITY A	INVESTMENT	YEAR 1	YEAR 2	YEAR 3	YEAR 4	YEAR 5	NPV	IRR
PRICE EACH		500	500	500	500	500	BASED ON	
COST EACH		250	250	250	250	250	5% INFLATION	
BEST CASE VOLUME	−2000000	1000	2000	3000	4000	5000		
MOST LIKELY CASE VOLUME		600	1200	1800	2400	3000		
WORST CASE VOLUME		200	400	600	800	1000		
GROSS PROFIT BEST CASE	−2000000	250000	500000	750000	1000000	1250000	$1,087,237	20%
GROSS PROFIT MOST LIKELY CASE	−2000000	150000	300000	450000	600000	750000	($109,563)	3%
GROSS PROFIT WORST CASE	−2000000	50000	100000	150000	200000	250000	($1,306,362)	−22%

PRODUCT OPPORTUNITY B	INVESTMENT	YEAR 1	YEAR 2	YEAR 3	YEAR 4	YEAR 5	NPV	IRR
PRICE EACH		500	492.5	485	477.5	470	BASED ON	
COST EACH		250	240	230	220	210	5% INFLATION	
BEST CASE VOLUME	−1500000	1000	2000	3000	4000	5000		
MOST LIKELY CASE VOLUME		600	1200	1800	2400	3000		
WORST CASE VOLUME		200	400	600	800	1000		
GROSS PROFIT BEST CASE	−1500000	250000	505000	765000	1030000	1300000	$1,640,903	31%
GROSS PROFIT MOST LIKELY CASE	−1500000	150000	303000	459000	618000	780000	$413,113	13%
GROSS PROFIT WORST CASE	−1500000	50000	101000	153000	206000	260000	($814,676)	−16%

PRODUCT OPPORTUNITY C	INVESTMENT	YEAR 1	YEAR 2	YEAR 3	YEAR 4	YEAR 5	NPV	IRR
PRICE EACH		500	525	550	575	600	BASED ON	
COST EACH		250	285	320	355	390	5% INFLATION	
BEST CASE VOLUME	−1000000	1200	2000	3000	4000	5000		
MOST LIKELY CASE VOLUME		600	1200	1800	2400	3000		
WORST CASE VOLUME		200	400	600	800	1000		
GROSS PROFIT BEST CASE	−1000000	300000	480000	690000	880000	1050000	$1,775,064	46%
GROSS PROFIT MOST LIKELY CASE	−1000000	150000	288000	414000	528000	630000	$656,875	23%
GROSS PROFIT WORST CASE	−1000000	50000	96000	138000	176000	210000	($415,962)	−10%

Figure 5-8. Comparative return on investment

products being evaluated, to give the reader a sense of the potential impact of volume and pricing on the return estimates. For Product A (Figure 5-8a), as for the other two, the revenue is based on a price of 500 dollars and a cost of 250 dollars. In these examples, the cost and price are held constant over the five-year product run.

The investment/return analysis for Product B is shown in Figure 5-8b. In this example, product cost decreases with volume, and the price in the marketplace also tracks downward. In this case, the cost decreases at a greater rate than the price, with the result that margin dollars are enhanced. Also, the initial investment for product B ($1.5 million dollars) is less than for product A.

The next example, Figure 5-8c, is for Product C. The initial investment is $1.0 million dollars. The product price increases with each year, and the product cost also increases; however, gross profit declines because the cost increases at a greater rate than the price, thus causing erosion in the margin. In this case, the combination of the product, profit, and investment generate a more favorable picture. A three-product comparison chart is available for your use in the Toolbox.

CRITICAL UNIT VOLUME DURING AMORTIZATION

1. AMORTIZATION

"Yeah, all these plans are nice, but show me the orders!" If this has been said to you recently by a senior manager or by your banker, it generally instills an unsettling feeling in the product manager. Whether we like it or not, this statement is rooted in fact and experience that reemphasize the critical need for a new product to become a viable business by generating unit volume. It is the volume that initiates and supports progress.

Volume means that a strong sales effort is occurring; it drives the unit cost down to where it has been targeted. It demonstrates the effectiveness of a marketing plan and establishes momentum. Fundamentally, volume creates the funds flow into the organization to begin to offset the investment in product development. A portion of each revenue element is used to pay back the organization for investment and therefore amortizes the development. There simply is no other means to offset these costs.

2. DON'T MISREPRESENT PROJECTED VOLUME

The fact that unit volume is the initiator of pay back is its own check and balance. If the development team overestimates the sales forecast, the organization will be unable to meet anticipated volume, and consequences detrimental to the product will occur. If the forecast is too low, the organization will not approve the expenditure for the initial development.

A common danger in new product development and the team's zeal in qualifying the program is to overstate volume to "sell" the program. This practice can be a career breaker for the team members. The best approach is to evaluate the opportunity realistically and forecast based on the most likely scenario. If the program is viable, it will be approved and eventually will be successful. Remember that both management and the banking community looks for promises kept, not promises made.

3. EXAMINE ROI: UNIT VOLUME PAYS FOR MANY ERRORS

"There is nothing wrong with the organization that more orders can't cure." How often have you heard this statement? It, too, has a basis in business experience. A healthy amount of orders often can mask inefficiencies within an organization. In point of fact, a downturn in sales forces organizations to redesign and streamline their processes. There is, however, no amount of cost reduction, trimming of expenses, or austerity measure that can compensate for lack of orders, even though the common knee-jerk reaction to lack of orders is simply to cut costs. The key to a well-run organization is to keep healthy programs going, trim off poor performers, and keep a focused eye on order input to keep from having to operate in an austerity mode.

To reinnforce this concept, look at the sample income statement (Figure 5-9a). It shows the sales input side of the income statement. Net sales are at full value in this example. The material labor and burden are factored in on a normalized basis across the series of five examples of decreasing revenue and profits. The five scenarios represent a shrinking business.

The material cost is factored at 50 percent of sales, while labor is factored at 6.5 percent of sales. For simplicity's sake, the burden is 2.5 times the labor dollars. The sales, general, and administrative expenses are shown as typical percentages of sales, leaving a net profit of 27.25 percent of sales. The expenses scale down as the business scales down, keeping the same percentage. This results in a fixed percentage of sales for the new profit with a decreasing dollar amount. To preserve the dollar amount of profit at $92,500 without decreasing the gross margin (as a result of volume reduction), the sales, general, and administrative expenses would have to decline disproportionately. In the example given in the figure, the scaled down expenses would be $117,000. To preserve the profit, however, these same expenses would have to total $84,625—a significant difference. The figure 5-9b shows how this disproportionate relationship exists, and how the impact of such a level of reduction can cripple a company.

The type of cost cutting just described affects staffing significantly. If the expenses are headcount-related only (an oversimplification for the sake of this example), the reduction rate will cause the enterprise to lose too many people, cutting staff below th levels needed for critical mass and for positioning the company for a strong development position in the future. In this example, we assume that every $100,000 of sales requires an additional staff member. Conversely, every reduction of $100,000 of sales revenue results in a reduction of staffing by one. At a one-million-dollar sales company, there are 10 people. When sales decline to 6.5 million according to this model, it would be necessary to fire 35 percent of the workforce!

INCOME STATEMENT

		CASE 1	CASE 2	CASE 3	CASE 4	CASE 5
SALES		1000000	912500	825000	737500	650000
RETURN ALLOWANCE		0	0	0	0	0
NET SALES		1000000	912500	825000	737500	650000
MATERIAL	@50% OF SALES	500000	456250	412500	368750	325000
LABOR	@6.5% OF SALES	65000	59312.5	53625	47937.5	42250
BURDEN	@ 2.5 TIMES LABOR	162500	148281.25	134062.5	119843.75	105625
GROSS PROFIT	@27.2% OF SALES	272500	248656.25	224812.5	200968.75	177125
SALES EXPENSE	@ 10% OF SALES	100000	91250	82500	73750	65000
ENGINEERING EXPENSE	@ 4.5% OF SALES	45000	41062.5	37125	33187.5	29250
ADMINISTRATIVE EXPENSE	@3.5% OF SALES	35000	31937.5	28875	25812.5	22750
SALES		1000000	912500	825000	737500	650000
TOTAL EXPENSES NEEDED TO OPERATE		180000	164250	148500	132750	117000
TOTAL EXPENSES REQUIRED FOR PROFIT OF $92,500		180000	156156.25	132312.5	108468.75	84625
REQUIRED REDUCTION PERCENT		0.000	4.928	10.901	18.291	27.671
% REDUCTION OF SALES			0.088	0.096	0.106	0.119
% REDUCTION OF EXPENSES TO PRESERVE PROFIT			0.132	0.153	0.180	0.220
NET PROFIT		92500	92500	92500	92500	92500

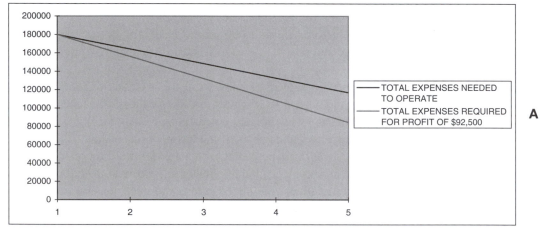

A

PEOPLE REQUIRED TO OPERATE	10	9.125	8.25	7.375	6.5
	WHOLE	FIRE 10%	FIRE 17%	FIRE 25%	FIRE 36%

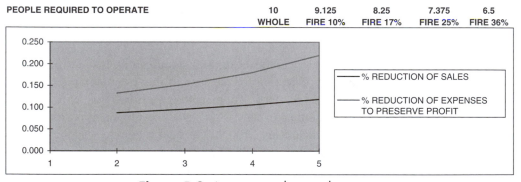

B

Figure 5-9. Income vs. volume and expenses

In an actual company, eliminating headcount as drastically as done on the spreadsheet example will create irreparable damage to the organization. Observe once more that little can compensate for unit volume.

4. THE IMPORTANCE OF AN ACCURATE UNIT VOLUME FORECAST

There is another, more self-motivated reason for the accurate assessment of volume forecast in new product development. The organization will be less than kind to you if the forecast is skewed to the high side; especially if initial numbers are not as aggressive as originally anticipated. In addition, whatever mentoring might be in place for you with senior management will soon fade in the wake of an overrated forecast.

The forecast of volume and the profits realized from them are factored into the budget. These funds will be used for other programs, so they must be achievable. If the funds are not realized, the budget process falls apart, and follow-up investment will suffer. Worse yet, management's confidence will be shaken to the point that follow-up funding will be more difficult.

5. FEEDING ON SCRAPS RATHER THAN OWNING THE MARKET SEGMENT

In a previous chapter, we discussed the need for timeliness in new product introductions. The market dynamics dictate that the first company to market usually is in the best position to reap the rewards of that market. If you miss the timing, you could end up serving a segment of the market you have no interest in serving, as shown in Figure 5-10.

In this illustration, the initial target is for the premium market segment, represented by the solid arrow pointing to this premium segment. If development is lethargic, the dynamics of the marketplace keep the market opportunities moving, The market in effect moves vertically past the manufacturer's target if the manufacturer's development is late. In such

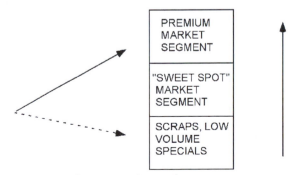

Figure 5-10. Market segment

a case, the dashed line represents the manufacturer ending up with low volume specials, completely missing the more desirable segments.

6. INTRODUCING A NEW PRODUCT WITHOUT VOLUME IS LIKE BUYING A HOME WITHOUT INCOME FOR THE PAYMENTS

Previously in this chapter, we discussed the concept of amortization as the streams of payments required to offset an investment. Ideally, the stream of payments, factoring in the temporal value of money, should exceed the original investment and incrementally add to the retained earnings. A company must have a stream of profit-generating revenue to amortize the development investment. Ensure that all of the expenses, costs, and investment are coordinated with the revenue and profit to guarantee that the program is on sound financial footing.

GENERATING CASH AND PROFIT

1. IMPACT OF ROUTE TO MARKET

The route to market can have a significant impact to the overall program. As a new product manager, your job is to maximize the profit for the company. Consequently, you must evaluate the tangible exchange that will take place between the company, the market, and the route to the final customer. The worth of the roles they play and the value given for those roles must be evaluated.

Pinpointing the risk and the party who endures the burden of warranty both enter into the analysis. A route that can actually deliver customers and volume sales that could not be delivered otherwise is worth more of the profit margin than a route that is merely in the opportunistic place at the appropriate time. Figure 5-11 shows the impact of the route on profitability.

In the example shown in the figure, the product has a normalized cost structure as follows for all three cases. In case 1, the channel cost is 15 percent of the sales price of $120, or $18. This leaves the manufacturer with a gross profit of $30.75. As the channel cost increases, the gross profit for the manufacturer declines to a crossover point, as shown in case 2. A further reduction causes an inverted situation in which the manufacturer is making less on the product than the channel taking the product to market. This is undesirable, as the manufacturer bears most of the risks, considering the warranty and the initial investment.

2. IMPACT OF ECONOMIC CYCLES

The impact of economic cycles is significant to the development of new products and new business. The ebbs and flows of the larger economy affect the profitability and cash

IMPACT OF CHANNEL ON PROFITABILITY

SALES PRICE	MATERIAL	LABOR	BURDEN	FAC COST	CHANNEL COST	CHANNEL COST	GROSS PROFIT
120	50	6.5	14.75	71.25	15%	18	30.75
120	50	6.5	14.75	71.25	20%	24	24.75
120	50	6.5	14.75	71.25	25%	30	18.75

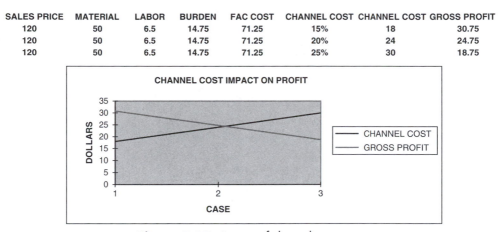

Figure 5-11. Impact of channel on cost

movement of a company. A well-run program can be affected by poorly managed corporate finances. In addition, there are several points in a cycle in which the company is cash rich and others in which it is starved for cash.

Figure 5-13 illustrates this point. In an economic downturn, sales of products (normalized as shown in the index values) are lower than previous months (T1–T8). This is a situation in which the company operations are throwing off cash, as shown by the values of the row labeled Cash (Activity). The values in this row are computed by subtracting the T values in the Index (Normalized) row from the previous T value, yielding a measurement of the cash flow. Accounts receivable is collecting for previous shipments at the higher rate. Meanwhile, accounts payable is paying out at a reduced rate. Cash is plentiful in this section of the cycle. At the lower level of business, management has throttled back on expenses and programs to maintain some level of profitability at the reduced levels.

As the business increases again, coming out of the downturn in the cycle (T9–T15), inventory must be purchased at a higher cost than the lower steady state, which increases accounts payable. Products are built at a higher cost, and accounts receivable is growing because of the increased shipments. This dynamic drains cash, as shown by the negative numbers in the Cash (Activity) row. The cumulative cash flow is shown in the third row of the table.

The graphic representation of what is occurring also appears in the lower part of Figure 5-12. This is not a pure representation of a cash flow example but is merely intended to stress a point about economic cycles and the company's ability to fund programs during specific economic phases.

CASH MOVEMENT IN ECONOMIC CYCLES															
	T1	T2	T3	T4	T5	T6	T7	T8	T9	T10	T11	T12	T13	T14	T15
INDEX (NORMALIZED)	10.5	9.5	8.5	6.5	5.5	5.5	4.5	3.5	3.5	3.5	6	8	10	11	11
CASH (ACTIVITY)	0	1	1	2	1	0	1	1	0	0	−2.5	−2	−2	−1	0
CUMULATIVE CASH	0	1	2	4	5	5	6	7	7	7	4.5	2.5	0.5	−0.5	−0.5
INDEX (NORMALIZED)	10.5	9.5	8.5	6.5	5.5	5.5	4.5	3.5	3.5	3.5	6	8	10	11	11
CUMULATIVE CASH	0	1	2	4	5	5	6	7	7	7	4.5	2.5	0.5	−0.5	−0.5

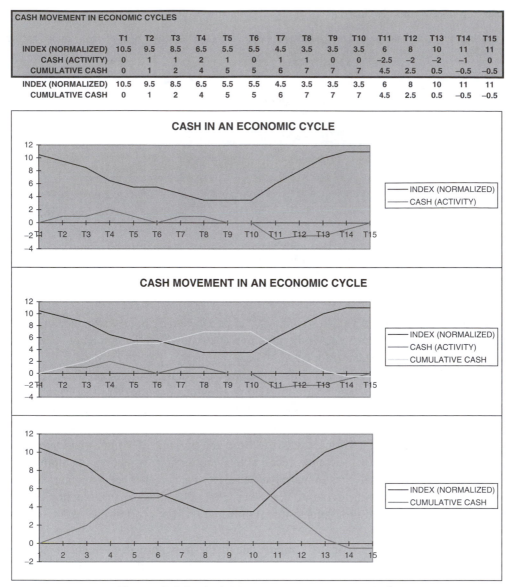

Figure 5-12. Cash Movement in an Economic Cycle

3. INSOLVENCY AFTER A RECESSION

As you view the charts in Figure 5-12, it is easy to see how recovery out of a recession or even a growth spurt can strap a company for cash. The finances of the company are most at risk during these times, and a watchful eye will focus on expenses. Many firms become insolvent during these times because of cash requirements.

If, however, the company has the financial strength and an opportunity worth the risk, the bottom of the cycle or a point approaching the bottom (depending on the length of the development cycle) is an excellent time, strategically, to prepare for the rise out of the downturn. Your company will be prepared to meet the market needs with new product, while your competitors struggled to survive the downturn. This is illustrated in Figure 5-13.

4. READ MANAGEMENT WELL TO SEE HOW FAST IT REACTS

In order to be an effective manager of new product development, you need to be able to read senior management's reactions to these financial challenges in economic cycles. If the company can tolerate the financial stress, investment in a downturn can pay off big dividends. You need to assess the management's appetite for risk, its ability to manage during a downturn, and its staying power in decision making.

5. WATCH THE HEALTH OF THE BUSINESS

Like it or not, the new product manager needs to keep a watchful eye on the general health of the business during the project. The financial pressure points in a project, (e.g., appropriation requests and equipment procurement) need to be handled within the framework of the finances of the business as a whole. For example, do not approach the senior management for funds in the same time frame in which the company is reporting a loss. You can improve your chances by letting the bad news of a loss wear off for a few days and present your request later, when it might be more palatable.

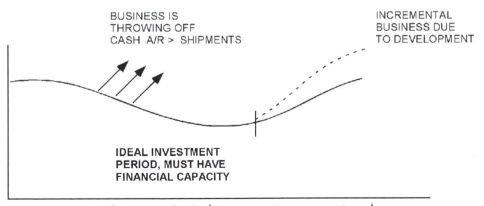

Figure 5-13. Cash movement in an economic cycle

6. CONSIDER NEW PRODUCT DEVELOPMENT AS A DISCRETIONARY EXPENSE

We discussed the overall strategic plan of the business and the need for new product development to be an integral part of the operation. In most companies; however, new product development is considered a discretionary expense. It is viewed this way because of the tendency toward shorter-range focus on profitability, which makes development subject to the nuances of quarterly reports. If there are going to be reductions in program expenditures, it is important to point out to senior management the impact of these reductions. Often these decisions are made in a meeting, and a vacuum of previous assumptions is made. The impact of previous promises is not readily apparent. Furthermore, the budget might not be updated to account for the reduction in future income due to the current reduction or delay.

It is also important to understand how management reacts to financial pressure within the framework of a product development. Is management prone to terminate expenses unilaterally at the first sign of bad news, or will it tend to follow through on commitments previously made to a new product?

7. IMPORTANCE OF ECONOMIC ANALYSIS AND COMPANY REACTION TO EXTERNAL FORCES

There are two factors that management needs to correlate in a downturn. The first is the general health of the industry and the economy in general. The second, and more important, factor is the company's performance, which encompasses its plans for the future. The two factors do not necessarily follow each other, as shown in Figure 5-14.

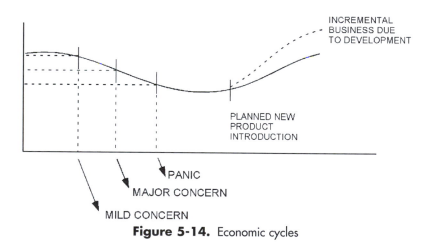

Figure 5-14. Economic cycles

Does management react with mild concern at the first sign of a downturn, or does it move directly into the panic mode? If panic is the first reaction, the impact of lost revenue will be more dramatic. It is management's job to react with prudence as the economy fluctuates; more important, management should react to both the health of the economy and that of the company itself. It is the program manager's job to be an advocate for project continuity.

8. THERE IS NO SUCH THING AS ONE BAD QUARTER

Every company has a certain amount of inertia associated with it. The communication between the financial reports and the field sales personnel is not direct, nor is it unfiltered by third-party interpretation. This is why initiatives take some time to pass through an organization.

The reason that there is no such thing as a bad quarter is because of this phenomenon. By the time the first report of bad financial news reaches the management of the company, the company in aggregate cannot react instantly to effect a change in the trend. Management can initiate action, but the action takes time to execute. Field sales personnel appointments are set up, focus is directed, and the final effect of the focus—the purchase order—does not occur immediately. There is generally a lapse of time between sales cultivation and the securing orders. Therefore, while this "redirection" is occurring, the next quarter is passing the company by.

9. LAUNCHING IN THE EYE OF A BUYOUT, LARGE VS. SMALL

A buyout is one of the most exciting aspects of a new product development manager's job, and it occurs in rare cases. There was a film some years back about the grand prix racing circuit. In it, the lead character was interviewed about his success in racing. He described how he wins races: "When I see an accident off in the distance, I accelerate as a first instinct, because others decelerate as their first instinct". It is not a very nice way to win, but it sometimes gets results.

There is a parallel between this racer's strategy and launching a new product initiative. While news of a buyout or change in ownership of a company circulates, a paralyzing effect occurs among the people involved. In the interim, very little progress is made; time is spent commiserating and guessing. This is the perfect opportunity to differentiate your team to the new ownership. The new management will be looking for this type of leadership.

In a larger company, maintain steady progress and continue with the plan. If your management is paralyzed, take time to explain the impact of no action. In a small company, if you are in control, accelerate your team's efforts—it will pay off in the long run.

10. HOW TO PULL IN YOUR HORNS

After the new ownership is in place and the dust has settled, there could be some cost cutting. You might be placed in a position to requalify and rejustify your program. Assumptions made before the change in ownership might no longer be valid to the new management, or even to the same management under new direction. Be prepared to fight even harder than before to preserve the program. Get in tune with the new ownership's direction, and focus and tailor the program's presentation accordingly. Also, prepare alternatives to the program; the purpose of such alternatives should be only to enhance the original plan or to align it more closely with new company objectives. The important point is to move ahead according to the plan in concert with the new ownership rather than rethink the tactics or strategy midstream to appease new management. If you fall prey to this latter temptation, you will get caught short in delivering previously stated goals with new assumptions overlaid on them.

PROFIT IN BACKLOG

1. MONITOR THE 90-DAY AND 180-DAY BACKLOGS

Keep tabs on the business level by watching out for the 90- and 180-day backlogs. These two numbers, along with the calculation of profit in backlog, can give you a sense of the general health of the business.

The profit in backlog is a detailed calculation of the various products in the backlog, their cost structures, and the amounts of gross profit they generate. If the accounting system is set up with allocations for sales, general, and administrative expenses, then an S, G, and A expense percentage can be entered in the appropriate space for each product category. The funds for development that you require in the six to nine months that follow will come from this backlog calculation.

Figure 5-15 shows an example of a profit in backlog calculation. The 90-day or 180-day totals are placed in the chart for analysis in addition to specific product information.

The chart in Figure 5-15 shows the total backlog numbers for each of five different products. Also shown are the prices for each product along with their respective cost structures. Sales, general, and administrative costs are entered as a percentage of sales, and net profit is calculated and totaled. This type of analysis can be used for the comparative analysis also. The sample expanded version of this calculation for 20 products is included in the Toolbox for your use.

2. USE AS PRESSURE CYCLE FORECAST OF THE GENERAL IMMEDIATE HEALTH OF THE COMPANY

The 90- and 180-day comparisons with previous backlog figures on a rolling basis can provide insight into trends or into critical changes to the business. Depending on the lead

PROFIT IN BACKLOG						
PRODUCT CATEGORY	PRODUCT 1	PRODUCT 2	PRODUCT 3	PRODUCT 4	PRODUCT 5	TOTAL PRODUCTS
BACKLOG UNITS	30	23	56	76	87	
PRICE NET EACH	6000	7500	4500	9000	1275	
BACKLOG DOLLARS	180000	172500	252000	684000	110925	1399425
MATERIAL	0.275	0.375	0.3	0.325	0.275	
LABOR	0.085	0.09	0.075	0.065	0.05	
BURDEN	0.2125	0.225	0.1875	0.1625	0.125	
FAC COST	0.5725	0.69	0.5625	0.5525	0.45	
SG AND A PERCENT	28.5	27	32	34	30	
PROFIT	25650	6900	29610	73530	27731.25	163421.25

Figure 5-15. Profit in Backlog

times, the 180-day backlog can be significantly smaller than the 90-day backlog; however, the comparative between the present 90- and 180-day backlogs with the previous measurements (for example, 30 days previously) can foretell certain trends. The pressure cycle forecast is available for your use in the Toolbox.

COST, VOLUME, PROFIT, AND BREAK-EVEN

1. BREAK-EVEN ANALYSIS

In any business, there are multiple costs associated with operations. Variable costs are a function of the volume of business, and fixed costs are a function of the business-incurred costs without reference to volume. There is a point at which the profit of the enterprise is equal to the fixed cost. This is known as the break-even point or zero-profit point. This is a crucial point to know even though no one ever wishes to operate in that mode.

As shown in Figure 5-16, the behavior of the fixed costs and the profit is similar to the equation for a line with a y intercept; for example, $y = m(x) + b$, where y is the profit, b is the fixed cost, m is the per-unit profit, and x represents the number of units. Assuming that the fixed costs are b, then the marginal profit per unit times the number of units equals b at break even. To find out the break-even point, determine the value of x, the number of units, where the total marginal profit equals the fixed costs.

In the example in the figure, the fixed costs are $150,000 dollars. The cost per unit is $275 dollars, and the sell price is $525 dollars. The break-even occurs at 600 units. This is the point at which the incremental profit of $250 dollars per unit multiplied by 600 units equals the fixed cost of $150,000. Copies of the chart and spreadsheet are available for your use in the Toolbox.

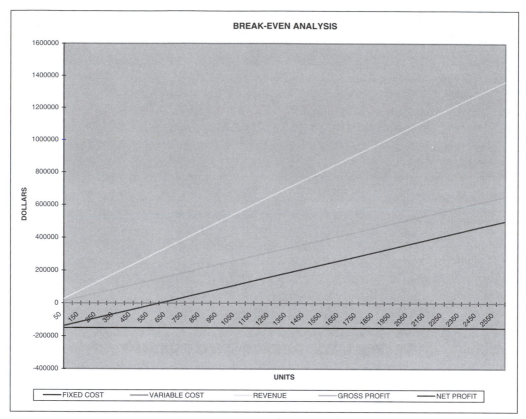

Figure 5-16. Break Even Analysis

2. SINGLE-PRODUCT COMPANY VERSUS CONTRIBUTION TO PROFIT MODEL IN LARGER COMPANIES

There is also a difference in the treatment and evaluation of profit between a large company with several product lines and a small company with one product or only a few products. The profit motive is much more intense in the small company and is critical in the single-product company.

We think and evaluate in terms of contribution to profit in the context of a large, multi-product company. The absolute dollar profit value is less important than its overall contribution, pull through sales, and strategic direction, even though it remains important to evaluate these products and the absorption of costs of their profits.

In the smaller company, there is necessarily little interest in percentages or global strategy; the issues are survival and profit dollars. The senior management and ownership of a small company generally do not respond to the discussion of what, to them, are abstract issues.

3. EVERY PROGRAM FLIES ON ITS OWN

Regardless of the size of the organization, the key to long-term success for the product, the team, and the individuals involved is to have every product be profitable in its own right. In addition, ensure that the accounting system is set up to evaluate the various products uniformly and that the indirect and S, G and A costs are absorbed according to their respective organizational activity bases.

There are many instances in which historical changes have been made in the treatment of products by accounting. Hidden costs or allocations levy undo organizational burden to higher-margin products in order to improve the financial "look" of others. If your organization is predisposed to this kind of treatment, costs could be posted to the new product to offset existing poor performers.

Each product should stand on its own merits in terms of profitability, and each should be managed by the correct financial data. Without these assumptions in place, incorrect conclusions can be drawn from the data, and product decisions are not well considered or executed.

4. INVENTORY: THE HIDDEN COSTS

We discussed the issue of accounting data being massaged to enhance the position of certain products. There is another issue in organizations—inventory—that affects the overall performance of a product line. Inventory is an indirect cost which, if not managed properly, can represent a significant burden on the corporation.

Most corporations are organized along functional lines, with product managers controlling specific products. Inventory is customarily considered to be the responsibility of manufacturing; however, this operational control generally defers to the product manager to facilitate the launch of the new product. If the product line is not well considered, or if the market shifts somewhat, the impact on inventory can be devastating.

The product manager needs to participate in inventory control and to operate within the framework of corporate goals given to manufacturing. In addition, product changes in response to problems can generate scrap, both in material and in labor.

FINANCIAL MODELS FOR SALES TRANSACTIONS

Shifting our focus to external factors in new product financial matters, there are multiple models for sales transactions. These models are representative in nature and are not meant to cover all the possible scenarios, but they do give an indication of how funds flow with the sales transactions.

1. DIRECT SALES MODEL

In a direct sales model, a sale is made to a customer. The factory purchases material and furnishes labor to build a product. The sales cost is incurred as the salesperson is securing the order. Other general and administrative costs are also incurred as a result of operations.

The product is shipped, and an invoice is generated and sent to the customer for payment. Payment, normally due in 30 days, can stretch out to 45 days. The company has carried the costs of this transaction for the 45 days; which helps explain why it requires cash for a product line to grow. As you increase sales, the cost to ramp up increases also. Figure 5-17 shows the cash flow for the direct sales model.

2. REPRESENTATIVE SALES MODEL

The representative model is different from the direct sales model in that the bulk of the sales expense that is normally associated with the direct sales force is not incurred prior to sale. The representative advances these costs by virtue of being an independent businessman. The factory issues a check for the commissioned sale, generally after receipt of net payment, to ensure that representative-entered orders are real and collectable. Figure 5-18 shows the financial model for the representative sales model.

3. DISTRIBUTOR OR VALUE-ADDED RESELLER SALES MODEL

The value-added reseller model is different from the previous models in that an entire transaction occurs between the distributor and the customer. The manufacturer builds the

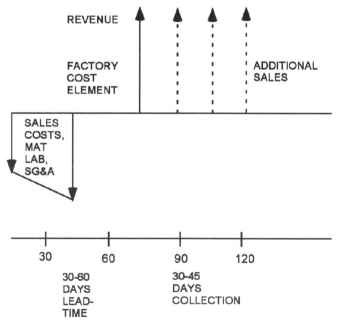

Figure 5-17. Direct Sales Model

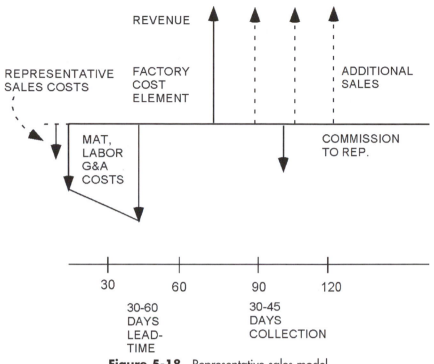

Figure 5-18. Representative sales model

products to a forecasted order and ships the product to the distributor. There might be other incentives between the distributor and the manufacturer, such as floor planning or rebates, but fundamentally, the distributor or Value Added Reseller (VAR) takes receipt of the product and resells it at a margin to the end user.

The manufacturer receives payment on a predictable basis, as manufacturer and distributor work together throughout the term of their relationship so that uncertainty of credit with the customer is eliminated or reduced. The distributor then must rotate the product (sell, ship, and collect payment) through its organization at a rate of inventory turns that exceeds the bank's finance levy for business. The distributor then ends up with the margin for profit, as shown in Figure 5-19.

FINANCIAL IMPACT OF LACK OF CONTINUITY

1. THE DAMAGE CAUSED BY LACK OF CONTINUITY

There is a cost impact due to lack of continuity within a product development program. A certain degree of efficiency is developed in a team's joint effort. Constant stops and restarts cause delays and also increase costs, and inefficiency settles into the team. This

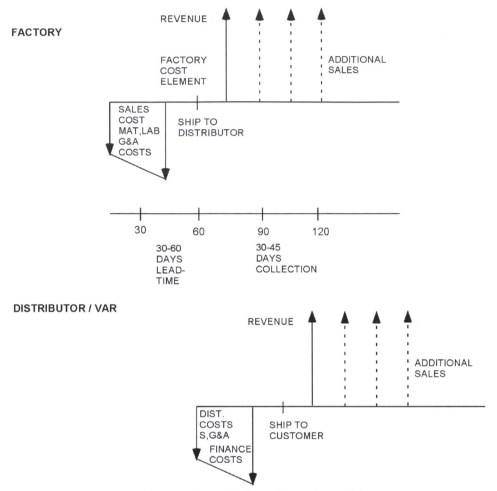

Figure 5-19. Distributor/Var. sales model

inefficiency manifests in two ways. The first is a tendency for the team to slow its progress by working at a measured pace. An attitude of "don't exert too much effort, they will change their minds again anyway" becomes commonplace. The second is due to refocusing the team on the latest project at hand, a process that is similar to the setup time for fixtures that must occur before progress can be made.

2. ENGINEERS ARE NOT LIGHT SWITCHES

In addition, development personnel are not light switches. They do not start and stop programs as swiftly as the senior management might desire. They will generally complete a certain phase of a program and document it, even though it might have been placed on hold. This is to force a natural break in the program, so that the program can be picked up at a later date.

As you work with more and more people and more projects that involve technical personnel, it will become evident that continuity is the best thing a manager can provide to the team. It is the manager's job to clear pathways and provide consistent direction for the development team.

3. MOMENTUM ONCE LOST IS DIFFICULT TO REBUILD

The issue of project management will be discussed in Chapter 7, and in it we will see that time lost in a development effort, given the constraints of the development program, cannot be made up, worked around, or reorganized around. It simply is lost. We can delude ourselves or pacify senior management by restructuring the program, shortening task times, or adding more personnel to execute tasks; however, if the program was laid out well initially, lost time cannot be recovered. Don't lose it in the first place! An effective new product development manager will foresee the potential for time loss and act to prevent it.

4. LOSS OF TRUST BETWEEN DEVELOPMENT AND MANAGEMENT

An often-underestimated effect of redirecting priorities for development people is the loss of trust or confidence that can occur between the development team and senior management. Too many swift moves can cause the development team to question management's sincerity, decisiveness, and commitment to the programs. If you are a senior manager, keep this in mind, as you might be the one causing problems regarding lack of focus.

5. LOSS OF CONTINUITY: SLOWER RESULTS AND POORER QUALITY

A final thought on continuity and the impact of its absence on cost is that in corporations, decisions are often made in a linear fashion with justification rooted in equations, relationships, and schedules. Unfortunately, we as managers forget that it is people who execute programs, it is people who make progress, and it is people who implement our visions. It is human nature that lack of continuity causes poor quality of work product from those people who are affected. A program that is interrupted several times will have poorer quality than one that is pursued from start to finish. There are details missed during the transitions, and frustration can lead to inaccuracies. Over the long run, it pays to establish a pace and preserve consistency of purpose throughout the program.

IMPACT OF HALFWAY EFFORTS

1. DON'T FALL FOR THE HALFWAY COMMITMENT

An often-used means for management to deal with the uncertainty of new product development is to commit to a program only partially. Half-hearted commitment gives management a false sense of security about the program because it deludes them into thinking they are mitigating risk somehow. Fundamentally, there is no way to mitigate risk after the scope of a program has been defined and the market opportunity has been identified. A halfway commitment merely places the risk of failure with the development manager. If you are this manager, take care in accepting this assignment, as senior management has severely hampered your ability to succeed. Furthermore, the halfway commitment is not a fair treatment of the diligence your team has performed in identifying the opportunity and scope of the product. It is a prescription for failure. This prescription is not only limited to the development effort; it includes you as well.

2. DEALING WITH MANAGEMENT'S FEAR OF THE UNKNOWN

The real issue in the application of the halfway effort or resource allocation is management's fear of the uncertainty of new product development. There are several reasons for this uncertainty. One reason is that management might once have funded a poorly researched program, which then failed. Now management has become very conservative. Another reason is that your program is off strategy, too expensive for the financial condition of the company. If you have followed the precepts presented in the previous chapters, however, you have accounted for these issues.

The final reason is that senior management might be risk adverse. In such a case, there is little that can be done, other than to represent the program with more reassurances. The risk aversion might be so strong that management is paralyzed, in which case, there are more severe problems in the enterprise than funding of the new product development.

3. NO SOLACE IN "WELL, GET STARTED, AND WE WILL EVALUATE LATER"

How often have you heard, "Here is some funding to get started, and then we will see how it goes."? This might sound comforting if you are in a financial firefight to get funding for a program; however, do not settle for it. When the going gets rough, in the middle of the program, your senior management might not be committed to going the distance to complete the program. There can be no solace in this type of funding, as your acceptance underscores and provides tacit approval of management's reluctance to make a command decision and see it through.

4. A GOOD PROGRAM WELL DOCUMENTED AND RESEARCHED DESERVES TO BE FUNDED PROPERLY

If there is a single-minded thought you need to take away from this chapter, it is that a well-researched program that is on strategy within the constraints of the business deserves to be funded properly. It is unfair both to the involved management personnel and to the development team to string them along or signal a false sense of commitment. If you cannot garner full support at this critical juncture, what does that say about the prospects if the program runs into trouble?

STRATEGIC IMPACT OF HALFWAY EFFORTS ON THE BUSINESS

1. LOSS OF THE COMPETITIVE EDGE

The arena of new product development is a race against both time and competitors. It is a race against time from the perspective of cost and return of funds to the corporation and a race against the competition to reach the market first so as to retain market share. When qualifying programs for the development of new products, a loss of competitive edge can occur if a firm financial commitment is not made. These halfway efforts serve to water down the programs and lead to their ultimate failure. While your company is vacillating financially over a program, others are pursuing programs to completion with increasing efficiency. Your company will get left behind. This situation arises frequently as a result of management wanting to accomplish too much in the way of development. Too many programs are started, and none end up with the support to execute them.

2. MORALE: DON'T DISCOUNT IT

Another circumstance to beware of is the loss of morale of a development team that has been supported only halfway. A team needs to complete a program as a group. If the priorities are shifting and the finances are a constant source of roadblocks, the team will lose heart in its efforts. It is very difficult to rebuild this enthusiasm with the same management in place. The team's morale must be nurtured during transition times to preserve the energy required to navigate the technical obstacles of the program. It is the role of the manager to insulate the team to some degree from the dynamics of the day to day operations and priority shifts that take place.

3. ACHIEVING CRITICAL MASS OF EFFORT

There is also the issue of critical mass of effort. The easiest way to allow failure is to fracture the team's effort. To achieve steady progress and momentum, a certain critical mass of effort must be maintained. It is what causes the team to progress because each member of the team is mutually interdependent for results. If there is no critical mass, individual team members take on more and more peripheral activities to the point where they think they are handling the bulk of the program. It is at this point that the process of momentum building breaks down. The members begin to work at their own paces because they perceive that they are responsible only to themselves, not the momentum of the team.

4. TIMING LOSS

The loss of momentum causes time loss in the project and all of the dangers associated with a late product introduction. The accounting function can actually create failure within the company by trying to control expenditures too tightly. The key is to create balance, select the programs to execute, and fund them properly. Don't let management start too many programs and then allow them all to fail.

5. THE FALLACY OF "IF IT'S A GOOD PROGRAM, IT WILL SURVIVE"

The notion that "if it's a good program, it will survive" projects a certain romanticism about product development. The most appropriate goal, however, is to create and foster camaraderie within the team to meet deadlines and market needs, not to place them in competition with other teams. An approach that places each team in competition for survival with the others, misdirects the focus inwardly into the organization and away from the marketplace and customers.

GETTING THE APPROPRIATION REQUEST APPROVED

1. THIS IS THE CRITICAL JUNCTURE FOR A NEW PRODUCT DEVELOPMENT ACTIVITY

The appropriation request marks the point at which conversation ceases and commitment begins. At the beginning of this chapter, we outlined a philosophy of viewing accounting and financial departments as partners rather than adversaries in the new product development process. The appropriation request is the milestone for making that partnership work.

The issue of funding cannot be ignored, given a lower priority, or diminished in importance in any way. If the tone of the rhetoric sounds like brinkmanship, to some degree it is. All of the previous effort, research, investigation, and planning lead to this point, so your team has the right to drive the decision to load the accounts with funding.

Any action short of funding at this point is merely conversation. It is your role as development manager to navigate the distinction with senior management.

2. MANAGEMENT DISPLAYS COMMITMENT HERE

Because funding marks a pivotal point in the approval process, management's true intentions become apparent at this juncture. This statement is made not to encourage confrontation or animosity between development managers and senior management, but rather to drive decision making and back up that commitment with corporate resources.

3. HOMEWORK DONE WELL WILL PAY OFF AT THIS POINT

The degree to which you prepared the research and planning is the degree to which you can be vocal and aggressive in securing funds. If your team did a poor job, you shouldn't expect too much if the first attempt at financing fails. If, however, you have prepared your case well, and if the opportunity is real and fits with the direction of the organization, be vocal and passionate in securing funds. This is not to say that you should be abrasive, but you should demonstrate to senior management that you are the program's champion and have the commitment to see it through to completion.

4. WHAT TO DO IF A PROGRAM IS NOT APPROVED

If you do not get the financing you are seeking, what do you do about it? There are some alternatives. You can scrap the program, broker the idea to someone else with the corporation's support, or finance it by alternate means. In each case, the onus will be on you to resell the opportunity to the appropriate party.

The idea can be brokered to another company; however, be aware that the other company might need to secure some of the people on your team. The other company should outline a quid quo pro arrangement, as you are turning over an opportunity that your firm spent considerable time in researching and planning.

The alternate financing approach might be best served by creating a joint venture or brand label arrangement with another company. This would allow your team to drive the development or at a minimum to participate in it. It allows a way to get started in the marketplace and gain experience with the business.

5. LEADING THE NEXT FINANCE STAGE

There is also the issue of securing financing as the program progresses. At each stage, the work of securing funds for the next stage of the project begins. All too often, the work of securing financing is left up to the last moment in which the decision actually takes place. The work of securing the funds needs to be done long before the decision making event. It is the role of the program manager to retain management's interest and support throughout the entire program, not just at the funding stage. To that end, if you are in a program presently, funding for the next stage must be planned for now.

6. CREATE THE FINANCIAL MOMENTUM

The goal is to create the financial momentum to keep funds flowing at a rate equivalent to the funds currently being expended. This process needs to be financially correlated to the progress of the development to obtain management's support.

SUMMARY

This chapter was designed to provide you with a basic background in accounting and finance to create understanding about the driving forces and pressure points in accounting. In other words, we looked at a project from the accountant's perspective.

A variety of information and specific tools for analysis and planning were developed and presented. These tools are included in a generic form on the companion disc for you to customize and use for your project-specific needs.

A financial model for a product sale was presented, along with definitions pertaining to finance and accounting. A review of cost systems and how to estimate an entire program was also included. A discussion on project and cost estimating, factors associated with the timing of future sales, and a sample of a comparative analysis for projects were also presented.

Following these items were discussions pertaining to the generation of cash and profit and the financial cycles of a business.

The channel to market financial impact was discussed next. Finally, we included a discussion on management's financial commitment to and funding of a project.

With a new product concept, a development plan, a business plan, and the funds to execute it, you are now ready to start out on the venture. This is the focus of Chapter 6: Starting Out.

STARTING OUT

ABSTRACT: You now have secured funding for the new product development program and are ready to start out. Now is a good time to select the team, right? Wrong. Ideally, the team should be selected long before the program is funded. By preselecting the team, you're positioned to begin development immediately upon funding. If the team is not already in place at this point, precious time early in development will be spent on putting a group together rather than prosecuting the development. In this chapter, considerable discussion will be devoted to the development team.

The development team is a highly important element of the development process, as it is people who generate progress and results. The other crucial elements that we will discuss in this chapter are team management and communications—including communications within the development team, laterally within the organization, and externally to the organization.

A STATEMENT ABOUT TEAMWORK

1. TEAMWORK

The term *teamwork* is an expression often misused as a rallying cry for a disjointed organization. It is a euphemism that senior managers tend to overuse when the goal is to coalesce and galvanize a diverse group of people to achieve a common objective. In reality, a team is a work in progress, in which the objective focuses the actions of the individual. In a team, the needs of the corporation outweigh the needs of the individual. A team is a forum for individual contribution to an aggregate goal.

2. MOST MANAGERS DON'T UNDERSTAND THE CONCEPT OF A TEAM

Most managers fail to understand the concept of a team as a work in progress. They give the "teamwork speech" and expect to have a disjointed group of people with differing agendas suddenly perform as a cohesive unit. The concept of teamwork, however, needs to be nurtured, reenforced and supported on a day-to-day basis. It is integral to the entire program, not just to its beginning stages.

3. TEAMWORK DOESN'T HAPPEN NATURALLY

Teamwork doesn't happen naturally. Groups are comprised of people who often have differing agendas and different perceptions of urgency. The manager must convert the energy of the individuals into directed energy focused on group progress—a very difficult task. The key to achieving this goal is to foster group progress with individual recognition and group rewards.

4. LEADERS GALVANIZE THE GROUP TO PERFORM

The development manager must shepherd the group to perform. This means consistent refocusing of individual efforts to the stated objective. The leader of the group must be respected by the group, not solely for technical capability but also by virtue of being the guiding force in prosecuting the development. The leader must provide the tools, the resources, and the uninterrupted time to allow progress to occur. The leader must verify that the progress made is appropriate to the initial objective and also ensure that the original objective is still valid. The group leader must foster enthusiasm, provide continuity during trying times, and protect the momentum of progress toward development.

5. ONLY THE LEADER IS TO BLAME

If you are the leader of a development group, *you* hold the primary responsibility for success or failure of the group. You (along with the team) will take the credit for success, but any failure will be yours alone. It therefore is in your best interests to understand the dynamics of a group of people and to become proficient at managing them toward a goal.

IDENTIFYING THE REQUIREMENTS

1. IDENTIFY THE RESOURCES

What are the resources required to initiate and complete the program? What is the mix of internal and external expertise that will be needed? What are the capital resources needed, and when will they be required? Is there a learning curve for the capital equipment? All of these essentials must be sequenced and coordinated.

As we will discuss later in this chapter, the program must be laid out and the resources identified and sequenced based on program activities. It is best to start with the project activities, sequence them, and generate the list of requirements to accomplish the task.

2. PLANNING THE REQUIREMENTS

Alterations in corporate finance philosophy have forced changes in the procurement of equipment and resources. Without external controls, most managers probably would spend all their capital expenditures in the first quarter after the start of a new fiscal year. This tendency could be due to shortsighted senior management and its changing preferences for expenditures and programs. There is a natural tendency to gather the resources even before they are needed in order to ensure availability.

In the ideal situation, not all the resources are required on day one of a program. In fact, however, they need to be scheduled to minimize financial impact and to allow the corporation and the team to absorb them and use them effectively. To the extent that the company can make a commitment and see it through, then, plan and execute the resource procurement accordingly.

3. ORGANIZE THE PERT CHART

There is only one systematic means to identify the resources and equipment needs for a program: Break the project down into its component stages and identify the subtasks for each stage. Each task will require personnel resources and, possibly, equipment resources. Note the requirements for each type of resource under the task. If you have laid out the program properly and the sequence is correct, the resource and timing will also be correct, as illustrated in Figure 6-1.

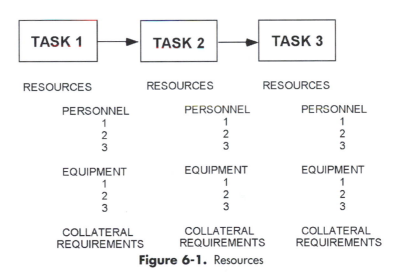

Figure 6-1. Resources

4. GRAPH THE DEMAND

The next step is to graph the manpower and resource requirements during the run of the development. For example, Task 1 might require three people during its duration, whereas Task 2 might require five people. The team needs to be staffed accordingly to meet the demands of the tasks given the time frame established. Supplying fewer personnel than are required will result in extending the execution time of a given task.

In a similar manner, the equipment needs to be sequenced, and procurement and commissioning schedules can be put in place. Integrating the equipment might also require its own personnel or simply part of the development team. All of these elements need to be considered for assessing the requirements of the program.

5. DETERMINE TOOLS REQUIRED

One way to keep the program on track is to determine the tools that the development personnel need early in the program. Often these tools are decided upon only as the development progresses, which results in building inherent delays into the program. Procurement delays, funding delays, and integration delays can occur.

A frequent excuse for the delay in the selection of these development tools is their continuous evolution and the desire to be on the latest platform or have the latest version of equipment or software. In reality, the delay incurred by not planning ahead often far outpaces the loss in efficiency due to using less than state-of-the-art equipment. Consequently, have the team identify the equipment and resources and procure and integrate them into the organization, so that they are truly available at the time they are needed.

6. DEFINE THE LEVELS OF EXPERTISE REQUIRED

In a similar manner to determining the tools required, you need to determine the level of expertise required of the team members. Be sure to get the requisite talent in place as a minimum before starting out; otherwise, you could end up looking for talent when you should be executing tasks with that talent, hence delaying the program.

7. ESTABLISH PATHWAY FOR TOOL DEVELOPMENT

One part of the identification and procurement process is a plan to integrate and diffuse the new technologies and methods into the development team. Regardless of the specialty, each member should be on a pathway toward certification in the technologies brought into the company. Following this strategy will go a long way toward giving management the flexibility to assign personnel to various tasks. Eventually, every member should have a broad repertoire of skills. Figure 6-2 illustrates the pathway.

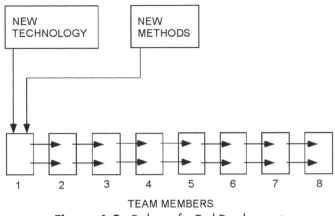

Figure 6-2. Pathway for Tool Development

The new technology and methods are taught to team member 1. After the work content for the task is complete, the team member transfers the technology to team member 2 and so on, until all of the team members are conversant with the technology and methods. This strategy provides the manager with the utmost flexibility.

8. REVIEW PROCESS QUESTIONS FOR PLANNING

There are several planning questions that must be considered as part of the process of identifying resources. At this point in the development of the program, most elements of the business plan should have addressed these issues; however, the following points can be used as a reminder for the program manager.

A. Write the planning statement.

What is the desired result or outcome? Be very specific about this. What are the measurements of success? What do we want to accomplish? Who, What, When, Where, How, and Why?

B. List the action steps in the plan, identify timing, and assign responsibility.

Develop the chronological list of action steps to be accomplished. Who is accountable for each task, and what is the realistic starting and ending time for each task? Also, list and correlate the interdependent tasks at this time.

C. Review each step in the plan and identify the critical areas.

Which action steps are the critical impact tasks? Where is uncertainty associated with a task or group of interdependent tasks? Where are there potential gaps or overlaps in responsibility?

D. Identify and list likely causes of key potential problems.

What are the specific influences or events that can cause these potential problems? Test each problem listed, to determine whether it represents a cause-and-effect relationship with the corresponding potential problem.

E. Develop preventative and facilitative actions.

What can be done to prevent each likely problem from occurring? What action should be taken now to facilitate the opportunity or to prevent the obstacle that could stand in the way of the opportunity?

F. Design contingent actions.

What contingent actions should be planned for each potential problem so as to minimize the impact? If the potential problem occurs in spite of preventative action, what must be done to mitigate the circumstances?

G. Identify and list contingency alarm points.

This is probably one of the most important elements of planning—to understand clearly the points at which you are in trouble and to set alarms for them so corrective action can be taken. Often, a program gets in trouble and the team is deeply immersed in the issues. If proper alarm points were used, precious time and energy could be preserved to focus on corrective action at an earlier stage.

ASSEMBLING THE TEAM MEMBERS

1. CANVASSING AND INTERVIEWING TEAM MEMBERS

The technique of putting a team together to suit the needs of a program is an acquired skill. It takes several years of operating in a team environment to understand the potential pitfalls and the ingredients for success. The elements discussed in this section are not totally inclusive of all issues in canvassing and interviewing team members; however, they do represent key points to keep in mind.

Personnel assigned to a development program are measured by the product of two attributes: intellect and energy. Intellect is the knowledge to create, design, analyze, and synthesize solutions. Energy is the human drive to apply that knowledge to effect progress. Either attribute by itself is worthless in a development person, but the product of the two is what makes things happen. A person with intellect and no energy to apply that intellect will not move a program along. A person with only energy and not enough intellect doesn't have the "tools" to do the job.

In our experience, however, energy, interest level, and drive can offset many shortcomings in experience and intellect. Never underestimate the capabilities of a person who wants to do something, and never count on a high intellect as a sole means to an end.

To assess accurately a potential team member's ability to execute a program, you need to conduct performance-based interviewing. Performance-based interviewing is the interviewing technique whereby the candidate is given the opportunity to draw parallels between his or her past accomplishments and the proposed program for which he or she is interviewing. It focuses on specific skill set requirements of the position and looks for those skills demonstrated in the candidate's previous work.

Performance-based interviewing is an objective, dispassionate method for assessing skill level and measuring it against the job requirements. It allows an objective selection between two individuals based on their anticipated contributions to the program.

2. MAKEUP OF THE TEAM

The makeup of the team is critical to the team's success. Certain personalities can enhance success or contribute to failure. Members with exceptionally strong personalities can undermine the team leader. Infighting can erode sense of purpose. The selection and complexion of a team must be geared toward the program at hand.

Each team needs a blend of creative people and pure implementers. There should be individuals with enthusiasm and also naysayers. Each type complements the other to create balance in the decision-making process. If the program requires navigation of uncertainty, focus the team orientation toward creative people. If critical time and cost issues are the focus, make use of implementers to deliver results.

3. ALIGNING THE TEAM

The alignment of the team is crucial to get off to a productive start; however, teams do not arrive in place and aligned. In fact, all teams when first assembled are disjointed, fragmented, and somewhat ineffective. They are a loose collection of people with differing goals, agendas, biases, and allegiances.

It's the manager's responsibility to focus the group on the goal and accomplish the result. Along the pathway toward the result, the manager needs to reinforce the interdependency between the team players and within the team as a whole. As this goal is attained, remind the team of the basic tenets that made them successful. The manager needs to drive the team culture from these tenets, which is why it takes time to construct an effective team. In some cases, several projects might be required to cement the team together.

Any senior manager who proselytizes about teamwork and who expects a full-functioning team in place on day one probably has never put a team together.

4. HOW A TEAM IS MADE

There is value in cementing team consciousness by team success; however, nothing cements team relationships as much as the process of the team overcoming a problem or an adversity together. For the manager, keeping a team together during these times can be a especially challenging, but the manager always must overcome the desires of the individuals and place the needs of the project above personal preferences.

Although members might want to dissociate themselves from the program, the manager must force cooperation that produces results. Later, the manager needs to remind the team of what works and what fragments a group, and how fragmentation can destroy a program. At the successful completion of the project, celebrate with the team and reinforce that commitment pays off.

5. DEVELOPMENT ENGINEERS APPLY THE TECHNOLOGY

The role of the development engineer involves a dichotomy in required personality traits. In one sense, the development engineer needs to be a creative, freethinking individual to overcome the uncertainty inherent in a program. In another sense, the engineer needs to be structured in implementing the solution to the problem. Most engineers are quite comfortable in the creative role, as they view resolution of the problem as a challenge. Mechanizing it so that it can be absorbed by the organization is something that many engineers are averse to. For many engineers, this task simply pales in excitement and satisfaction compared with the problem-solving aspect of the project.

These two apparently opposing personality traits are very difficult to find or cultivate in one individual. Many times, it is easier and more expedient to implement these dual functions with two people, each of whom possesses one of the required traits.

The needs of the program at any point in time define which traits are required in the development people. Often, the need for free thinking and creative solutions to problems is greatest at the beginning of the program, while the more rote activities of implementation take precedence toward the end. The development people need to flex from one process to the other seamlessly; when this is not possible, the program must be structured so that different personnel take over as the program changes.

6. GET REQUISITE TALENT

As a basic tenet of assembling a team, be sure to secure the requisite talent at the beginning of the project. A common theme throughout this book is that time is a competitive weapon. This being the case, you want to minimize lost time by securing talented people at the onset. Otherwise, you will be spending precious time getting the personnel up to speed instead of making progress on the project.

Many times, senior management attempts to influence the selection of team members as a means of containing expenses for salaries; however, this practice sets the project up for delay. The team becomes comprised of "trainable" personnel rather than experienced personnel. You have learning curve issues and experience curve issues to contend with. To understand the impact of this type of team selection process, consider a mistake that could be avoided by the selection of experienced people. In the long run, you will save expenses by not committing the mistake of selecting "trainable" personnel, and you will save even more valued time.

7. OBTAIN TEAM MEMBERS WITH DRIVE

Secure personnel with drive, and it will pay off well in the long run. This strategy also allows the manager the freedom to manage the program rather than expend energy toward incentives for the group to perform. We have found that it is quite helpful to select personnel who have something to prove in life. The challenge for the manager is to channel that desire and make it consistent with company and program objectives.

Team members with something to prove are driven to perform. They break down barriers and navigate uncertainty better than most people because they are focused on the result. All too often, teams are comprised of personnel who are not driven and who have few personal goals that are linked to corporate goals. The problem is that these people do not contribute energy to the group and actually cost energy to extract minimal performance.

8. TEMPERAMENT OF THE TEAM/PLAYERS

In new product development, remember that results are the only things that count. Senior management is appreciative of all the "storms" the team might have gone through, and your navigation skills might earn you recognition or a better assignment someday. Ultimately, however, management is only interested in whether you "brought the ship in." Results are the only tangible measurable; consequently, the team must perform.

Developing a new product places the team under pressure, and tempers can flare occasionally. This is why it is a good idea for the team's manager to understand the temperament of the team.

Individuals respond in a variety of ways when placed under pressure. It is important to recognize this fact and to mitigate the effect of conflicting personality profiles within the team. Members of the group need to be reminded that time, not their colleagues, is the adversary and that the needs of the program outweigh individual needs. There are many ways of profiling personalities and many recommended methods for managing them. Our focus here is not on the mechanics of people management; rather, reemphasize the objective of the team and manage toward that goal—to execute a program in the best way possible, to meet the needs of the corporation in the time frame allotted while cultivating an atmosphere of mutual respect and cooperation, and to provide a vehicle for individual personal and professional growth.

There can be occasional conflict; that is acceptable as long as the conflict is constructive toward the program. The team leader must manage conflict. Each member can have a dissenting view. The views may be discussed, voiced, and argued. In the final analysis, however, it is the manager who needs to make a command decision. It might not be popular, but it is the group decision. In addition, the manager's job is to monitor behavior and immediately dismiss a member who engages in destructive conflict for the purpose of delay or for political motives. This doesn't mean dismissing a dissident, the bearer of bad news, or someone who is passionate about his or her work. It does mean removing personnel who knowingly do not operate in the best interests of the program. If this activity is taking place, action must be taken immediately. Failure to do so will undermine the manager's leadership and delay progress for the whole group.

9. DYNAMIC INTERPLAY

The management of a development team is a dynamic endeavor. Assumptions change, situations arise, obstacles present themselves, and attitudes change. The manager needs to watch diligently, listen, observe, and redirect and reinforce behaviors to facilitate the completion of the project.

ORGANIZATIONAL FORM

1. TYPES OF ORGANIZATIONAL FORMATS: FUNCTIONAL, PROJECT, MATRIX

The issue of which organizational format to employ in a program is a critical one. Given ample resources and reasonable commitment, almost any of the three commonly accepted formats can work; however, in actuality, there are never enough resources or time to execute a program ideally. Therein lies the importance of which format is used and how to maximize the use of the development personnel. The three basic types of project organization are as follows:

- Line or functional form
- Pure project form
- Matrix form

In the line or functional form, a group of projects is assigned to a functional manager. The members of the team report to the manager in the standard hierarchical structure. This form of project organization is illustrated as follows in Figure 6-3a:

The pure project form of organization is one in which each project is self-contained and resources are assigned to a single project for execution. Members do not work on anything other than the project. The traditional skunk works is the ultimate in pure project format. It is illustrated in Figure 6-3b.

In the matrix form of organization, development personnel are located in a resource pool. Assignments are given to the pool, and talent is assigned on availability. Section managers in charge of specific technologies or skill sets provide line management. This format can cause conflicting priorities between section (line) managers and project leadership. This form of organization is illustrated in Figure 6-3c.

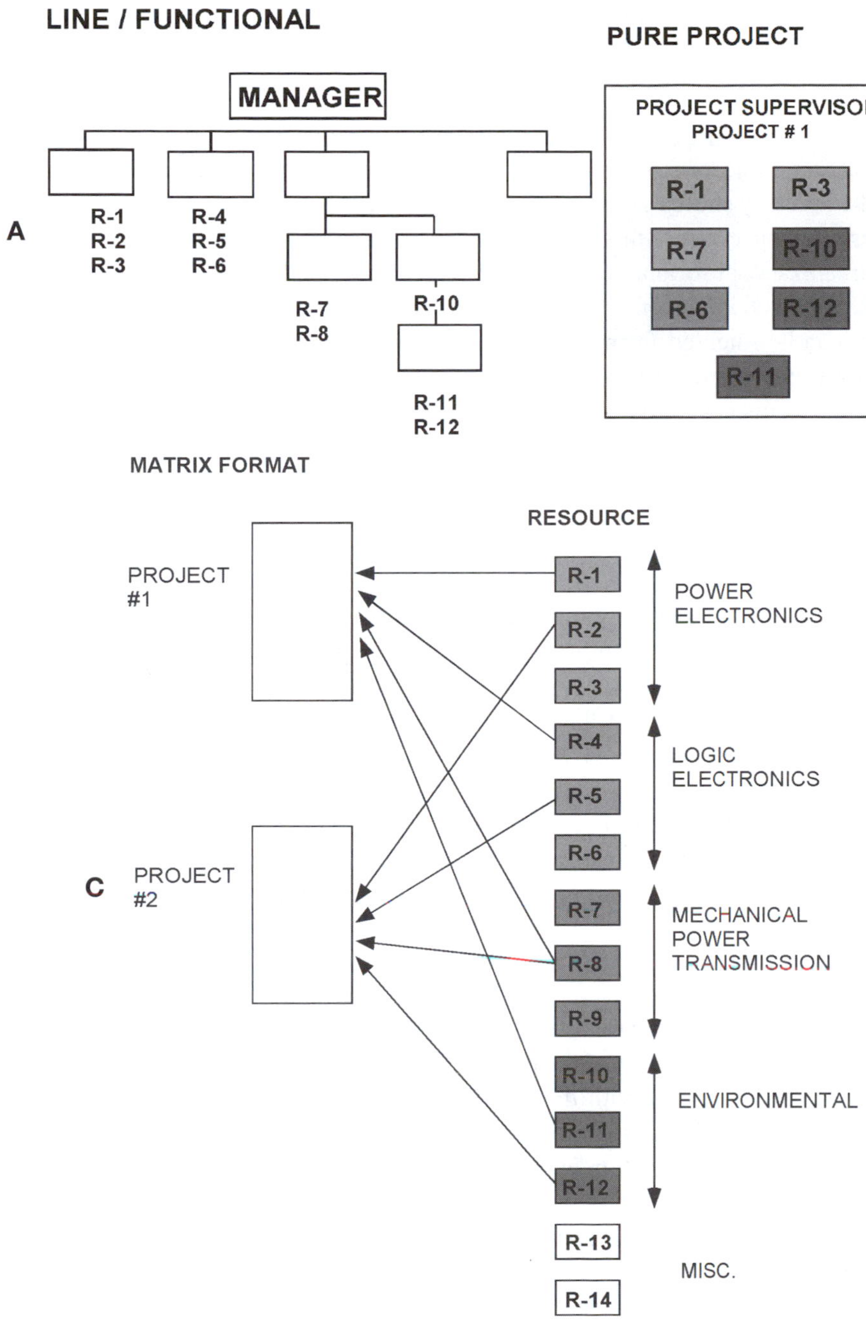

LINE / FUNCTIONAL

PURE PROJECT

A

B

MATRIX FORMAT

C

Figure 6-3. Forms of Project Organization

2. PERSPECTIVES ON ORGANIZATIONS TYPES

Line Organization Format

The line or functional form of organization is probably the next most effective in that one manager for all the projects directly applies the resources to the programs. Compromises will have to be made by strategizing on the manager's part; however, if the members of the department are flexible enough, this can be effective also.

The line organizational format is another relatively good method for applying resources to a program. There is, however, a drawback in this form of organization, whereby engineering talent can be shielded from the customers and corporate strategy by virtue of the bureaucracy of a line organized department. Often referred to as monolithic, this organizational form can execute programs with proper management. From a personnel perspective, it is easier to "lose" someone within the organization, hence the use of personnel resources is not as good as in the matrix form.

Pure Project Format

The pure project form is probably the most efficient organizational form for executing a program. It focuses the development horsepower directly on the task at hand with no compromises. There is single-minded focus, which helps a team be highly effective.

The pure project format is one of the best formats to employ when program execution is critical. This format places resources exactly where they are required to ensure the application and management of resources with respect to project timing and results. In terms of allegiance and ownership of the program, this format is one of the best choices. It has disadvantages in terms of flexibility of resources, which is simultaneously part of its appeal. Flexibility in terms of the application is in conflict with this type of format.

Matrix Format

The matrix form of project organization is the most comfortable for senior management to support throughout a program because the ratio of projects to people seems better. Senior-level managers have a tendency with this form to stuff projects into the pool at will, with little regard for the effect. Resources seemingly are endless. Unfortunately, this format is used when resources are actually thin and there are many projects that have to be executed simultaneously. Consequently, programs become resource starved through reprioritization, and infighting between project managers who are competing for resources and expertise. The matrix form is a very commonly used format nonetheless, even if the company's formal structure shows something else.

The matrix form of organization is sparing in terms of the use of resources. It makes use of the proper personnel for each segment of the project. Typically, personnel are used to their full potentials. The weakness with the matrix form manifests in two ways, both stemming from the same root phenomenon. Program allegiance is generally weak in these systems, as personnel become "experts" in a specific discipline and choose their favorite program to concentrate on. It is thus difficult to measure the application of talent in uncertain situations and to assess their diligence. Accounting for their time is difficult, as is focusing horsepower on the tasks at hand. Project contention issues also arise, making multiple projects management far from easy.

No single format is optimal for all types of development. One must evaluate the enterprise philosophy, the temperament of the resident talent, the program needs, and the ability to apply resources to the various programs.

3. SELECTING THE RIGHT FORM

If you have the flexibility to design an organizational format, select the one that best fits your needs and the organization's level of flexibility. If you cannot change the format, understand the weaknesses with the one you must live with, and bolster those weaknesses to serve the interests of the project. You might want to take the time to project potential conflicts in a program according to each organizational format and to generate different scenarios. By so doing, you can gauge the impact of organizational structure on a program.

Sometimes it is very difficult to change entrenched organizations from one pure form to another. Enter the dotted-line relationship. A dotted-line relationship is a sign of two things generally. First, it is a sign of a weak organization, and second, it is usually a sign of an organization in transition. If your company needs to transition from one format to another, you might want to make use of the dotted-line relationship, which also assesses behavior of the individual reporting to two people. Keep in mind, though, that it is not fair to the individual to let the format go uncorrected indefinitely.

4. MAKING DO WITH EXISTING ORGANIZATIONAL FORMATS

Sometimes there is no simple or practical way to modify the existing format of a team. Remember an earlier observation: Almost any system or format will work as long as it is managed properly. If the format cannot be changed, work within the parameters at hand. Deliver demonstrable results to senior management first, then negotiate organizational changes. This strategy is especially important to follow if the manager is new to the organization.

5. MEET THE PROGRAM NEEDS, NOT ORGANIZATIONAL PREFERENCES

One final thought on the subject for organizational formats: Let the needs of the project drive any organizational format changes, not the other way around. If the project requires a certain format, modify the organization to that format. Do not try to meet the political needs of the organization with the program; the program simply cannot be compromised.

APPRENTICESHIP AND MENTORING

1. FOCUS ON PERSONNEL DEVELOPMENT

Traditionally, the two practices of apprenticeship and mentoring have represented the means to transfer technology from the experienced, sage, and wise members of the workforce to the younger, inexperienced, and naïve members. This strategy has worked historically in industry and could be effective today if not for the rapid change in technologies. Even mature members of the workforce need to expend a portion of their day learning the new technologies and methods. The focus has shifted to training new members on basic requirements for a lifetime of continuing education. If apprenticeships historically involved a fixed time to practice and learn a fixed body of knowledge, then today, the body of knowledge is ever increasing. Consequently, we need to give the new members the tools to learn and then get them into the productivity race.

On the professional side, mentoring is a traditional method to augment a new member's education with the wisdom of the "elderly statesman" of the corporation. Mentors generally provide advice regarding career development, pathways to success, and elimination of roadblocks. Mentoring is effective today, even in the wake of corporate hacking away at the higher-level managers during cutbacks. Unfortunately, in some organizations, the use of mentoring has been replaced with self-preservation; in a healthy situation, however, it remains an effective means for educating the younger members of the workforce.

2. GROWING THE YOUNG WORKFORCE

The growth and assimilation of the young workforce is almost a sacred corporate trust. Through mentoring and the transfer of wisdom, the culture and values of the organization get passed from management to the new hires. If a company is going to grow, the integration of the value system and the culture must be spread throughout the organization. If this job is left undone, the enterprise will eventually be destroyed, as the basic value system and frame of reference system are corrupted. As in life, take care in growing the "young"; the results will have tremendous positive impact.

3. CROSSFERTILIZING THE EXPERIENCE BASE

Previously in this chapter, we discussed the assimilation of new technology by different members of the team. It is important to have a coordinated plan for crossfertilizing the technology and skill sets of the members. The ultimate test of crossfertilization of knowledge and skill sets is to be able to reassign members in midtask and to have the members still be able to complete those tasks seamlessly. Pursuing this strategy yields the maximum flexibility of the team and, equally important, creates loyalty and enthusiasm within the group, as no one remains vocationally stagnant.

4. CONCEPT APPLIES TO ALL DISCIPLINES

The practices of apprenticeship and mentoring are valuable for all disciplines of a development group. They apply to professional, vocational, and technical disciplines. Consider how apprenticeship evolved historically, as the transference of knowledge to young employees who had to be taught the procedures and methods for a task or series of tasks. It was instruction on the mechanics of the job. After the apprenticeship period, the individual was expected to demonstrate a certain level of skill, which was further enhanced throughout the individual's career.

This concept is still valid today for all disciplines. Individuals new to the organization need to be taught the values and mechanics of the job to which they are assigned.

In contemporary business practice, enterprises tend to give new hires a small dose of instruction and then send them on their way. More popular in the U.S. than in Europe, this practice can generate waste and a frustration on the part of both individuals and management. Managers simply need to dedicate the resources to cultivate new individuals in their professional and vocational roles.

The practice of mentoring is less about mechanics and more about work style. The workplace is increasingly diverse, and the style of management can affect program outcome more significantly than in the past. In current practice, turnover in the managerial ranks increasingly points to the need for mentoring. Mentoring imbues the individual with the value system of the corporation. It teaches the individual how to conduct him- or herself within the constraints of the values of the organization. Properly implemented, these values are tremendously more powerful than any energy a single manager can expend.

5. THE CASE FOR BOTH APPRENTICESHIP AND MENTORING

In the contemporary development environment, there is a practical need for both apprenticeships and mentoring to cultivate new personnel in the professional and vocational areas. The objective is to create a group of team members out of individuals who are well trained, who operate within the boundaries of the company mores, and who are motivated to improve.

MANAGEMENT OF THE TEAM

1. SYSTEMS FOR PROGRAM MANAGEMENT

Many systems for program management are offered in various forms. Fundamentally, they are based on two basic types of project tracking methodologies, namely:

- Gant chart
- PERT (Program Evaluation and Review Technique) chart

The Gant chart is represented by a list of activities along the Y-axis and a timeline along the X-axis. The X-axis indicates the start and stop time for each activity.

The PERT chart shows a sequential connection between the separate tasks. Each task has a point of connection for starting and a point of connection for completion. The interconnection of the points represents the logical sequence of activities. The PERT chart is the more accurate method of program management, as it accounts for task listing, timing, and interconnection of events. Figure 6-4 shows an example of a Gant chart. In it appear the tasks listed on the Y-axis and the timeframe listed on the X-axis. A start date and an end date are established for each task. In this chart, there is some logical organization of the tasks, but not to the degree that is found in the PERT chart.

The PERT chart outlines both the tasks in a project and the interconnections between task start and end dates. This constitutes the logical sequence of tasks for the program. In addition, the critical path is defined—the shortest pathway from start to completion without compromising the integrity of the logic. An example of a PERT chart appears in Figure 6-5.

As the figure shows, each task is interconnected to other tasks to formulate the logical sequence of task completion. In this example, the second and third tasks can be executed in parallel time frames with the tasks immediately below them. In addition, the last task cannot be started until the previous tasks are complete.

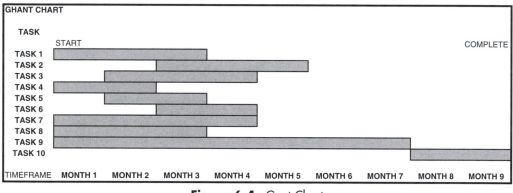

Figure 6-4. Gant Chart

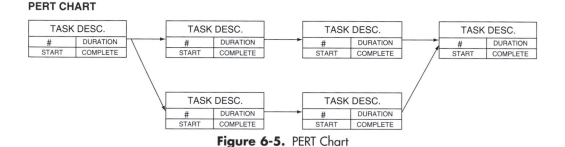

Figure 6-5. PERT Chart

2. MODEL FOR PROGRAM MANAGEMENT

The management of a program is an interdisciplinary, widely diverse activity. There are issues to manage that range from creative specifications to communication and security (both internal and external). Figure 6-6 is a model showing the diversity of management issues involved in a project, in a format that can be analyzed and optimized.

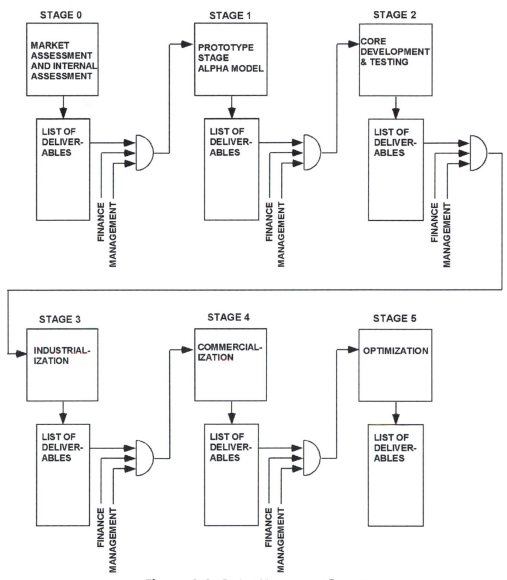

Figure 6-6. Project Management Format

The overall project is comprised of six basic stages. They are as follows:

1. Market assessment and internal assessment

2. Prototyping

3. Core development and testing

4. Industrialization

5. Commercialization

6. Optimization

The first stage is the assessment phase, which involves internal introspection of company and direction and an external look at product opportunities. The next stage is the prototyping stage. It leads to the physical embodiment of a product concept as a starting point. The third stage is the core development of the product. This is the stage in which the product is developed and testing takes place.

Next is the industrialization stage where the product is made manufacturable and the manufacturing processes are developed. The fifth stage is the commercialization stage where the product is rolled out to the marketplace and units are placed in the hands of consumers. The final stage is the optimization stage, when feedback and improvements are incorporated.

Throughout the entire program, each stage has all the required deliverables, finance, and managerial approvals for proceeding to the next stage. Program management is the coordination and the execution of all of these tasks in an effective and timely manner.

3. INTERNAL, EXTERNAL, AND COMBINATION RESOURCE MANAGEMENT

Not all of the resources to be managed are internal to the organization. Besides vendors and advice gathered from others, there could be an arrangement by which outside resources are required because they are not available within the constraints of the company. In such cases, there must be a contractual means for obtaining the resources.

The program manager must have the power to effect results for the program. Simply stated, this influence needs to extend to all areas having an effect on the progress of the program.

4. MANAGING UNCERTAINTY IN A PROGRAM

Oxymoron is the term that first comes to mind when referring to the management of uncertainty. By its very nature, uncertainty is "uncertain" in terms of degree, duration, severity, and extent. To suggest that it can be managed in some deterministic, scientific manner is somewhat naïve. Rather, the idea is to mitigate the impact of uncertainty and all that it brings with it. Take care at the outset of a program to identify the items that are known and understood and those items that are fraught with uncertainty. Channel energy to understand and clarify the uncertain items as soon as possible.

5. THE DYNAMICS OF SCHEDULES, TIMING PROGRESS, AND SLIPPAGE

The impacts of schedules and slippage in the schedules can devastate a program. What seems like a small slippage in a schedule can accumulate to huge delays in product introduction. Unfortunately, many teams feel that slippage in a schedule can be made up as the program progresses. This is rarely the case. In addition, there could be a tendency to try to make up for lost time by changing the logic of the program to work around delays. Although this goal is admirable in terms of trying to preserve program timing, this practice can introduce oversights and poor compromises.

To help you gain a sense of the degree to which slippage can affect a sample program, Figure 6-7 shows a sample project consisting of six stages. Each stage has a scheduled duration.

As the project exits stage 0, the team incurs a one-day delay. In a 20-day activity, this represents a 5 percent delay. Because this is the first stage, the cumulative delay is also 5%. The use of the cumulative measure is the percentage of that stage's total duration that represents the accumulated delay.

As the project exits stage 2 (another 20-day duration), two days are lost in delay. This represents a 10 percent delay for that isolated stage and a cumulative delay of 15 percent dragged into stage 2. Progressing through the example, a few days here and a few days there add up to a staggering 46 percent delay dragged into stage 5. This means that 46% of the time originally budgeted for stage 5 of the project is delay. It is impractical to catch up within the stage you are operating in and get back on track for each stage. To bring the program back into alignment with the original schedule, this amount of time would have to be added to budgeted time planned for stage 5. Looking at the project as a whole, the overall delay represents 8.22 percent, or 12 days. The point to emphasize is that a project simply cannot catch up on a delay. Once the time is gone, it's gone!

6. MANAGING FOR RESULTS

What are the practices that the group and the company need to follow to become an effective development team? There are distinct practices that will result in distinct goals for the company. The summary chart is presented in Figure 6-8.

As shown, one key practice involves the training and use of a common vocabulary throughout the enterprise. This assists in yielding a corporate goal and in cultivating effective and uniform understanding of goals and issues with the development project and the business. It also assists in linking the goals of each department.

The next practice is the development and use of common methodologies for evaluations and decision making. Working to reduce misunderstanding and false starts and progression

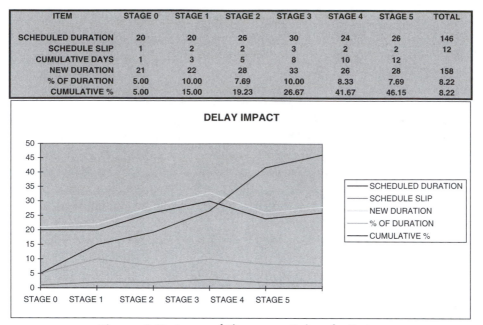

ITEM	STAGE 0	STAGE 1	STAGE 2	STAGE 3	STAGE 4	STAGE 5	TOTAL
SCHEDULED DURATION	20	20	26	30	24	26	146
SCHEDULE SLIP	1	2	2	3	2	2	12
CUMULATIVE DAYS	1	3	5	8	10	12	
NEW DURATION	21	22	28	33	26	28	158
% OF DURATION	5.00	10.00	7.69	10.00	8.33	7.69	8.22
CUMULATIVE %	5.00	15.00	19.23	26.67	41.67	46.15	8.22

Figure 6-7. Impact of Slippage on Delay of a Project

	EFFECTIVE AND UNIFORM COMMUN-ICATION	INTERDEPART-MENTAL LINKING STRUCTURAL OBJECTIVES	DEVELOPMENT PROGRAM TIMLINESS	OPTIMIZATION OF CORPORATE RESOURCES
COMMON VOCABULARY	x	x		
COMMON METHODOLOGIES FOR EVALUATIONS AND DECISION MAKING	x		x	x
PROCESS FOR EST. PRIORITIES		x	x	x
DECISIVENESS AND CONSISTENCY OF PURPOSE		x	x	
REALISTIC PLANNING AND EXPECTATIONS			x	
TEAM STRUCTURE AND AUTHORITY	x		x	

Figure 6-8. Managing for Results

down the wrong pathways, this practice establishes a uniform basis for decision making. This has a net downstream effect of promoting program timeliness. The use of corporate resources is also a collateral benefit, as common values and goals tend to encourage placement of resources to the proper tasks.

Establishing and supporting a process for determining priorities is the next desirable practice. It assists in the structural linkage of objectives by ensuring that each department has the correct priorities. This practice encourages development timeliness and the most effective use of corporate resources.

The next desirable practice is the corporate operational trait of decisiveness and consistency of purpose. This has tremendous benefit in cementing interdepartmental linking objectives and the timely execution of a program.

Next, realistic planning and expectations bring about benefits in terms of helping the organization be timely in their pursuits. Unrealistic expectations and planning done through rose-colored glasses create intrinsic delays in the execution of the program.

Finally, team structure and authority have a direct impact on both corporate-wide communications and program timeliness. If the team is a functioning unit with effective management and clear goals, the organization will be plugged into the program, and the product will have a much better chance at a timely introduction to the marketplace.

Taken individually, no single one of these practices is necessarily a driving force effecting the desired goal; however, much of the work of the development manager and senior manager is concerned with the proper alignment of the organization and the removal of obstacles. Implementing these practices within the enterprise helps smooth the path for management personnel. Constant vigilance and management talent are still required to bring a program to successful completion.

7. FOLLOW-UP SYSTEM

With any program, there is an enormous number of details associated with specific tasks, and these details require follow-up. So numerous are these details that it would be impossible to schedule all of them on the program's Gant or PERT chart. In addition, there are sub-tasks associated with the management of the project that might need investigation. Items such as scenario development, "what if" investigations, logic and schedule manipulations, all require follow-up.

To best keep track of these details, implement an action item follow-up system. This follow-up system also can serve as a permanent project record for future use in demonstrating diligence or resolution of issues. With the use of computers and mechanized methods, any number of implementations can be used. Regardless of which one is used, be sure that the selected system includes the following components:

- Item number
- Description
- Action required
- Assignment date
- Responsibility
- Completion date
- Running status

An example of a format for a follow-up system is shown in Figure 6-9.

The example system represents only the information to be recorded and maintained. To be effective, this needs to be managed every day. As each item becomes complete, it should become part of the permanent record of the project. Follow-up dates should be adhered to strictly to prevent loose, unresolved details from bogging down the project.

ACTION ITEM FOLLOW UP SYSTEM								
ITEM	DESCRIPTION	ACTION REQUIRED	ASSIGNMENT DATE	ASSIGNED INDIVIDUAL	COMPLETION DATE	RUNNING STATUS	FOLLOW UP DATE	COMMENTS
1								
2								
3								
4								
5								

Figure 6-9. Follow-Up System

CULTURE OF THE GROUP

1. SET THE PACE—SENIOR MANAGEMENT AND TEAM

As discussed previously, not every team is fully functional at the outset. It is even less functional if the members never have worked together before. Consequently, the manager needs to establish the culture of the team. This means imbuing the group members with program values. It is best to establish this project culture as early as possible in the development cycle, as time is at a premium and time should be spent in execution rather than in group alignment.

The pace of the program should also be set with the cooperation of both senior management and the group. In effect, the program manager needs to push the team for performance and also to urge senior management for funding and resources. Each faction needs to be urged by the other's commitment. For example, the team should work diligently, as it is spending corporate resources. Senior management should take note of the commitment and performance of the team and release funds in a timely manner.

2. WATCH THAT FIRST OBSTACLE

When a project is first starting out, actions speak volumes. Resolution of that first obstacle is crucial to setting the tone for subsequent activities. Raise the expectation level high and resolve the first obstacle as soon as practically possible. Do not address it with complacency, as everyone will then view the group as ineffective and lacking commitment.

The group's test when encountering the first obstacle and every one thereafter is to overcome it without delay. An oft-quoted expression is applicable here: "An obstacle is what you see when you take your eyes off of the goal." In program management, time and complacency are the enemy.

3. VARIABLE WORK EFFORT APPLIED TO VARIABLE WORKLOAD

What we have said up to this point does not imply that development should be slavery. Rather, the concept is to create a dynamic performance within the group that is variable. As the needs of the program dictate either by schedule or adversity, the demands of team rise to meet the program's needs. This philosophy needs to become part of the operational characteristics of the team.

Because the project is under some sort of project management control and the demand for personnel's time is identified initially, the demand variable should be left to overcoming the obstacles.

Another variable factor is time off, which can include vacations. Plan to work around personnel with critical skills during these times. As mentioned previously, the idea is to reduce the effects of uncertainty as best as possible.

INCENTIVES FOR THE DEVELOPMENT GROUP

1. INCENTIVES

This section does not delve into the means for compensation or incentive programs for employees, other than to endorse some means for encouraging the development group to bring the project to completion within the prescribed parameters. Members of the team must set aside their personal agendas to become an integral part of the team. Therefore, it is only fitting that the team be rewarded as a unit for their performance. Salary and benefits get the personnel in place and reporting for work each day. Incentive for group performance will drive them to achieve results on time and within budget.

2. RELATE TIME AND QUALITY OF WORK TO INCENTIVE

The incentive should comprise two components for development personnel: personal and professional. This means that one component should specifically reward the individual members of the group in monetary fashion. The other component is to be spent on new discretionary equipment, training, or processes, to assist the members in professional growth. This practice serves to tie performance to both personal gain and professional improvement.

3. RECOGNITION OF PLAYERS

Special care needs to be exercised in rewarding key performers. In many cases, these are more valued members of the group in terms of contribution; however, care must be taken so as not to disenfranchise the other members. Fundamentally, most professional people realize the contribution capability of higher-level individuals, and to some extent, the incentive can be a driving force for their own improvement. The use of the second component has a leveling effect in that all of the personnel may have access to the discretionary equipment.

4. ELIMINATING PRIMA DONNA ATTITUDES: WE WIN OR LOSE TOGETHER

Finally, the key to long-term effectiveness of the team is to eliminate prima donna attitudes among the top performers. People who know their value to a team can begin to become prima donnas, and these attitudes must be discouraged. They serve to demoralize the balance of the group and over the long term will destroy any leverage that might be created.

The value that must become part of the team is that the group wins or loses together, regardless of how talented one or another member might or might not be. The group wins, or the group loses and some other group wins. This is the law of the product development jungle.

MANAGEMENT REPORTING

If the development team is doing a great job in executing a project and senior management doesn't know about it, is the team still doing a great job? Sounds like the example of the tree falling in the forest with no one to hear it; does it make a sound?

Senior management has a right and a need to know status of a program and rates of progress. The development manager has the responsibility to keep senior management properly informed. This current section will focus on setting up systems to accomplish just that.

1. SET UP A GOOD, ACCURATE, EASY-TO-USE REPORTING SYSTEM

Program management is not about sophisticated management reporting systems; it's about getting results. Therefore, set up a management reporting system that is easy to use and effective at communicating essential information. Make it specific enough to be accurate and have depth. Make it general enough for a senior management overview. Have audited data entered, and make it as automated as possible. The manager's job is to manage the development, not to "reinterpret" data and present it to senior management.

Although many systems generate their own reports for status, this section will present a format that consists of the data in which most senior managers are interested.

2. LAY OUT THE PROGRAM IN DETAIL

Begin by laying out the project in total. Starting with stage 0, include the following list in detail. By laying out all this data for each stage, the requirements for the program are readily apparent. Figure 6-10 represents an example of this process.

1. Funding

   This is the stage's financing package. Detail the budgeted expenses and allocate them to tasks, performance, and deliverables. In this way you will have a good idea what the deliverables are costing the project.

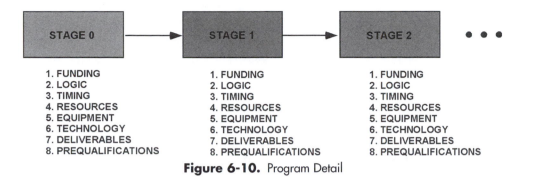

Figure 6-10. Program Detail

2. Logic

   Lay out the project in PERT format showing interconnected tasks and milestones. Take care to identify the connective logic, as this ensures diligence and quality. Haphazard redefinition of the logic is a sign of a program out of control.

3. Timing

   Lay in estimated times for each activity and task. The use of development personnel's estimates is the best choice here. Any other estimate won't be "bought into" by the team, as the estimate is not theirs.

4. Resources

   Identify the required resources, and schedule them on the timeline.

5. Equipment

   In a similar manner, identify all of the capital and expense equipment required for the project. Specify dates on the timeline when they will be required, and also factor in training time if it is required.

6. Technology

   List and schedule the technological requirements for the project. Make sure to include collateral requirements to integrate the technology into the company. Also allow time for absorption of the technology by the team members. This technology transfer must take place for the project to be successful in the long run. This is because the licensing of technology is for one purpose, which is to get your company into the game initially. *Staying* in the game is the responsibility of the company.

7. Deliverables

   Schedule and lay in the list of deliverables for the stage. This is the strict list of deliverables that must be completed in order to consider the task of the stage complete.

8. Prequalifications

State clearly on the schedule (now multidimensional) the prequalifications needed for the next stage that must be tended to now. An example of this type would be to start campaigning for funds needed for stage 3 during stage 2. Another example is the recruitment of talent required in the future, so that personnel are in place and functional when they are required.

3. DEVELOPMENT TIME MANAGEMENT AND ENGINEERING LOG

The next area for consideration is the development time and engineering log. This is a tool to manage the engineers' time and to ensure the project is getting the proper time allotted. Development engineers' time is used in many ways, even when they are assigned to a specific program. As will be shown in more detail later, there is always danger of misallocation of time to nondevelopment activities, which delays programs and makes the group progress lethargic.

An example of a time reporting system is shown in Figure 6-11 a truncated format for review. A complete monthly chart is available for your use in The Toolbox.

As shown in Figure 6-11, the use of an engineering time log for tracking development time spent generates the percentage of time used on various aspects of development activities. In this example, the actual development time represents only 47.4 percent of the hours expended by the enginer. Orders required 20.7 percent of the time.

Engineering changes, which have little contribution to value, represented 10.1% of the total. Through the tracking log, the nondevelopment time can be accounted for. If the manpower loading was allocated at 100 percent for development, it is apparent how the project will be significantly delayed with only 47.4 percent being expended. Moreover, this represents only one week. Left unchecked, one week dovetails into another, further reducing the effective development time.

4. REGULAR STATUS MEETINGS

The use of regular status review meetings is beneficial to maintaining the program on track. The status meeting accomplishes several things:

- It establishes consistency in progress assessment.
- It establishes accountability for the individual team members.
- It assists in identifying upcoming issues.
- It provides a forum for discussion on resolving problems.
- It focuses attention on critical path tasks.

Depending on the type of development project and the group's level of experience, the optimum time for these review sessions can range from every week to every two weeks. These are "nuts and bolts" sessions that are designed to assess progress and navigate around

ENGINEERING TIME LOG

		MON	TUES	WED	THUR	FRI	SAT	WK-1 TOTAL
ORDERS								
	1	4	4	4	5	6	7	30
	2	1	6	1	1	1	1	11
	3	1	2	2	2	2	2	11
	4	1	1	4	4	4	4	18
	5	1	1	5	5	5	5	22
SUB TOTAL		8	14	16	17	18	19	92
DEVELOPMENT PROJECTS								
	1	7	7	7	7	7	7	42
	2	7	7	7	7	7	7	42
	3	7	7	7	7	7	7	42
	4	7	7	7	7	7	7	42
	5	7	7	7	7	7	7	42
SUB TOTAL		35	35	35	35	35	35	210
ENGINEERING INVESTIGATIONS								
	1	1.2	1.2	1.2	1.2	1.2	1.2	7.2
	2	1.2	1.2	1.2	1.2	1.2	1.2	7.2
	3	1.2	1.2	1.2	1.2	1.2	1.2	7.2
	4	1.2	1.2	1.2	1.2	1.2	1.2	7.2
	5	1.2	1.2	1.2	1.2	1.2	1.2	7.2
SUB TOTAL		6	6	6	6	6	6	36
ENGINEERING CHANGES								
	1	1.5	1.5	1.5	1.5	1.5	1.5	9
	2	1.5	1.5	1.5	1.5	1.5	1.5	9
	3	1.5	1.5	1.5	1.5	1.5	1.5	9
	4	1.5	1.5	1.5	1.5	1.5	1.5	9
	5	1.5	1.5	1.5	1.5	1.5	1.5	9
SUB TOTAL		7.5	7.5	7.5	7.5	7.5	7.5	45
MISCELANEOUS								
	1	2	2	2	2	2	2	12
	2	2	2	2	2	2	2	12
	3	2	2	2	2	2	2	12
	4	2	2	2	2	2	2	12
	5	2	2	2	2	2	2	12
SUB TOTAL		10	10	10	10	10	10	60
TOTAL		123	135	139	141	143	145	443

Figure 6-11. Engineering Time Log

and through barriers. They are not strategy meetings, nor commiseration sessions. They have the sole purpose of getting things done.

There is another type of progress meeting that may be held every two months. This is a management review meeting to discuss overall progress issues with senior management. Specifically, the idea is to not get pulled into details on each project but rather to inform management properly on progress and to resolve linking objective issues.

Two-month intervals are not magical by any means, and the frequency can vary based, again, on the project; however, this series of meetings is designed to evaluate development operations and programs for effectiveness.

5. PRODUCT MAINTENANCE VERSUS DEVELOPMENT

The issue of product maintenance versus development is an interesting one. Consider, for example, a case in which a fixed number of development people develop product number one. As the product ages, there will be changes to the product driven by external factors to the organization. The support of these changes is referred to as product maintenance. As more products are added to the portfolio, the demand for development engineers' time begins to grow. Time reporting will show, as in the previous example, that development time for new products does not get expended on development. It gets parsed out to all of these other activities, which results in severe project delays that are not recoverable. Figure 6-12a illustrates how this can happen and how it appears on a time report.

Figure 6-12a shows three scenarios to illustrate the loss of development effectiveness because of the effect of dilution due to product maintenance. Each scenario is for a development program requiring 6000 hours of development time. The available hours are the total hours available for the group. The maintenance hours are the required hours needed to support the product previously introduced. The development hours are the hours that can be expended on the development project. The remaining hours are the hours left to complete the project after the month's development hours have been expended.

In Scenario 1, for example, the program requires 6000 hours. With no maintenance requirement, the 2000 hours available each month can be expended on the development program. It is complete in exactly three months. In Scenario 2, there are maintenance hours, which dilute the development effort. This results in a 4.28-month time frame for program execution. Scenario 3 consists of maintenance hours for two previously introduced products, which dilute the engineering effort even more. At some point in assembling a portfolio of products, maintenance can exceed development given a fixed number of people. The crossover is shown graphically in Figure 6-12b.

6. PERIODIC MANAGEMENT REPORTS: A MODEL

A must for the objective communication of product development progress is the use of a standardized management report. There are three sections that comprise the report, namely:

A. Product Status Report

The product status report is a short synopsis of the progress toward achieving functionality of the product and other specification items such as performance, size, form, and fit. The entire functional specification should be outlined in a tabular format, with check marks for parameter achievement. A gradient may also be used to grade

PRODUCT DEVELOPMENT / MAINTENANCE TRADEOFFS

ITEM	MO. 1	2	3	4	5	6	7	8	9	10	11	12
	SCENARIO 1			SCENARIO 2				SCENARIO 3				
PROJECT HOURS REQ'D	6000			6000				6000				
AVAILABLE HOURS	2000	2000	2000	2000	2000	2000	2000	2000	2000	2000	2000	2000
MAINTENANCE HOURS	0	0	0	600	600	600	600	1160	1160	1160	1160	1160
DEVELOPMENT HOURS	2000	2000	2000	1400	1400	1400	1400	840	840	840	840	840
REMAINING HOURS	4000	2000	0	4600	3200	1800	400	5160	4320	3480	2640	1800

3 MONTHS

4.28 MONTHS

7.28 MONTHS

A

DEVELOPMENT VS MAINTENANCE

AVAILABLE HOURS
MAINTENANCE HOURS
DEVELOPMENT HOURS

HOURS — MONTHS

B

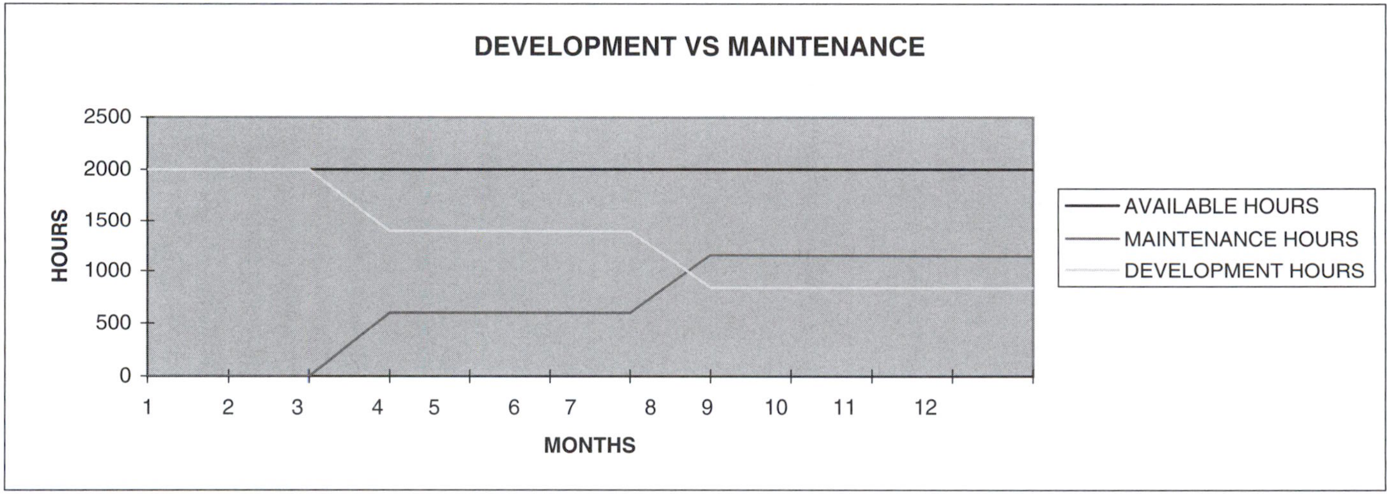

Figure 6-12. Development Hours

achievability for each item of interest. Also in this section of the report, the original target manufacturing cost is compared with current estimated projections.

B. Project Status Report

The project status report summarizes the investment and exposure of the company in funding the program and compares the sunken investment to the total and remaining investments. It summarizes personnel investment, hours, and dollars.

C. General Business Condition

Finally, the general business condition report gives senior management an idea of the outlook of the business in terms of forecast sales, incoming order rate, shipments, and profitability.

A model for this format, which includes a place for narrative in each of the three sections, is shown in Figure 6-13.

A sample management report format is presented for your use in the Toolbox.

7. TEAM PUSH/MANAGEMENT PULL

As a final note to this section, the use of management reports has as one of its aims the creation of a synergistic exchange between the team and the senior management group. The team pushes the project through the process by effort and deliverable results; the management pulls the project by setting high expectation levels and removal of barriers for the team. When the two forces work together, they create a fireball of energy and enthusiasm for completing the project successfully.

COMMUNICATION SYSTEMS

1. SET UP COMMUNICATIONS

The communications system set up at the beginning of the project will determine the effectiveness of the project. A communications system is not necessarily medium dependent. It is simply the unidirectional and bidirectional exchange of key information to move progress along.

By definition, the communications system must effectively do the following:

- Transfer information without distortion.
- Acknowledge and confirm information.
- Maintain a certain amount of security.
- Allow a channel for meaningful feedback.
- Create understanding in the receiver intended by the transmitter.
- Act as a document and retrieval system.

MANAGEMENT REPORT FORMAT

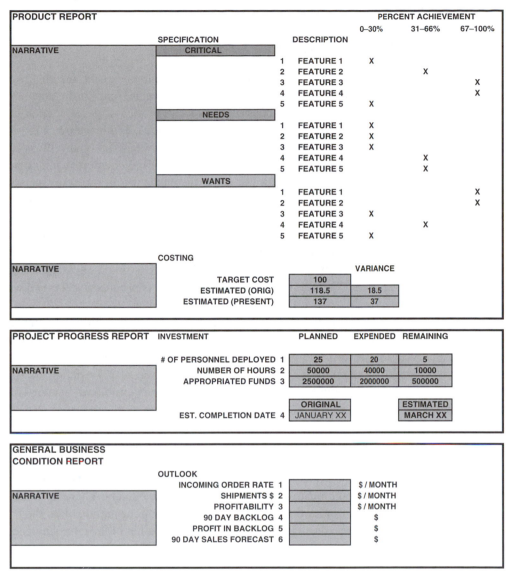

Figure 6-13. Management Report Format

These items require a fair amount of set up and security to work in a real development situation. In addition, they are multidimensional in that communications occur both internal and external to the company. The importance of a properly implemented communications system cannot be overstated. It will prove to be invaluable as the project progresses and as details become more prolific and harder to keep track of.

2. A MODEL FOR COMMUNICATIONS

Communications within a company and outside of the company occur on all levels. There are communications within the company, through the channel, and to the customer. At all levels and in all directions, the transmission and receipt of messages, data, and decisions are opportunities for misunderstanding, inflated expectations, and inaccuracy.

We all have experimented with the game of "pass the message." This game illustrates how a message can change its meaning or emphasis as it passes from one person to another. Consider communications between companies in which the elements of time delay, personalities, and focus differ. These factors complicate the job of accurate receipt of transmitted data, as Figure 6-14 illustrates.

Depending on the stage of the transaction, there can be diagonal communications, such as sales personnel of the manufacturer communicating with the engineering or finance personnel of the customer. Also consider how many individuals a given message passes through and the degree of interpretation that can take place as it is progressing the pathway from transmitter to receiver.

Another element to consider is the security of communications both within the company and outside of the company. The heavy black lines on the illustration can serve to illustrate firewalls that prevent the breach of communications security. As we will see in the next section, the use of network communications has improved the speed of information transfer tremendously. It has also introduced a whole new set of security problems—such is the nature of progress.

3. WANs, LANs, AND PEOPLE (NETWORKED COMMUNICATIONS)

With respect to protecting intellectual property, trade secrets, and interests of the business in general, network communications cannot protect your company. There is no software, protocol, or arrangement of hardware that can make up for poor judgment on the part of team members. In order to use these tools effectively, each member of the company needs to operate like a businessperson. This means being savvy enough to know what and what not to disclose to personnel outside of the company.

If the company personnel understand the business, the threats, and the opportunities and are truly interested in furthering the business, they will conduct themselves in your best interests. If they do not understand the issues, train them quickly, or eliminate them as possible sources of information leaks. In chapter 2, we discussed the spheres of influence, which showed how information could pass from one person in the company through people interfacing with them to others outside of the company.

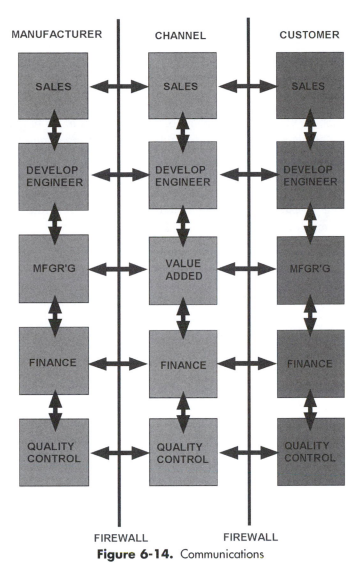

Figure 6-14. Communications

Implement internal communications controls to be able to determine whether sensitive information is being made available to competitors or to the general market, where it can fall into the hands of competitors.

As a final note, secure people you can trust in every aspect of the company. There are no systems (electronic or otherwise) that can provide security as good as trustworthy people who have a vested interest in the business and are streetwise enough to perform their jobs and protect the company's interests.

THE PROGRAM INITIATION

1. ESTABLISH THE START AND END DATES

As important a milestone as the end date of a project is, the start date is equally important. The beginning date establishes that point where planning, conversation, and forecasting diminish in relative importance to actual execution. To that end, the organization and the development group need to understand that development is in progress, and the single-minded purpose now takes precedent: completing the program successfully.

Establishing the end date also sends the right message to the organization and the development group about the expectations for completion and general timeliness. Fundamentally, however, the handling of the first obstacle will set the tone.

Setting the end date as relatively inviolate eliminates the use of excuses for inaction in getting back on schedule. The dates and deliverables drive the activity.

2. MAKE USE OF TRAINING AS PART OF THE PROGRAM

Each day brings new learning, and so too with product development. Although there are high expectations for the development group and the company population in general, there is no need to expect that the team is at capacity in new learning. The manager needs to balance the needs of the project and the need for continued development of the team members.

If such training and new learning do not take place, the group will fall behind in capability with respect to the market. This, in turn, will reduce your competitive position for the next program. The operative lesson is to integrate training and new learning wherever possible.

3. ONE STEP AT A TIME

Execute the program one step at a time. If the proverbial "journey of a thousand miles begins with a single step," then remember that the thousand miles is covered by countless steps, one after the other in relentless succession. This is also the case with product development. Daunting at first, the program schedule might seem insurmountable, given all the tasks, subtasks, and activities. Take each task, however, and execute it one step at a time until all steps are complete.

4. HAVE FUN!

To place the program in perspective, the new product development activity can be the hardest, most difficult work a person can do. But lighten up, and have fun. This can be the hardest job you will ever love! The challenge of technology, its application, and the excitement of completing it within the allotted time frame can make the job truly enjoyable.

As with anything else in life, look two steps forward, and take the next step toward the goal. Plan the contingency, and keep the group's eye on the goal.

5. LET'S GO!

If you feel that product development is like a race, you are probably accurate. The kick-off is the formal beginning of a long process. Mark the time, pull the trigger, and let's go ahead!

SUMMARY

This chapter is a transitional chapter. It describes the point in the development program at which planning gives precedence to execution. Once the starting point is established, the balance of the effort is in producing tangible results.

In this chapter, we reviewed several of the elements that comprise the formulation of the development team and the tools for making them effective.

Initially, the issue of assembling the group and transforming a loose collection of people into a functioning team was discussed. This led to the discussion of organizational formats and how to match the format to the specific project under development.

Mentoring and apprenticeship are more important then ever before as we move forward in new product development. It can be a means for imbuing new members of the team with the values of the company as well as with technical know-how.

Management of the group was also discussed, with respect to generating the results effectively. Tools for tracking and managing the time of the group were presented.

An important element in developing a product is to organize and lay out the project in total. This structuring of the project will yield big returns if the schedule should become compromised in any way. A clear listing and interconnection of the tasks make planning and execution more manageable.

Also discussed was the culture of the development team and the company. The culture is developed over time but must be established as a baseline immediately. The team needs to be action oriented, and obstacles must be navigated around or eliminated. Develop a culture based on a sense of urgency.

Finally, we discussed communications used in the development process. Communications are a key element for successful programs; they can be a program's downfall or make it successful.

With the completion of Chapter 6, the group is now ready to start the execution of the development project with the information in Chapter 7.

EXECUTING THE PLAN

ABSTRACT: This chapter discusses the actual execution of the new product development project. We will review the basic tools required and the basic structure of running an organized program. The time for planning is at an end; the effort is now focused on the execution. Once the program is in motion, it involves an interdisciplinary watch over timing, functionality and product cost. The manager of the program must now advance on all of these fronts and make progress toward the successful introduction of the product.

A major thrust of the chapter will be problem solving. Problems will befall the development if the program demands that new ground be covered. It is in the best interest of the program to deal with them effectively to enable on-time delivery of a successful product.

Another focus of the chapter is the issue of documentation. A vocational hazard when absent, documentation is often overlooked, but this omission often comes back to haunt the team in some way or another. Properly executed documentation protects the company in the event of product liability.

We will also discuss quality management systems to provide continuity across several developments. Finally, we include a section on patent protection and strategies for the use of patents within the context of the new product development process.

MECHANICS OF PRODUCT DEVELOPMENT

1. FLOWCHART DEVELOPMENT

As discussed in Chapter 6, the flowchart of the program is an important step in executing the development on time. There are many software programs to accomplish the task of documenting the logic, keeping track of the tasks at hand, and tracking the timing of the tasks. Regardless of which software program is used, discipline in enforcing the program tasks and schedules is of paramount importance. The software can only keep track and prompt. Management of a successfully executed program requires the investment of human energy to stay on track.

2. ORCHESTRATING TIMING, RESOURCES, AND CREATIVITY TO MEET THE OBJECTIVE

Only human energy can assess, revise, and navigate obstacles. This occurs by amassing horsepower to overcome or work around those obstacles. In either case, timing, resources, and creativity must be orchestrated to meet the objective. Failure to employ these qualities will cause endless delays and corrupt the timeline of the program. The new product development manager must keep a watchful eye on all dimensions of the program. Figure 7-1 illustrates the philosophy of management:

As Figure 7-1 shows, the manager must look ahead in the project timeline to anticipate future details as well as keep track of documenting and completing previous details. By looking ahead, the degree of impact of uncertainty can be minimized. By mopping up details, nothing is left to chance, and progress is truly in the forward direction, with little danger of past baggage haunting the program. Analogous to a symphony conductor, the development manager must coordinate the team resources with reference to the scheduled tasks.

3. PROGRAM MANAGEMENT

The issue of program management is complex and pervasive, involving the entire company. One needs to think of program management as having two loops of control. The inner loop is dedicated to the specific project, while the outer loop coordinates several projects that contribute to the total activity within the corporation. This can be represented as in Figure 7-2.

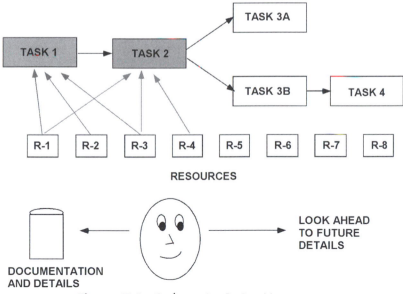

Figure 7-1. Orchestrating Project Management

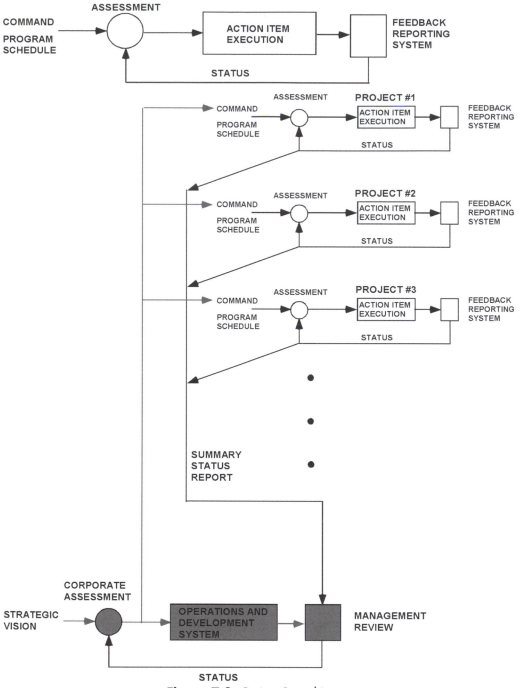

Figure 7-2. Project Control Loops

The Inner Loop

The inner loop of project control is comprised of a classical feedback system in which the schedule, analogous to a command, is input to the development system. The development system executes action items to complete tasks of the program. The means for feedback to advise of status is analogous to the reporting system of a program. The status is compared with the schedule (command) in the assessment node, and an action is directed to the individual development system.

The Outer Loop

The outer loop of control correlates the individual programs to the larger corporate strategy and operations. The overall corporate strategy is the command signal for the corporate business operations and development system. As depicted at the bottom of the illustration, the operations and development systems execute the business strategies and development. The output of the assessment node serves as the command for the individual programs. A management review evaluates the output performance of operations and feeds information back to the corporate assessment node, completing the outer loop. Additionally, summary status reports taken from each individual project input to the management review. This review serves as an entry point for feedback on all of these programs and extends to the higher-level corporate assessment node.

The corporate assessment node evaluates the difference between business strategy and operations performance and then generates appropriate action items and tasks to be executed by the operations and development system.

4. PROJECT MANAGEMENT SHOULD BE ROTE, NOT HEROIC

Senior management, in its desire to accelerate an individual program, often lends a bit of romanticism to a project, calling for heroic efforts on a regular basis and injecting an *esprit de corps* into the team. This approach can serve to accelerate a program for a time; however, a steady diet of this behavior can burn out talent rapidly or even encourage cynicism. The idea is to make the process of project execution almost rote—not to the exclusion of enthusiasm or creativity, but to establish a pace of performance and results expectations that challenges personnel and delivers results to management.

MANAGING PEOPLE

1. PROJECT MANAGEMENT IS PEOPLE MANAGEMENT

Forget the claims on the box covers of project management software. Software cannot "manage" a program.

There is more to program management than keeping track of the details of logic, timing, and resources. Although a fully integrated system of software based project management can be quite valuable in executing a project, once the details are tracked, project management is essentially people management. It involves the motivation and guidance of very talented personnel to accomplish a specified result in a scheduled time frame.

Therefore, development managers should become as conversant with people skills as they are with the requisite technical skills. The accomplishment of things to exacting specifications is a difficult endeavor. Accomplishing them through others elevates it to a further level of difficulty, and accomplishing them through others within a specified time frame elevates this to a unique challenge suited for only a few people.

2. THE UNPREDICTABILITY FACTOR IN PERSONNEL

There is an unpredictability factor in development personnel that must be recognized. Progress might not be linear; therefore, the schedule must be tied to actual patterns of progress. For this reason, it is helpful for a team to have completed several projects together previously to refine the estimating and performance skills of the group. Team members have both "good" days and "nonproductive" days. Progress is a pathway, and the manager needs to guide the way.

3. MANAGING TECHNOLOGICAL RISK

Managing technological risk is an often-oversimplified practice. Often, managers look at risk and uncertainty, make an immediate assessment, and assume that overcoming it is part of the development. Risk is, however, a significant barrier to a program, and as in the case of resource allocation, simply redoubling efforts will not be sufficient to prevail. In fact, risk should be quantified in each stage of the project. As the project has a Pert chart to execute tasks, there should be a scaled uncertainty measure underlying each task. Uncertainty should be measured on a relative scale, in which the most uncertain element represents 100 percent, as shown in Figure 7-3.

As Figure 7-3 shows, uncertainty is charted along with the talent levels of resources assigned to the task. While this is only a sample representation, not the actual form that might be used, it does serve to point out that uncertainty changes in the project and varies by task. The manager needs to amass the intellectual talent and bring it to bear on formidable tasks to resolve uncertainty in a timely manner and prevent delays. Simply stated, this means preparing for Task 3 while the team is still completing Task 1.

EXECUTING THE PRODUCT DEVELOPMENT PLAN

1. ANY PLAN WILL WORK—THE KEY IS ADHERENCE

A plan is a pathway toward the completion of a program. The key to completing the program is adherence to the plan. There are several elements that must be coordinated as part of the product development plan, namely:

ITEM	TASK 1	TASK 2	TASK 3	TASK 4	TASK 5	TASK 6	TASK 7
DURATION	5	7	35	23	12	16	25
RESPONSIBILITY	MAA	TSB	PCR	GHT	DFT	FTD	PQR
LEVEL 1-10	4	3	8	6	7	3	7
UNCERTAINTY FACTOR	10	15	100	35	40	23	80

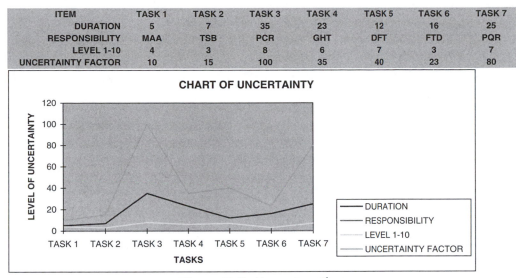

Figure 7-3. Uncertainty Chart

- The plan itself with tasks and logic
- The schedule
- Chart of uncertainty
- Resources
- Work content
- Management reporting
- Measurement of performance

All of these elements must cohere throughout the development process. The manager's role is to give careful consideration to all of the elements and juggle them to serve the needs of the project in total. This might mean trading off elements for expediency; although this practice is not recommended, the manager might need to exercise such an action to move a program out of the doldrums.

2. A LOOK AT TASK MECHANICS

All tasks should be handled in the same manner. Standardization serves to establish a common denominator for scheduling a project. To further define the task execution, every task should include the following five integral aspects:

A. Assessment

Assessment is the analysis and breakdown of the tasks into its component parts. It results in the identification of the deliverables that constitute completion. It can also be thought of as task scope.

B. Strategy

The strategy is the plan for implementation of the task. It specifies the means by which the end occurs. It can be a detailed representation of the substeps needed to complete the task.

C. Execution

Execution involves the actual implementation of the substeps and components of the task. It is the core development activity. Many development people jump into it without taking the necessary initial steps.

D. Verification and Metrics

Verification is another often-overlooked item. All too frequently, the verification step is waived in anticipation of the design review. There are a few problems with such a practice, however. The first is that verification is an integral part of the task process. It ensures that the start of a successive task is built on a solid foundation of the previous task. Second, precious time can be lost by not verifying, by starting the next step only to find that previous steps must be retraced and corrected first. Third, a design review is simply an overall review that is not intended to catch mistakes or serve as a comprehensive recheck of previous work. Metrics should be part of the verification step; test data prove that the verification has legitimacy and is tied to the original intent.

E. Documentation

"No job is done until the paperwork is complete." This age-old saying is no longer medium-dependent on paper; however, it does speak with relevance to every task, subtask, and component of the project. Documentation of the elements of the task show diligence, chart progress, establish points of reference with data, and serve as a reference library for future queries.

3. TASK BREAKDOWN STRUCTURE

Each task of the project must be developed within the context of a task breakdown structure. The task breakdown structure is an organized method for identification of the technical and administrative disciplines involved in that task. Each subtask component takes on the same format of assessment, strategy, execution, verification, and documentation. Time is allotted to each activity, and talent codes are also assigned. The various disciplines engage the task and execute it. Figure 7-4 illustrates the task breakdown.

As Figure 7-4 shows, the assessment and strategy for each discipline should be started and completed at the same time to ensure that the task is initiated with all disciplines in communication and at a common starting point.

TASK BREAKDOWN STRUCTURE

TASK 1 CODE	DISCIPLINE	ASSESSMENT	STRATEGY	EXECUTION	VERIFICATION	DOCUMENTATION	TOTAL
M1	MECHANICAL	4	4	45	16	16	85
P1	PACKAGING	4	4	80	12	12	112
E1	ELECTRICAL	4	4	16	12	18	54
E2	DIGITAL ELECTRONICS	4	4	60	40	40	148
S1	INTEGRATION	4	24	72	40	40	180
T1	MANUFACTURING DESIGN	4	4	40	24	40	112
	TOTALS	24	44	313	144	166	691

Figure 7-4. Task Breakdown

It is acceptable for the separate disciplines to conduct discrete investigations for optimizing a solution, so it is not absolutely necessary for the investigation start times to be concurrent; however, the activities should be. As shown in the example, the integration activity showing a strategy planned for 24 hours might cause a delay, depending on how other tasks interface. If this is the case, the program manager might elect to initiate this activity in advance so that execution can be simultaneous with the same phase of the task as performed by other disciplines.

With the basic metrics in place, it now becomes necessary to stage execution by the individual disciplines in a sequential fashion based on the interdependence of the requirements of a given task. The truncated Gantt chart shown in Figure 7-5 represents this staging; with the sequencing in place, all of the tasks and subtask activities are interlinked.

The resource requirement used in the execution of each task can be identified, and the requirement for each period of time is known and scheduled (in days, in this example). This process can be made as specific as needed to manage the project effectively. Fundamentally, however, the Gantt chart emphasizes that given the program timing, the

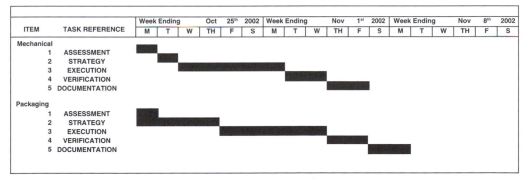

Figure 7-5. Gantt Chart

task breakdown, and the logic, the required resources must be available and supplied, or the timing and, eventually, the sequencing of the program will degrade, causing nonrecoverable delay.

4. LOGIC VERSUS TIMING TRADEOFFS

There could be a need to trade off timing and logic occasionally. This practice is ill advised, however, from the standpoint that sequential revision of a project plan only serves to confuse participants. Such a practice also tends to degrade the integrity of the logic and qualification, as subtasks will be missed or overlooked.

The other point to remember is that revision of the sequencing disturbs the resource allocation, further confusing the scheduling. It is best to expend the energy required to be on time the first time.

5. SETTING UP A DEMAND-PULL SYSTEM

To manage the development process effectively, construct a situation in which the market side of the business pulls development along. This can be done quite simply by focusing on dates of completion and sales lost when dates are missed. By focusing the group on dates, a sense of urgency is created, and participants in the subsequent task create the demand-pull on participants in the previous tasks. The manager's role is then to enforce quality and consistency in the completion of the tasks. A demand-pull system is shown in Figure 7-6.

To further complete the presentation of demand flow, the personnel's first step as part of Task 2 is to verify the completion of Task 1. This process ensures that a program does not forge ahead with poor results from an isolated task. Building intermediate checks into the execution process also ensures a higher degree of quality for the program overall.

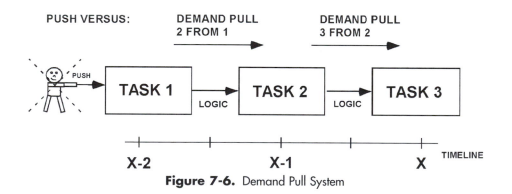

Figure 7-6. Demand Pull System

PRODUCT DEVELOPMENT PHASES

This section of the chapter addresses the core mechanics of the product development phases. Whereas the previous section discussed program charting and sequencing, this section focuses on the stages of the program and the accompanying deliverables.

1. THE SIX STAGES OF PRODUCT DEVELOPMENT

There are six distinct stages of product development. These stages outline the maturity of a product, from the inception of an idea all the way through to an optimized product. To repeat the stages originally presented in Chapter 6:

- Stage 0: Market assessment and internal assessment
- Stage 1: Prototyping
- Stage 2: Core development and testing
- Stage 3: Industrialization
- Stage 4: Commercialization
- Stage 5: Optimization

Let us now examine each stage and outline the deliverables. The presentation of these deliverables represents only a general order for executing the tasks required for the deliverable. The actual sequencing of the tasks and deliverables could be different in an actual project because some activities will be done in parallel. A checklist of deliverables is available for your specific project in *Toolbox*; the narrative is also included.

STAGE 0: Market Assessment and Internal Assessment

The first stage is the assessment phase. This is an internal introspection of company and direction and an external look at product opportunities. The goal of this stage is to obtain a clear idea of how the product opportunity fits with company goals and objectives. The following is a list of deliverables for this stage.

A. MARKET OPPORTUNITY ANALYSIS

The market opportunity analysis is a summary of the opportunity for a new product or product line in the marketplace. It is the germ of a business idea where the product concept starts. This study is cursory in nature but must result in identification of a clear opportunity.

B. COMPETITIVE ANALYSIS

The competitive analysis is a summary of the current activity being conducted by the competition, who could be traditional or from nontraditional competitors. The analysis should be specific enough and accurate enough to outline the current products available, because the product your firm will design must leapfrog competitors' products with features advanced enough so that when it becomes available for sale, the offering is still competitive. This analysis should also consider the life histories of competing products and technologies.

C. PRODUCT CONCEPT

The product concept is generated from analyses of the opportunity and the competitive position and from a projection into the future. It outlines the product to be developed. It should use whatever medium is appropriate to describe what the product is and how it will be positioned in the marketplace to capture share.

D. FEASIBILITY STUDY

Given the product concept and the company capabilities, this study yields an answer to the question of whether the intended pathway is feasible for the company to follow. Although some products are significant opportunities, their execution might not be feasible for the company. Time is the other important factor to take into consideration at this point in the evaluation. Is it feasible for the company to develop this product concept in the allotted time frame?

E. COMPANY STRATEGY

How does the product concept execute the overall company strategy? This position statement can describe a strategic initiative satisfied by the product opportunity or describe how the product will allow the company to further its long-range goals.

F. FITABILITY STUDY

The fitability study is a refinement of the strategy study. If the strategy summary is a broad-overview look at the pathway provided by the product, then this is a detailed analysis of how the product fits into the existing scheme of products and offerings from the company.

G. TIMING

All things being equal, products are opportunities for a limited time frame, and as such, they need to be developed and marketed within that time frame, or the opportunity is lost. This section of the summary outlines the overall time frame for the program. It identifies the program in terms of priority, funding, and resources.

H. MARKET DYNAMICS ASSESSMENT

What are the market dynamics like? Is the market rapidly changing and difficult to read, or is it more stable and predictable? Is this market similar to others that you have seen to the extent that you can fit a pattern to them and predict the next trend? All of these considerations contribute to this assessment of the general activity of the market.

I. GO OR NO-GO DECISION

Given all the data in the summary and the experience base of the people involved, document the "go" or "no-go" decision and the circumstances and assumptions surrounding it. By documenting the decision, analysis for future programs can become more informed.

J. MARKET REQUIREMENTS SPECIFICATION

For a "go" decision, the market requirements specification must be prepared. This is the document that connects the customer's and the market's needs for the product specific to the company developing it. The specification represents the customers' voice on what they want to have in the product and serves as the base specification for the product on which all later versions are built.

STAGE 1: Prototyping

The next stage is the prototyping stage. It is the physical embodiment of a product concept as a starting point. This is the stage at which the product concept evolves into a piece of hardware that is functional. It is desirable to deliver this prototype to the customer's attention as rapidly as is practical to secure early feedback, while still maintaining integrity of intellectual property. The deliverables for this stage are as follows:

A. PRELIMINARY DESIGN SPECIFICATION

The preliminary design specification is development's response to the functional specification, which represents the voice of the customer. The design specification is the best assessment of what can be accomplished.

B. INTELLECTUAL PROPERTY REVIEW

At this point, it is a wise idea to prepare a patent review to determine whether there are any patent issues outstanding or whether there is a potential conflict with existing patents. This review should be performed as early as possible to allow alternate approaches if necessary. This deliverable should identify opportunities for securing patents, trade secrets, copyrights, and trademarks.

C. DESIGN STRATEGY

The design strategy is the development approach to "inventing the product." It deals with issues such as selection of technology, degree of risk, product reliability needs, and cost. It also has implications for development time (to the extent that one approach might be more expedient than another) and for possible allocation of resources. This issue can have far-reaching effects if not considered properly, so it is best to take care in this initial evaluation.

D. TEST STRATEGY

The test strategy follows the design strategy and is part of the overall production strategy. It is a plan for how the new product will be tested. It correlates closely with the design strategy, as both are technology dependent.

E. MANUFACTURING STRATEGY

The manufacturing strategy is a plan that outlines the philosophy of manufacturing— for example, three circuit boards fastened to a frame with collateral components bolted down, versus three-dimensional circuitry molded into the product itself. Is the product to be a throwaway item, or is it repairable? Answers to these questions constitute the manufacturing strategy.

F. QUALITY PLAN

The quality plan is the master plan for the quality claims of the product throughout its life cycle. To differentiate the issue of quality from that of reliability, the quality of a product is strictly a measure of how accurate and repeatable the final product is coming off of the production line, and how accurately it matches the drawings and documentation used to describe its manufacture.

G. RELIABILITY REQUIREMENTS

Reliability is a measure of how long and how accurate a product will perform its intended function. A description of product reliability has its own mathematical relationships. Most consumers are interested in this measurement, although they refer to quality. This is a measure of the value in use that the consumer will obtain from the product.

H. SAFETY REQUIREMENTS

The safety review is a comprehensive analysis of the issues of safety and how they affect both the product and the user under various scenarios of use and misuse. This review needs to be a structured and somewhat exhaustive effort to demonstrate diligence on the part of the manufacturer. In more recent trends, this effort needs to affect product design, building prevention of misuse directly into the design rather than relying on the use of warnings.

I. COST ANALYSIS WITH SENSITIVITY

As part of the initial product planning exercise, the product's cost needs to be paramount. From this point onward, most development and manufacturing activities tend to add cost. Therefore, the starting point estimate of cost must be low enough to support all of the incremental costs associated with developing the product.

J. STATUTORY AND REGULATORY APPROVALS

Most, if not all, industries have a means for regulation of that industry. Products involving electrical circuits, for example, could come under the regulatory control of Underwriters Labs, Canadian Standard Association, CE compliance, VDE, and many other organizations. The need at this stage of the program is to articulate the plethora of standards the product needs to meet in order to be accepted by the marketplace.

K. VALUE ENGINEERING STUDY

This study is the means by which the needs, wants, and desires of the marketplace are organized, evaluated, and compared with implementation costs. These implementation costs affect manufacturing costs, development cost, and time.

L. AFTERMARKET PLANS

Consideration of the aftermarket for the product must be factored into the overall product plan. If the product is disposable, the issue is minimal. If, however, the product is repairable or needs to be maintained in some way, an entire business infrastructure will be required to support the aftermarket plans. In addition, the eventual disposal of the product should be planned by the manufacturer and communicated to the consumer.

M. PRELIMINARY RETURN ON INVESTMENT

To obtain management approval or determine whether the program is even worth doing, a preliminary return on investment calculation needs to be made. As simple as this calculation is, many companies fail to perform it and consequently lose a dispassionate means to evaluate one opportunity against another. Companies can then embark on what might be the least expedient pathway toward their strategic goals.

N. DESIGN REVIEW RESULTS

Alongside all of the previously mentioned studies related to product and technology, there should be a crossfunctional design review to ensure that the pathway toward the new product is correct. The design review should check that all of the studies have been performed and that they were performed accurately and against some standard, not just glossed over.

O. PACKAGING OPTIMIZATION

If the product is packaging dependent, it is a good idea to attempt to optimize the package at this point. It is difficult to optimize when the team is immersed in the technical challenges and functional requirements of a program. Hence, by constraining the packaging needs early in the program and enforcing them throughout the program, the packaging requirement can be met without undue strain.

P. CAPITAL REQUIREMENTS PLAN

The capital requirements plan needs to be outlined at this point in the program from the perspective that it is related to both the manufacturing approach and the design strategy. By doing the capital requirements plan at this stage, misunderstandings about future expenditures and potential refocusing on a different design approach can be avoided.

Q. TECHNOLOGY INTEGRATION STUDY

As products become more complex and several technologies are used in the implementation of a product, it becomes necessary to plan how the development team will integrate the technologies into the product. This plan is necessary, as test plans and design verification will require it.

R. FINAL DESIGN SPECIFICATION

With all of the fact-gathering and assumptions complete, it is time to prepare the final design specification. This will be the permanent plan of record as to what the new product has been determined to be at this stage of the development. There can be subsequent changes; however, this is the plan of record and can be used as a measure of the extent of future changes.

S. ALPHA UNIT DEVELOPED

The final deliverable for this stage is the production of an alpha prototype. The alpha is the designation for the engineering prototype. It may be made in a lab, without tooling or specialized parts. It could have a form that will not be the final form. It could fail to have the aesthetics of the final unit; however, it does demonstrate the functionality and the performance of the final unit. It is suitable for showing to a prospective customer to obtain feedback.

STAGE 2: Core Development and Testing

In this stage, the embodiment of the development is tested by design verification and design qualification. It is the stage at which the product matures from the physical embodiment of an idea into a product that can be produced with predictable performance. Its list of deliverables is as follows:

A. CRITICALITY ANALYSIS

The criticality analysis is an exercise in determining the degree to which one failure in the product could prove critical. It draws a relationship between the failure mode and the effect within the application. This can also serve as a basis for further examination during the safety review. A design failure modes and effects analysis may be included here.

B. SAFETY REVIEW

The safety review is a summary of the general effects of the safety issues on operation of the device and the personnel. It must show due diligence on the part of the manufacturer such that each issue was identified and accounted for in the design so as to mitigate potential damage.

C. PRELIMINARY FORECAST

The preliminary forecast is important from two perspectives. The first is from that of the return on investment calculation. The second is from a commitment perspective. The marketing function must commit itself and its sales force to deliver the forecasted numbers. Requiring this forecast as part of the project plan helps ensure commitment.

D. BETA UNIT

The beta unit is, for all practical purposes, as close to a production unit as is possible without the manufacturing setup being complete. In fact, there can be circumstances in which the beta unit will be made from production tooling.

E. CHANGE CONTROL SYSTEM

Once the product has matured from the laboratory stage, an engineering change control system is required to enforce some means of product configuration control. This is needed to prevent changes from creeping in without the team agreeing to them. The engineering change control system is often the only means for documenting and tracking and enforcing product configuration control.

F. DESIGN VERIFICATION TEST RESULTS

The design verification results show that the product is designed to meet the requirements of the design specification. These tests document where the product meets, exceeds, and might fall short of the design specification. The evaluation is through the use of metrics.

G. DESIGN VALIDATION RESULTS

The design validation test results show that the product meets the intended use by the customer as defined in the functional specification. It documents where the product will meet, exceed, or fall short of the functional specification.

H. STANDARDIZATION TESTS

As part of the qualification process for meeting certain standards, the test results must be documented. Test results could show that changes are required for the product in order to meet the standards. This deliverable is helpful in managing the transition from a nonqualified unit to a qualified unit.

I. PILOT RUN PLAN

The pilot run is the first practice production run of the product. It "test drives," so to speak, the manufacturing documents, the design documentation, and the processes. The logistics and the details of these are found in the pilot run plan. A process failure modes and effects analysis may be used here.

J. LIFE TEST RESULTS

In order to substantiate the claims made for a new product, life testing must be initiated and completed. This life testing will surface product weaknesses and allow the development group to make specific changes. If the claim is made in the marketplace, then value in use must be integral to the product.

K. DESIGN REVIEW RESULTS

This is a formal design review process in which all of the plans and certification efforts undergo a peer review. This is a healthy exercise, as it allows dispassionate review of a product and also extends knowledge about the product, approach, and technology to others outside of the development group.

L. SIGNIFICANT PERFORMANCE AND APPLICATION FEATURES DOCUMENT

If the functional specification is the voice of the user or consumer and the design specification is the voice of the development group, then the significant performance and application features document is the voice of nature. This is a development-initiated document that defines the absolute boundaries of product operation.

M. PILOT RUN RESULTS

The results of the pilot run must be documented and any anomalies resolved. A development environment is much more forgiving than a manufacturing environment; problems could surface in the pilot run that will require attention.

N. FINAL PRODUCTION STRATEGY

The final production strategy is created from the results of the pilot run. It incorporates the solutions and/or changes that arise as a result of the pilot run.

O. FINAL TEST STRATEGY

In the same manner, the final test strategy is driven by the changes from the final production strategy. These are specific test issues that might need to be added or deleted to accommodate the changes in the pilot run.

P. FINAL QUALITY PLAN

Also included in the list of changes of a production nature is the final quality plan.

STAGE 3: Industrialization

Next is the industrialization stage, in which the product is made manufacturable and the manufacturing processes are developed and put in place. At this stage, the product has matured from a development environment to a manufacturing environment. Development involvement takes place primarily from a support and integration function. Changes and enhancements are limited.

A. CAPITAL EQUIPMENT AND SETUP

With the new product going into manufacturing, all of the capital equipment needed for production must be specified, set up, calibrated, and made functional. This deliverable is necessary to eliminate capital equipment as a variable in the manufacturing process.

B. MANUFACTURING SETUP

Concurrent with the capital equipment setup, the balance of the manufacturing line must be set up to accept the new product. This includes all documentation systems and procedures.

C. TOOLING AND EQUIPMENT QUALIFICATION

If any tooling and equipment are required for the product, this is the time to set them up and also eliminate them as a possible cause of variability.

D. FINAL PILOT RUN

Sometimes an additional pilot run is necessary to test the changes resulting from the first run or any subsequent changes to the product. The deliverable for this phase is final pilot run certification.

E. PRODUCT CERTIFICATION

At this point in the program, it becomes necessary to certify the product. This ensures that the product has been designed, configured, manufactured, and tested according to specifications and standards with the appropriate safety review input.

F. FIELD TEST DATA AND FEEDBACK

The beta test program is required to test the entire customer and user side of the product development. This means customer validation of the usability and completeness of user manuals, part numbering systems, the order entry process, commissioning, and startup if required. The results of this testing program verify acceptance. The feedback can also be used to modify the marketing plan used at the commercialization stage.

G. DESIGN FREEZE

A difficult deliverable to achieve, there comes a point in the product development where the design simply must be frozen and any improvements scheduled for subsequent release of the product enhancements or versions. The product does not change for non–safety related issues beyond this point.

H. CHANGE CONTROL SYSTEM ACTIVE

As part of the frozen product status, the product needs to be under strict change control. All product design and process changes occur under this control with a protocol for the release of the change.

I. PROCURE PRODUCTION MATERIALS

In preparation for the production run, and with the design frozen, this deliverable confirms that forecasted product quantities are ordered from suppliers and that items with long lead times will not impact availability for sale dates.

J. QUALITY DATABASE INTEGRATION

The quality database needs to be an integrated database complete enough to feed back field problems and detailed enough to effect solutions. If, for example, root cause analysis is required at the component level and a demonstrable corrective action must be initiated and completed, then the "hooks" for the data need to be implemented in the database.

K. FINAL COST EVALUATION

At this point in the development and manufacturing implementation, the final cost data needs to be prepared. It will be used for comparative analysis with original estimates as well as for preparing an assessment of the gross margin and market-level pricing. Occurring just prior to introduction, this evaluation will ensure that the program is launched with the correct pricing and that there is little misunderstanding regarding margins.

L. FREEZE MANUFACTURING, TESTING, AND QUALITY PLANS

With the design frozen, the manufacturing, test and quality plans also need to be frozen. These plans must likewise come under change control.

M. FINALIZE MARKETING STRATEGY

With the field feedback in place and the cost of the product known, the marketing strategy can be finalized. This must occur to ensure that the marketing effort is appropriate and directed to produce early results, rather than stumbling around trying to discover what works and what doesn't in the marketplace.

N. PREPARE LAUNCH PLANS

With the marketing strategy in place, the launch plans can be finalized. Action items, responsibility, and completion dates will summarize the required effort.

O. INITIAL STOCK IN PLACE

Depending on the route to market for the new product, there might need to be a significant amount of initial stock required to launch the product. Manufacturing orders must be initiated to build the stock. Marketing effort must be expended to place the stock where it can have the greatest impact on generating sales

STAGE 4: Commercialization

The fifth stage is the commercialization stage, when the product is rolled out to the marketplace and units are placed in the hands of consumers. This is the pivotal test towering in importance over all of the others because it demonstrates the final acceptance or rejection of the product by the consumer.

A. FINAL COMPETITIVE ASSESSMENT

To assist the marketing personnel in their activities, a final competitive analysis that takes into account all changes made to date needs to be created. This analysis shows how the product compares to competing products. It also serves as a basis point for the product evolution flowchart and for the timing of enhancements and improvements.

B. FINAL LITERATURE

If any material changes were made as a result of the industrialization phase and the field test program, the literature, in whatever medium, needs to be updated so that all documentation (both internal and external to the company) is consistent.

C. FORECAST FOR PRODUCT

Upon initial commercialization of the product, the marketing function in conjunction with the sales function and the input from the sales channel generate a usable forecast. This will be used to correct any accuracy problems with the previous forecast.

D. SALES CHANNEL INITIATION AND TRAINING

At the time of product introduction, it is critical to spread word of the new product as quickly and uniformly as possible. To that end, the sales channel needs to be trained and "turned on" to the product—oriented to the new product and focused on the sales of it.

E. CUSTOMER VISIT SCHEDULE

It is also a good idea to have the sales channel document its customer visit schedule as a deliverable. For use as a planning tool, this item also becomes a check on the ability of the sales channel to be effective in promoting the product.

F. NEW PRODUCT RELEASE

Finally, the deliverable of new product release is the formal release of the new product to the marketplace. This occurs for a fully tested product in stock at the right price and marketed through the most effective route to the user. In other words, no excuses—the product is ready in every sense of the program.

STAGE 5: Optimization

The final stage is the optimization stage, in which feedback and improvements are incorporated. The product has matured beyond initial release, and the traditional learning curve cost reductions occur due to this effort. As the marketing effort is scaled up, the cost is reduced, and the enhancements are incorporated; this is the stage at which gross profit is generally maximized for the corporation.

A. FIELD FEEDBACK PLAN

With the product having some longevity in the marketplace, it is necessary to compile and analyze field feedback. Such feedback will be used to generate the required enhancements in functionality and also to identify the areas in which cost reductions are necessary.

B. MANUFACTURING FEEDBACK

In the same manner, there needs to be manufacturing feedback to determine what product changes are necessary to facilitate improved or less costly manufacturing methods.

C. EXPANDED PRODUCTION PLANS

Depending on the ramp-up in sales volume, the use of additional manufacturing capacity—either through capital equipment or time shifting of manufacturing cycles—might need to be considered. In either case, the marketplace demand must warrant the ramp-up in capacity. This deliverable should be staged in advance of the market need but not so far so as to be underutilized.

D. VALUE MANAGEMENT

During the initial product planning stage, when the requirements specification was being prepared, there was a method used referred to as *value engineering*. Value engineering determines the value placed by the user on certain features, and the pricing/functionality matrix was generated according to it. Now that the product is on the market and the marketing and sales force has experience, the use of value management should be performed to recheck where the consumer places value in the product and what features are worth what portion of the price.

E. PRODUCT EVOLUTION ANALYSIS

The product experience will allow an informed product evolution flowchart to be updated. Given that the product exists and that the business will lead the company in some direction, the use of the product evolution flowchart is a useful deliverable in ascertaining the next form of the product.

F. COST REDUCTION

The value management exercise and the cumulative experience in the marketplace preferences and manufacturing capabilities should make possible a cost reduction in the product.

G. PRODUCT OBSOLESCENCE

As part of the product evolution flowchart, the product eventually must be retired. This deliverable will define the manner in which the retirement will take place. If it is a repairable product, how long will parts be on hand? If it is not repairable, what is to be offered in its place? These issues are part of the product obsolescence plan.

2. FIRST TIME THROUGH VERSUS EXISTING SYSTEM

There is an important perspective to be aware of in executing a new product development. There is a fundamental difference in human energy level required to implement a development within the scope of an established system of procedures versus starting from scratch. The job is significantly complicated if an organization must institute procedures and put them in place; it then becomes even more difficult to institute a system and make it prevail throughout the organization.

If an entire new system needs to be established, concurrent with the development, factor in more time to execute the program. If the organization resists the implementation, the chances of doing the program on time and getting procedures in place that will prevail are quite slim. Extra effort at the onset for the purpose of establishing authority will have far-reaching benefits for the program and the development group.

3. PUTTING PROCEDURES IN PLACE AND MAKING THEM STICK

Regardless of the circumstances or the expectations, the procedures and processes just discussed must be put in place in an effective manner. They must also be used as part of the program to cement them in place in the overall organization. Failure to do this will result in a program with a one-time success in spite of the lack of procedures and little future leveraging of talent, or at the worst, failure of the program and dismantling of the development group.

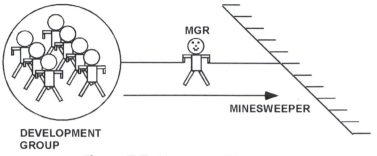

Figure 7-7. Manager as Minesweeper

4. THE MANAGER AS MINESWEEPER

With the team in place and functioning properly, the manager needs to act as a minesweeper for potential problems that may affect the program. The manager must lead the group through the potential conflicts and resolve the issues along the way. Figure 7-7 illustrates this point.

As shown in Figure 7-7, the manager needs to clear the pathway for the team by precipitating issues that can affect the group's performance adversely and then resolving them. Eventually, the group will become proficient at doing this as second nature.

TRACKING PERFORMANCE

There is an important distinction between tracking a project and freeing up personnel for optimum performance. The idea behind the tracking of performance on a project is to understand the progress and effect corrective action if delays occur. To be truly effective in a program, the manager must clear the pathway for maintaining steady and predictable progress to keep the program on schedule. Looking at it in a different perspective, the tracking of performance should confirm an on-time schedule.

Operationally, product tracking is like a control loop. The project plan is the command signal; the summing node is the management review; and the development team makes progress. A feedback reporting system (project tracking) summarizes progress for comparison to the plan at the summing mode or management review. Management intervention acts on the development group to effect changes in the progress. Ideally, as stated previously, proactive management should relegate the role of project tracking to that of confirming an on-schedule situation. This is illustrated in Figure 7-8.

1. THE PLAN DICTATES THE TYPE OF TRACKING

There are different types of plans and different means for tracking progress of the plan. The tracking mechanism must be commensurate with the plan type. If the plan calls for a

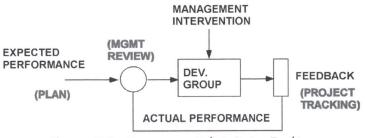

Figure 7-8. Management Role in Project Tracking

milestone of performance every six months, there is no need to track performance daily. If the plan calls for tasks to be completed in a two-week time frame, a one-week follow up is insufficient to be effective. With proper planning and proper follow-up frequency, the program can have a tracking system that will be effective without being overbearing.

2. ALL SYSTEMS WORK WITH DISCIPLINE

As stated previously, most systems will be effective as long as they have discipline. Discipline is the operative word. The type of discipline needed for development is not the "bull of the woods" type but rather the steady progress and follow-up to ensure results within an allotted time frame. Because it has been discussed that time lost cannot be made up, the unwavering need is to keep on schedule at almost any cost. Staying on schedule will occur through the steady disciplinary process of following up on action items, responsibilities, and completion dates.

3. WITH INTERDISCIPLINARY PROJECTS, USE PERT/CPM METHODS

One of the most important factors in program management is staying true to the logic of the project. The logic is the interconnection of activities and the order in which they are initiated and completed. In addition, the interdependency of the activities must be preserved to maintain integrity of the overall program. To that end, it is advisable to make use of the PERT critical path method of program control. This method shows the interdependencies as well as the longest or critical path through the program. With these two bases of information, the logic integrity will be preserved, and the critical path is the focus.

4. MANAGEMENT OF MULTIPLE PROGRAMS

This is all well and good for a single project run by a single manager of a single group, but what about the management of multiple programs? Herein lies the mysteries of effective management, as multiple projects have multiple priorities planned along multiple schedules. Coordinating these requires a level of management that is in a league of its own.

The projects must be broken down into their component parts, tasks, and subtasks, and talent must be assigned to the activities called for in the specific time period. Schedules cannot be missed under any circumstances, or the flow of progress will unravel like a sheet of music played out of timing.

Given the proper resources, such tracking ensures timely completion of all the programs. The key, however, is the issue of resources. Rarely are there enough resources to spread over the several projects called for in a senior manager's plan. Program management then becomes a juggling act to place resources with the tasks at hand, with eventual sloppiness about completing details. This process becomes self-defeating over the long term.

Senior management is interested in progress and completion. Be sure not to overload the development personnel or become a team so overloaded as not to be able to complete anything.

5. WINGING IT: WILDCATTING DEVELOPMENT

Not every company is interested in implementing a structured product development system. Some are interested in a more opportunistic approach, where the employees are directed to work on the "product opportunity of the day." Typically found in the smaller companies, senior management fails to reinforce a system for the generation of new products and fails to use discipline in executing a program in the wake of changing market conditions, lost orders, and competitive thrusts.

Unfortunately, this type of company exhibits a high cost of development engineering on its income statement, with little completed to show for the effort. There is also a significant chasm between the sales effort and the development effort. Materials are a high portion of manufacturing cost, as design-to-cost efforts are erratic. Rework is high because development is rarely completed on a product before it is released for production.

Despite all of the negatives, there is still a way for managing a development group immersed in this type of environment. The use of management by queues seems to work out well in such situations, although it offers no predictability about when a program will be completed. The queue is the list of projects assigned to an individual development resource. These projects are waiting for the resource to be applied to them. When the market condition changes, the priorities of the queues are reordered. The manager in such a scenario also must manage the lack of continuity caused by the changing conditions.

This environment requires a special type of product development professional. He or she must personally provide the continuity that is lacking in the system. Management by queues also requires the development personnel to have experience in the development arena and to have the initiative to stop and start and retain the pertinent data for a program.

6. FOLLOW-UP MEETINGS

To establish and maintain the continuity in tracking performance, the use of follow-up meetings is essential. The follow-up meeting yields the appropriate performance feedback, keeps all parties aware of project issues, and creates expectations and accountability among members of the group.

7. THE ROLE OF ACTION ITEMS

In Chapter 6 we discussed the use of action items in a program management venue. For effective program management, the action item is only one part of the equation; it is one link in the chain of hierarchical control. The other essential links are the assignment or responsibility of action items, and the completion date. As shown in Figure 7-9, the action item supports the subtask. The subtask supports the tasks, and the tasks support the stages of the project. The stages then support the project itself.

8. FOLLOW-UP SYSTEMS

There are many mechanized systems for following up on details. The important point to remember is that the details must be followed up on—each and every one of them, all of the time. If the manager is cavalier about details, the group eventually will become cavalier about the program. Using mechanized software-based systems can be helpful and can create automatic follow-up based on dates and times. These systems are only effective, however, when the manager obtains action *with* the follow up on a timely basis.

OBSTACLE REMOVAL

1. BECOME AT EASE AT PROBLEM SOLVING

Obstacles are a fact of new product development. They occur at every turn, and it is the group's and manager's responsibility to navigate these obstacles and bring the program to

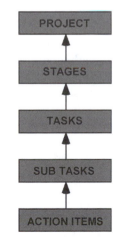

Figure 7-9. Action Items in Hierarchy

successful completion. Such navigation should become second nature to the group. Team members need to be at ease with problem solving and overcoming obstacles. Furthermore, they need to become at ease resolving these issues within a time frame.

2. DON'T JUST SOLVE THE PROBLEM FOR THE TEAM—TEACH MEMBERS HOW TO EVALUATE AND ACT

There is a saying that goes something like this: *Give a man food, and he will eat for a day. Teach a man to farm, and he will eat for a lifetime.* The same holds true for the management of development groups. Use the overcoming of obstacles as an object lesson to teach the group how to navigate for itself.

3. MULTIPLE PATHWAYS TO THE GOAL

The manager needs to develop a philosophy within the group that there are several ways to achieve the objective. Alternative means must always be available to get the program on track or to keep it there. The group needs to develop contingent strategies around these obstacles and exercise them without delay.

4. STRATEGIES FOR WORKING AROUND CONTINGENCIES

The Timing for coping with contingencies is crucial to minimizing lost time. All too often, the manager and group wait until they actually run into an obstacle before reacting to it. This approach wastes time and limits options. It is critical to recognize potential obstacles looming in the future and to take effective action earlier. Consider Figure 7-10.

Figure 7-10 shows two scenarios for engaging an obstacle. The first is the method generally used by default. The project rolls along until it encounters the obstacle. The management then scrambles to recover from the lost time and in the process uses precious time to determine a way out of the situation.

In the second scenario, management looks into the near future and tries to recognize potentials for obstacles or patterns that might contribute to them. Generally there are signs

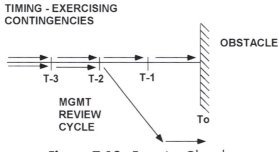

Figure 7-10. Engaging Obstacles

of impending obstacles that can be recognized well in advance of the actual impediment. The key is early recognition to and reaction to mitigate the damage of time lost. In the example shown in Figure 7-10, the recognition and the action take place at T-2, so that an agreed to approach is in place and workable at T-0.

If a problem is real and inevitable (or even likely), do not wait to exercise the option of working around the problem early.

KEY PLAYERS AND BACKUP

1. EVERY TEAM HAS KEY PLAYERS

Individual contributions can vary from one member to another. No one contributes exactly the same amount of value to the project each day, week, or month. In a coordinated sequence of scheduled events, however, the contributions of all are required to move the program forward. Furthermore, failure of one member, regardless of how little, can affect the program. Using the example of the racecar that loses the race because of an 89-cent part, the lowest-level contributor can take the program off schedule or affect the timeline.

The key contributors of a program can be recognized for their contributions but must be placed in perspective. No one player can be larger than the effort of the team or the program as a whole.

2. NOT GETTING HELD HOSTAGE

It is vital for a manager to prevent being held hostage in a program by the key players. Given their talents and critical importance to the program, it behooves the management to develop alternatives and strategies for dealing with these resources.

3. PRIMA DONNAS NOT ALLOWED HERE

Dealing with team members who have prima donna attitudes is a difficult and tedious management task. The management certainly can afford to absorb the effects of these attitudes when they are directed at them; however, the real danger is in the demoralizing effect these people have on the rest of the group. They have a tendency to bully their way around, discount the ideas of others, and align support along polarized positions, which are not necessarily in the best interests of the program.

Fundamentally, these people have to be dealt with in a firm and fair manner. If necessary, they might need to be removed from a program. Keep in mind that a single project does not make or break a company, generally. Because the main objective is to execute a program and to cultivate a functioning team, prima donnas have no long-term place in that team.

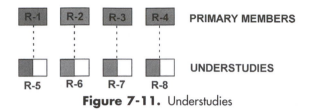

Figure 7-11. Understudies

4. DEVELOPING TACTICS TO COVER LOSS OF KEY PEOPLE

There are several tactics and strategies that can minimize the damage caused by a disruption in applied talent to a program. These include cross-training personnel to diffuse key knowledge among several players, using the key players as trainers of the balance of the group, and continuously evaluating personnel for upward mobility for training. Select "understudies" for each of the key members and ensure that they have at least half of the knowledge base of the person they are backing up, as illustrated in Figure 7-11.

The use of understudies will ensure some coverage in the event that a key member is lost. Such a strategy is also a good means for development planning of the team members.

Each member so trained is effective in reducing the dependence on the prima donnas and also sends a clear message to them that their behavior will not be tolerated for any period of time.

5. OVERCOMING THE CRISIS: TRADING GROUND FOR TIME

If for some reason you find yourself running a project with a high-risk key player, you might need to endure using that person until a backup plan is exercised. This means keeping this person on the team even though he or she might be disruptive and bad for the group. The manager needs to hold the group together until a replacement is found. Then the personnel action can be taken and the values of the group reset to the priorities of the project.

DEALING WITH SHIFTING LINKING OBJECTIVES

1. COMPANY CULTURE

The linking objectives are very important to a program as no project can be started and completed in the vacuum of other departments. Given this, senior management should not expect a group to operate without the timely and effective participation of the other departments of the company. To that end, it is a senior management issue and they are solely responsible for ensuring these objectives are linked and prioritized. If the culture of the organization does not allow for this, precipitate the clarification with management. If you are in a role to prevent it from happening, link the interdepartmental objectives. It will save a significant amount of effort later. Failure of addressing this will cause problems and these problems, generally, will not go away.

Some company cultures encourage the competition for interdepartmental resources amongst program managers. If this is the case in your company, decide if this is the situation you will put up with or if is intolerable. If you decide it is tolerable, then become proficient at it, in order to be effective in executing the program.

2. PROJECT TEAM FORM

To demonstrate the requirement for aligned and linking objectives, consider Figure 7-12. It shows several example projects to be executed within the company. The various departments need to contribute resources to complete the projects effectively. Project A is the highest-priority project, which should be on all departments' top priority list, not just those of Engineering and Development. In this example, the balance of the departments do not consider project A as a priority, and worse, manufacturing is focusing on Project D, the seemingly lowest-priority project.

At the right of the chart is a concurrence assessment. In this exaggerated example, there is no concurrence with any of the projects or departments, which indicates a severe senior management problem.

The business of running a company must be a balanced focus for the departments. The objectives require both execution of the normal business activities (as shown at the bottom

PROJECT PRIORITY		SALES AND MARKETING	ENGINEERING AND DEVELOPMENT	MANUFACTURING, MATERIAL PROCUREMENT, METHODS, MANUFACTURING SYSTEMS	ADMINISTRATION, FINANCIAL, ACCOUNTING AND ORDER ENTRY	CONCURRENCE OF LINKING OBJECTIVES?
HIGHEST PRIORITY	PROJECT A	B	A	D	C	NO
HIGH PRIORITY	PROJECT B	C	B	C	B	NO
LOW PRIORITY	PROJECT C	D	C	B	A	NO
LOWEST PRIORITY	PROJECT D	A	D	A	D	NO
NON NEW PRODUCT DEVELOPMENT VOCATIONAL ACTIVITIES		NORMAL SALES AND MARKETING ACTIVITIES WITH EXISTING ACCOUNTS AND PRODUCTS	PRODUCT MAINTENANCE AND ORDER PROCESSING	SCHEDULING, MANPOWER LOADING AND SHIPMENTS	TREASURY FUNCTIONS, CASH FLOWS, FINANCIAL STATEMENTS, BUSINESS ANALYSIS	

Figure 7-12. Linking Objectives

of the chart) and the project work associated with advancing the business through new product development.

3. COORDINATING ALL COMPANY PROGRAMS

The object lesson is that all company programs need to be organized and resources applied in a productive fashion in order for any of them to progress. A haphazard approach using freedom to shift resources might appear feasible on paper, but in reality this shifting causes loss of time on reassignment, start-up, and shutdown.

In program management, the one thing a company needs to do to maximize the development output for each dollar spent by establishing and nurturing momentum. Constant reassignment depletes any momentum that has been established.

4. ASKING FOR ADVICE FROM PEERS

In the absence of leadership, the program managers can get together to resolve the priority issues. The degree to which this can work is dependent on the each program manager's cooperation, maturity level, personal agenda, and commitment to the organization as a whole.

This is a worthwhile approach to investigate, as the managers can decide and present the agreed-to priority to management, asking management to either approve the priority or reorder it. Either way, senior management ultimately gets pulled into the process.

5. EXECUTING WITHOUT RESOURCES

There are occasions when a program must be executed in the wake of resource limitations and reordering of priorities and personnel shifts. Given the corporate hierarchical form, development managers are left with little choice but to consistently inform senior management of the resulting impacts. Often, an established pattern by managers is to bring the issue to brinksmanship in some meeting. The more effective way might be to create small brinksmanship episodes but more frequently, widening the scope of personnel involvement and knowledge. This tactic gets the word out to senior management in several different perspectives and also informs the rest of the organization of the impact of company decisions. Consider the approach as incremental brinksmanship: low-key but frequent emotionalism, with an ever widening audience.

6. IS SENIOR MANAGEMENT AWARE OF THE SITUATION?

If this is a severe issue, the development management needs to determine whether senior management is aware of the situation. If they are but do nothing to resolve it or at least to acknowledge it, the development team has a much greater problem. There must be a covenant between the management of development and senior management. If there is no corrective action, there is no covenant, and the team is potentially wasting its time. In these situations, when senior management fails to address these issues, the team must make a decision to continue under these circumstances or to precipitate a means for resolving it.

PROBLEM SOLVING

The issue of effective problem solving is one of the most important in new product development. The ability to employ effective problem-solving skills is a major requirement for the group. Fundamentally, little can be accomplished without this skill, and almost anything can be overcome with the skill. Consequently, this section focuses on the problem-solving process in a generic fashion so it can be applied in cases specific to your projects.

1. THE GOAL: FOCUS ON NEAR TERM AND LONG TERM SIMULTANEOUSLY

The development manager can be effective only if there is focus on both short-term and long-term goals of the company. The short-term goals are oriented toward the direct management of the tasks of the project. The long-term goals of the company are the various milestones and product introductions for the company. This requires perspective from different points of view and the ability to operate on various levels of details, as shown in Figure 7-13.

2. PROBLEM-SOLVING CATEGORIES

In applying the principles of problem solving, the first step is to determine what the present situation is. This is referred to as *issues review*. The objective of this review is to determine whether there is a problem to resolve and what type of problem it is. Most often, the problems encountered fall into one of three categories:

- Cause assessment
- Decision management
- Planning architecture

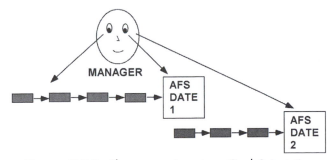

Figure 7-13. Short-verses Long-term Goal Orientation

ISSUES REVIEW

Let's examine the situation first. There are four basic areas of consideration involved in this review. These questions are sequential and are arranged to uncover what type of problem is being considered. The points for consideration are as follows:

1. EXAMINE THE PROCESS AND OBSERVE THE BROAD OR COMPLEX CONCERNS

List the concerns for examination.

2. SEPARATE THE CONCERNS INTO DISTINCT ISSUES

- How does the collection of concerns relate to each other and to the higher-level issue?
- Is there more than one issue involved?
- Will one action resolve or cause explain the issue?
- What evidence surrounds the issue(s)?
- How can you be more specific?

3. CATEGORIZE AND BREAK DOWN LARGER ISSUES INTO ROOT ISSUES

- What is the relative importance of each single issue?
- What is the impact/seriousness/urgency of the effects of each issue?
- How is each issue changing as time progresses?
- Based on impact, urgency, and trend, what is the hierarchical order of the concerns?

4. DETERMINE A STARTING POINT FOR ANALYSIS

- Is there a change from the standard or expected?
- What is the extent of the change?
- Under what conditions did it occur?
- Is the cause known?
- Is the cause relevant?

  If the answer to questions 4 and 5 is *yes*, then you need to determine the cause.
- Is there a choice to be made between two alternatives?

  If *yes*, a decision needs to be made. Use decision management perspectives.
- Is there a project or decision to be implemented?

  If *yes*, use the planning architecture.

By using this series of questions, an issue can be analyzed to determine what type of problem is occurring and how to proceed. It will be most helpful when the issues and data are written down and considered within a framework. Assuming that the initial issues review is complete, the following is a presentation of the three analyses that will be used.

CAUSE ASSESSMENT

The cause assessment is used when development managers need to identify root causes of problems to effect corrections and place the program back on track. This next series of questions can assist managers to determine the root cause systematically based on examination of facts and deductive reasoning.

There are several basic points for consideration in determining root cause of an observed problem. They are presented as follows:

1. SPECIFICALLY STATE THE ISSUES

- Specifically, what is it?
- Specifically, what is wrong?
- Specifically, what standard exists?
- How is that change from the expected?

2. DESCRIBE THE PROBLEM

List the examined facts:

- In what context is the issue observed?
- What exactly is wrong?
- State the location of the issue in the observed context.
- Where in the object/context does the issue occur?
- Time and date stamp examined phenomenon.
- When in the life cycle of the context was the first issue observed?
- State any correlations in the examined issue.
- Quantify how much of the context is defective.
- Quantify how many issues are occurring.
- What are the trends?

List the relational facts.

- On what similar circumstance might we expect to see this issue, but do not?
- What similar issues might we expect to see but do not?
- At what other locations might we expect to see the issue, but do not?
- Where else on the object might we expect to see the issue, but do not?
- When else could we have observed the phenomenon, but did not?
- When else in the object's/context's life cycle could the issue have occurred, but does not?

- Where might you expect correlations, but do not see them?
- How much of the object/context might you expect to be affected by the issue, but is not?
- With how many objects/contexts might we expect to have issues, but do not?
- What might the trend be, but is not?

3. IDENTIFY DIFFERENCES BETWEEN THE EXAMINED AND RELATIONAL OBSERVATIONS

What is different or unique about (Observation) compared to (Matching Relational fact)? Repeat for each set of observations.

4. LIST RELEVANT RECENT CHANGES

5. GENERATE LIKELY CAUSES

6. TEST LIKELY CAUSES

- If (state a likely cause) is the real cause, does it explain why it is (state an observed fact) and not (state the corresponding comparative fact)?
- How does this likely cause explain each set of observations and analysis?
- Can this likely cause be eliminated because it fails to explain the observations and analysis?

7. SELECT AND VERIFY THE CAUSE MOST LIKELY TO BE SUSPECT

How can we prove that this is the most likely cause?

DECISION MANAGEMENT

The next area of analysis is the decision management process. To make a decision is relatively easy in light of alternatives given or choices to be made. To make an informed decision that has basis and is supportable with options for correction and recovery is another matter. This section focuses on decision-making with the objective of managing the process for results. The decision-making process is broken down into several elements for consideration. They are presented as follows:

1. WHAT IS THE PURPOSE OF THE DECISION?

2. WHAT ARE THE CRITERIA FOR THE DECISION?

- What factors should be considered?
- What are the critical elements that should be included in the criteria?

3. CATEGORIZE THE CRITERIA

- Are there compromises to the criteria?
- Are the criteria clear and measurable?
- Will we be able or willing to accept the alternatives if these criteria are not met as stated?

4. GENERATE THE ALTERNATIVES

- What are the possible courses of action?
- What options are available or can be engineered?

5. COMPARE THE ALTERNATIVES

- How well does each alternative perform against the criteria?

6. IDENTIFY RISKS AND ASSESS RISKS

Engineer recovery from adverse risks.

7. DECIDE AND ACT

Taking in to account the alternatives and factoring the risk assessment into the analysis, select the best, most balanced alternative.

PLANNING ARCHITECTURE

The planning process is a structured method to analyze and plan. It consists of several major steps in the planning process. They are presented as follows:

1. State the deliverable of the plan.
   - What is the desired outcome. Who, What, Where?

2. Categorize and prioritize action items.
   - Identify prerequisites and dependent activities.

3. Identify critical pathways and elements and personnel.

4. Identify and list potential problems.
   - Consider probabilities.

5. Identify causes of critical potential problems.
   - What specific influences or events could cause each potential problem?

6. Develop preventative and alternate action(s).
   - Identify new alternate critical pathways based on circumstances.

7. Design contingent action(s).

8. Identify and list contingency trigger points.
   - How will we know when it is time to put the contingency into action?

As with any of these guidelines for consideration, they are simply guidelines for a thought process. They are not the magical means to solve all problems; however, they can assist analysis, and most important, they can assist in completeness of analysis—completeness in terms of thoroughly researching the issues and alternatives prior to drawing a conclusion. So often, companies rush to conclusions only to discover that further research and fact finding are required.

1. INNOVATION AS A PROBLEM-SOLVING TOOL

Most progress is gained by hard work, effort, and innovation. Innovation is a key component in problem solving. It also can be a means for generating unique approaches to marketing or capitalizing on an idea. Given a fixed resource budget, a fixed number of people, and a competitive arena, innovation might be the only mechanism for competing effectively. It also can be the means to level the playing field against an opponent with limitless resources.

There are several basic steps to the innovation process. They are presented as follows:

A. Develop the goal statement.

To (Fill in Specific goal).

Lead with an action verb, e.g., improve, enhance, reduce.

Structure statement in which the choice is open-ended.

Describe the end result desired in positive terms.

B. Develop the design attributes.

Develop single-issue attributes stated positively with the end result.

C. Select the key design attributes relating to the goal statement.

D. Expand the ideas from each key design attribute.

E. Select and combine ideas.

F. Evaluate ideas in relation to the goal statement.

G. Modify and develop solutions.

2. ROOT CAUSE ANALYSIS

The determination of the root cause of a problem or issue is critical to resolving it successfully. Often, companies immerse themselves in the details without determining the root cause. This tactic results in wasted time and effort. In addition, the participants end up trying to make a situation "fit" observed facts rather than "mining" the observed facts for the data and patterns of the root cause.

3. FISHBONE, PARETO, ET AL.

The classic fishbone chart is a means for determining cause by breaking down issues into component causes. It ensures completeness of an analysis and allows multiple pathways for multiple issues. The name *fishbone* derives from the appearance of the chart, in which each secondary issue or tertiary issue fans out from the higher-level issue that precedes it. Figure 7-14 illustrates a generic fishbone analysis. As shown, each cause is broken down into secondary or subcategory contributing causes. These secondary or subcategory contributing causes are then further broken down into tertiary causes, which contribute to the secondary causes.

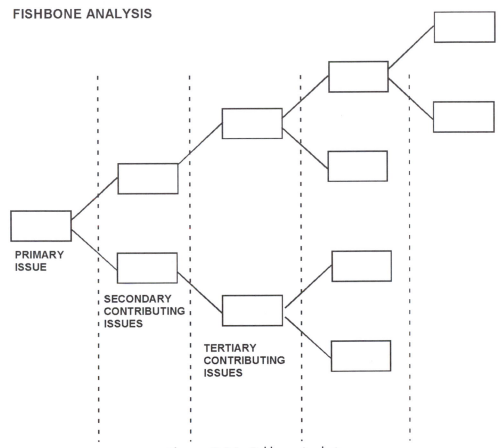

Figure 7-14. Fishbone Analysis

For the analysis to be valid, it is important to break down each possible issue into all of its contributing causes. Failure to miss one can mean a breakdown of the process.

The next area of consideration is the Pareto chart. The Pareto chart is a means for analyzing loose, disjointed occurrences or incidences or data and assigning them to like categories. These are then totaled by incidence type and numerically presented in a chart format. The highest number of incidences is presented, and the balance is presented in decreasing order. This ordering allows the loose disjointed data to be presented in an organized fashion for problem resolution. As shown in Figure 7-15, there are nine pieces of data assigned to four categories. Specifically, incidence reference numbers 3, 7, and 9 fit conditions in category 1. Incidence reference numbers 1 and 2 fit conditions for category number 2, and so on.

By charting the Pareto chart of the categories, we see that the category 1 incidences represent a more serious problem than the other categories, but not by much. There is much to be learned from the chart. There is the sheer number of incidences per category, the distribution of incidences among the categories, and the number of categories. In this way, a

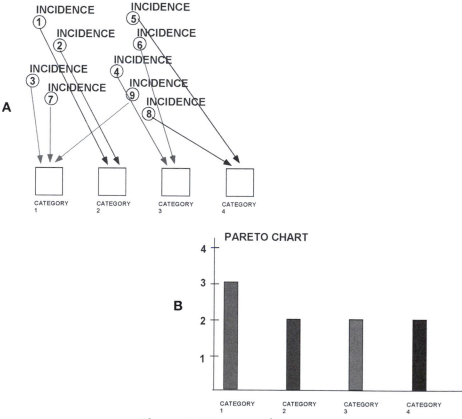

Figure 7-15. Pareto Chart

group can focus on the most severe problems first and also ensure that all problems are addressed in some way.

4. ATTACKING THE SPECIFIC OBSTACLE

There is one basic principle to guide you through problem solving—the use of overwhelming force to overcome obstacles. The disruption that an obstacle can cause is irrecoverable in many cases. Therefore, the objective is to minimize the time spent in overcoming it. Consequently, bring overwhelming force and persistence to bear on the obstacle.

5. PUSHING THE RIGHT AND WATCHING THE LEFT

There are times where some problems defy resolution. They simply are too complex and intertwined with other problems and therefore are difficult to understand. In these cases, it is a good idea to introduce causation of some known type and observe the reaction. The idea is to profile some type of transfer function or characteristic to understand the dynamics of the problem. By causing input and observing the reaction, it might be possible to minimize the problem to understandable components for quicker resolution.

6. SOLVING PEOPLE PROBLEMS

The issue of personnel problem resolution can be more involved than a static or dynamic problem. This is because people are human and have feelings, egos, and emotions. They are part of your team for several reasons, the sum of which might not even resemble the reasons you may have selected them. In spite of all this, the manager has a job to do and a program to deliver to senior management. The following represent several basic tenets in dealing with personnel issues:

A. SPEED: Do not delay; act decisively and swiftly, assuming the facts are known.

B. ACCURACY: Ensure that the facts used to make a decision are accurate; avoid guesswork.

C. QUALITY, FAIR PLAY: Administer discipline, advice, and appreciation in a fair, equitable, and consistent manner.

D. CHANGING PLAYERS: If players must be changed, determine actions and replacements, and act swiftly and decisively. If termination is required, act decisively with the individual and the group. Do not get persuaded into second chances if the best choice for the program is known.

E. DEALING WITH NONCONCURRANCE AND DISAGREEMENT: Encourage some disagreement and constructive confrontation. It is a healthy means for decision-making and alternative evaluation. Management can overrule certain objections where disagreement might occur or preferences might be voiced; however, if a nonconcurrence is voiced by a member of the group, take it seriously. This person is, in effect, "stopping the production line" because he or she feels so strongly about an issue. Personnel generally do not behave this way in a cavalier manner, so respect their nonconcurrence.

DECISION MANAGEMENT

1. DIFFERENCE BETWEEN MAKING A DECISION AND MANAGING A DECISION (DECISIVENESS)

Almost anyone can make a decision. This is not being decisive. Decisiveness is the ability to make informed choice and see it through to completion—to manage the decision. Decision management is easy if team members are goal oriented and diligent in their methods. If obstacles befall the consequences of the decision, be persistent and overcome the obstacles. If the situation is fraught with obstacles and cannot be made practical, cut the losses, reverse the decision, and move on with the project.

All too often, managers become so attached to their own decisions that they let pride stand in their way of working in the best interest of the program. In many cases, they can fail the program and themselves by not reconsidering the decision. One or more of these humorous statements might sound familiar: "My mind is made up, don't confuse me with facts," or another favorite: "Here is the conclusion, now go get me some facts to back it up!"

2. REVERSING A BAD DECISION

If reversing a decision is right for the program, do not hesitate to do it. It's that simple, that painful, and most important, that effective. Most people in management fear reversal of a decision; however, it will earn the respect of your superiors and subordinates. Your superiors will respect the fact that recognition was made and direction changed at your ego's expense for the good of the program, and subordinates will respect your leadership in discontinuing pouring resources (their talents and energy) into something that will not work out.

3. DECISION-MAKING IN UNCERTAINTY

The real test of decision-making is to be able to make and manage decisions in uncertain situations. This is somewhat different from the linear method, in which all or most facts are known. Traditional analysis methods allow the decision-maker to analyze, act, and implement. In uncertain situations, sometimes all that is available is intuition or an informed guess. Therefore, one needs to "feel" one's way in the decision management process. The differences between the traditional decision-making process and decision-making in uncertainty are shown in Figure 7-16.

As shown in Figure 7-16, the decision-making process differs in uncertainty, where information is sought actively as the decision is being made and managed. As the decision is made, the decision-maker monitors it closely at each step for confirmation or for early signs of failure or anomaly. Through such monitoring, the decision can be reevaluated and modified to suit the prevailing condition.

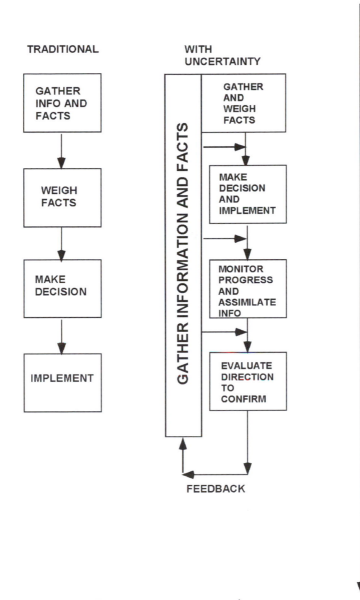

Figure 7-16. Decision-Making Process

4. REMEMBER: YOU ARE PAID FOR DECISION MANAGEMENT, NOT FOR DECISIONMAKING

"That's why you get paid the big bucks, to make the tough decisions." Have you ever heard that line? There is some basis of truth in the statement. Senior management does pay the development team and management to make and manage difficult decisions, to commit to a course of action and follow it through; it does not pay these individuals to make a decision and then walk away from it if it becomes inconvenient or difficult. This being the case, give senior management what they are paying for.

CONTINGENCIES

1. THINK AHEAD AND PLAN CONTINGENCIES

As the program plan is developed and implemented, there should be thought given to contingency planning. At critical points in the program, the team might want to think through contingency pathways. Figure 7-17 shows a contingency plan developed for a path on which a milestone cannot be reached. In this example, the pathway follows the line to the alternate pathway. These contingency paths ensure mitigation of the time damage to a program and allow for a more seamless transition to an alternate path. The team should plan these paths and crossover points in addition to the basic program flowchart. These paths can be thought of as bridges to an alternative flowchart.

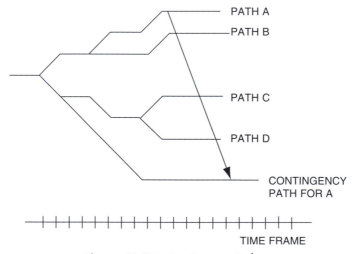

Figure 7-17. Contingency Pathways

2. EXAMINING DIRECTION AFTER A SETBACK

The true measure of a team's character and commitment is how it reacts to a setback. A setback in a program is one of the most difficult things to overcome. It generally follows a period when a tremendous amount of human energy has been expended in the attempt to overcome an obstacle. At this time, the team is already stretched and must endure the disappointment of the setback. This is the point, however, at which the maximum amount of energy and tenacity must be applied to overcome the problem.

Rather than redoubling the group's efforts, now would be an opportune time to re-examine the tasks and milestones currently in progress and refocus in the perspective of the overall program. This process is illustrated in Figure 7-18.

The "*field of view*" shown in Figure 7-18 is the manager's wide-range view of events and task requirements, involving the fact gathering that must take place prior to resetting the group's effort and getting the program back on track. The idea is to prevent another setback immediately following the first, which could demoralize the team.

3. RESETTING THE TEAM AFTER A SETBACK

If the program does suffer a setback, there needs to be a management effort to reset the group's effort. This is important, as the group's momentum has been compromised. Resetting the effort has two components: morale issues and vocational aspects. The morale issue must be resolved by the leadership of the manager. The vocational aspect results from the manager's sweep of the field of view, reinitiation of group activities, and driving progress.

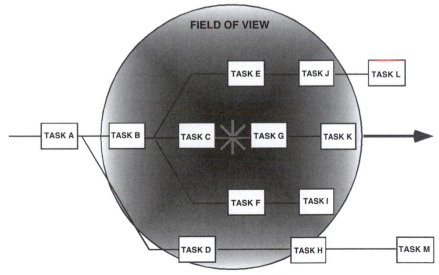

Figure 7-18. Setback

RECOVERY SKILLS

1. DON'T LET ADVERSITY SLOW THE PROGRAM DOWN

If it is a sure bet that adversity is present in an uncharted field of endeavor, then recovery skills are an imperative tool in the manager's portfolio. How fast a group can recover is an indication of how timely a manner the program will be completed. In this section, we focus on common-sense recovery moves to minimize the damage caused by a setback.

2. INCREASE PACE TO OFFSET LOSSES

To demonstrate the importance of keeping on schedule, the manager must increase the pace of the development effort to make up for lost time. At best, this tactic will keep the program from losing ground further. The increased pace sends a clear message to the team that delays will not be tolerated. In addition, team members will tend to operate much more diligently in the future to avoid further setbacks, knowing that the result will be an increased pace.

3. TEAM WILL OVERCOME BY ITSELF NATURALLY NEXT TIME

An important value to instill into the team is that the way to prevent setbacks is by operating in a manner that will avoid pitfalls. By extension, this means that the team will begin to look ahead for possible future problems and to plan contingencies—all of the skills we previously discussed. In this way, the group leverages the knowledge base of the manager.

4. "FAILURE IS NOT AN OPTION"—THE MESSAGE OF APOLLO 13

A line from the script of the film *Apollo 13* can serve as a motto for project development groups. At the darkest point in the effort to return the astronauts to earth, the comment was made that "failure is not an option." This slogan can be used to inspire a team to perform. Each individual is part of something larger than him- or herself. Team members are interdependent, and one member needs the performance of the others.

DOCUMENTATION

"I seem to remember something similar to that." How often have you heard that statement when someone is queried about substantiating data, test results, or justifications? The response indicates a less-than-adequate documentation system. If there is one element of the product development process that is crucial, it is an effective documentation system. This system must be accessible from the standpoints of both information going into the system and information retrieved from the system.

1. THE VALUE OF DOCUMENTATION

The value of documentation is tremendous from a program management standpoint. Documentation is required because human beings have memories that are finite and of questionable accuracy, with retrieval issues and interpretation issues. Therefore, documentation represents part of the communication process between parties involved in the development process. In addition to its library functions of storage, retrieval, and access, documentation serves to accommodate the changing personnel of a project and to orient newly assigned personnel.

The importance of documentation in the development environment cannot be overstated. This is not to say that every detail of the program's execution must be noted and categorized; such a practice would make sifting through the minutia for useful information quite difficult. Rather, the goals are to record and file pertinent information in a ways that makes retrieval logistically practical.

2. THE INCREASING REQUIREMENT FOR DOCUMENTATION

There is an increasing need for documentation in the business world today, thanks to three driving forces at work: more stringent standards and certification requirements of products, protection against litigation, and protection in a general atmosphere of mistrust. In addition, more and more collaborative programs are being done and the need for accurate and timely information is forcing the use of a documentation system to communicate between the parties of each company. To meet the demand for this increased scope and accuracy of documentation, therefore, the easiest approach is to maintain the documentation as the events are happening, the tests are being performed, and the program is being executed.

3. BASICS FOR ALL COMMUNICATION

A basic requirement of effective communication is that the "message," which is transmitted by the transmitter over some medium of communication, must be received and interpreted by the receiver with the same meaning with which it was transmitted. This is a rather difficult objective to accomplish in today's environment because of the different biases and perspectives of each transmitter and receiver. In addition, there is a steady stream of data inundating us all, but the true information is buried somewhere deep within it.

In practical terms, then, extra care needs to be taken to present and assimilate information as it relates to a development program. Data, test results, conclusions, recommendations, strategies, and approaches need to be interpreted so as to establish a clear understanding between the transmitter and the receiver.

4. THE TYPES OF DOCUMENTATION

In this section, the several types of documentation will be discussed, and a suggestion for documentation models is given.

A. INTEROFFICE COMMUNICATIONS

Interoffice communications between program team members are extremely important to document. They represent the professional discussion between colleagues that defines approaches and documents decision-making and problem resolution. There exists a host of mechanized methods for this type of communication and for its storage and retrieval. Almost any system can work. Each interoffice communication should include the following elements:

Title: A title is used for cataloging purposes. If the communication is in an electronic database, future sorts can be done by titles if needed.

Date stamping: The date stamp records when the communication took place. It can also be helpful in reconstructing events or decisions made in the past.

Time stamping: Similarly, the time stamping is helpful when a flurry of communication from several parties is occurring and research into the chronology of communications is needed.

Addressee: The name of the addressee is needed to identify the primary receiver of the transmittal.

Distribution: A distribution list identifies all the parties who have been given the information or data.

Technical data: With electronic communications, it is very easy to include technical data or reports in a file attached to the communication. When the data is referred to in the body of the communications text; it can be attached electronically.

Recommendations or conclusions: Most communications should have a conclusion or recommendation for best effectiveness. This element is missing in many communications, however.

Action plan, responsibility, and completion date: Interoffice communications should be purposeful, in that they should communicate by instruction, action plans, assignment of responsibility, and completion dates. All parties then know both their own duties and others' duties. These elements have an organizing effect and allow the interfacing required for members of the group.

B. EXTERNAL COMMUNICATIONs

External communications should include the same basic elements as interoffice communications; however, the distribution should be more carefully organized and controlled. Security also should be in place to prevent unwarranted ingress to the system and data corruption, and to keep unauthorized information from exiting the system to unauthorized parties external to the organization.

In an electronic implementation, terms such as "firewall" and "access" are relevant, and in nonelectronic systems, "security" and "need to know" are the operative terms.

Whether or not a communications system is electronic, security must be maintained for the protection of the company and the program. Whenever secure information leaves the company, reaching a vendor, an external resource, or (inadvertently) a competitor, your competitive advantage has been compromised. Take care to secure appropriate information effectively.

C. TEST DATA

As the product line is being developed, it is important to document the test data. These will serve well as reference data in the future if required for defense in litigation. All test data should be documented, including tests showing failures. Conclusions should be drawn and recommendations made, and retesting of the product showing that it passes where it previously had failed will chart the course of development for the record.

D. DESIGN CRITERIA

Design criteria documentation should show how the product is conceived in terms of design approach and implementation. The design criteria show the outer envelope of product scope and define the limits within which the design team will work. These criteria also show how the product matured from initial concept to final implementation and what the development group envisioned as the product operating scope.

E. S.P.A.F. THE OPERATING ENVELOPE

The S.P.A.F. is the Significant Performance and Operating Features document. This defines the scope and operating boundaries of the product as it has been designed and introduced into the marketplace. The actual claims made in the marketplace and in the specification might be somewhat more conservative than the S.P.A.F.; however, the existence of this document and its comparison to the claims and the operating boundary indicate the degree of conservatism in ratings, and the manufacturer's intended margin of safety and scope of use.

F. PROOF OF DUE DILIGENCE

The basic idea behind the documentation of test data and design decisions is to demonstrate proof of due diligence in the design and manufacturing of the product. In litigation, this might be the company's only instrument of defense. It is often remarked by development people that documentation "slows" the product development process. With contemporary advances in computer technology and office mechanization, however, proper documentation of a product's development has no material time impact on a program. Furthermore, there is much to be gained by having access to accurate information about the development process itself.

G. PROCEDURES

In the same manner, the procedures used in the development process should be documented. If the program is being executed with procedures being developed for the very first time, the use of format and mechanization will prove invaluable. As will be discussed later in this chapter, quality management system requirements dictated that procedures and results of actions be documented to ensure that the program and operations are conducted within accepted guidelines.

H. CORRECTIVE ACTION SYSTEM

As part of a total quality management system, any corrective actions need to be documented. For example: If the product were to fail a qualifying test and product changes were made, the document set should reflect these events and the design change. The documentation should also record field and customer experiences and resolution.

The corrective action documentation should have four basic components. They are as follows:

Initiation: This is the initial query or complaint that requires resolution.

Acknowledgment: The acknowledgment is the formal acceptance by the corporation of the query or complaint that must be addressed procedurally.

Action/implementation: This is the agreed to action that the corporation will implement to resolve the issue or address the query.

Close out: This represents the formal resolution of the issue with the time and date, along with the planned date of implementation.

PLANNING FOR PRODUCT LIABILITY

In this section, we examine the effect of product liability on the development process. Traditionally, there has been little connection between the design process and potential liability other than implementing good design practice. In actuality, however, there should be a direct connection—a means for corrective action and a documentation system in place as the product is developed.

The planning for product liability must be both proactive and defensive at the same time. It must be proactive from the perspective of attempting, through design, to minimize the chances of injury or property damage through use of the product. It must be defensive from the perspective of accumulating the documentation, indicating due diligence in the design and manufacturing process. Such documentation will be required in the unfortunate event of litigation.

1. THE PRESENT PRODUCT LIABILITY SYSTEM

The present product liability system in place in the U.S. is a system in change. Traditionally, product liability actions and results favored the plaintiff with use of strict liability laws. This has had a deleterious effect on manufacturers, the court dockets and the insurance companies. Large payouts have caused increasing costs of products and have had a dramatic effect on the design and development of new products.

For example, the U.S. Chamber of Commerce estimates that the cost of legal services and insurance premiums for fiscal defense of product liability is $100 billion annually. Part of the problem is that some of the products that have been in use for 20 to 40 years carry the liability forward to present time. This practice tends to encourage the proliferation of legal action and awards. In addition, the quest for reasonableness has given way to tactics and strategies for maximizing or minimizing the payout, depending on which side of the court one is sitting on.

Recently, however, legislation has been passed that reverses to some degree the historical pattern that has developed. The American Law Institute has issued Restatement of (Third) Torts, an analysis of current and future legislation that is driving the law away from a concept of strict liability to a negligence-based system. A negligence-based system differs from one based on strict liability by its requirement for reasonableness in the litigation. For example, when determining the effectiveness of a warning label on a piece of equipment, strict liability requires juries to determine whether a better warning label could possibly be produced (an easy assumption with the benefit of hindsight). The negligence-based analysis requires a decision based on whether the warning label supplied was reasonable in light of the equipment supplied, the intended use, and the intended user's level of knowledge related to the safe use of that type of equipment.

2. CORPORATE PROTECTION

Given the conditions in which manufacturers must conduct business operations, how can manufacturers protect the corporation? The answer lies in the methods and procedures used in the development process. The protection of a company is comprised of the two basic components, preventative and defensive.

The preventative actions take place long before an unfortunate accident and impending lawsuit. They occur through good design practices, a thorough understanding of the use of a product, the diligence of understanding the potential misuse of the product, and high-quality manufacturing processes.

The defensive posture and positioning take place after the accident and prior to or upon the initiation of the lawsuit. In such a case, records will be taken, fact finding takes place, and structuring of the corporate position occurs in a loss mitigation phase. The corporation is drawn into litigation either in court or out of court, and energy and funds are expended to fight off the suit.

There are three basic tenets in the prevention and defense of a product liability action, as follows:

A. Plan ahead

Adopt a realistic attitude toward product liability by clearly developing an understanding of the hazards associated with the products. Create and disseminate efficient and effective warning systems. Establish a solid information base, project hazards, make recommendations, and document the dissemination of safe use and handling of information.

In addition, seek out third-party assessment of the product design to double-check the company's design operations and provide unbiased opinion of design practice in the event of future litigation.

B. Demonstrate concern for safety in design operations

When conducting design operations, make user safety of paramount importance. In a potential litigation, the manufacturer's interest can be served by a demonstrable concern for safety. As issues surface, create a system for addressing them, resolving them, and documenting the agreed-to approach to market. The manufacturer must demonstrate a comprehensive understanding of the product's uses and limitations and take reasonable steps to advise and protect users. Juries tend to punish manufacturers for flagrant disregard for safety and also have a tendency to be understanding in cases where the manufacturer truly has made an effort to design safety into the product.

C. Exercise the duty to warn

The onus is on the manufacturer to continuously exercise the duty to warn users of potential harm through the use and/or misuse of the product. Appropriateness of warning labels and instruction will be evaluated by potential juries for reasonableness. If a lack of information makes the product unsafe, it is the manufacturer's requirement to warn. Design engineers should eliminate hazards where possible and specify collateral safety devices where appropriate. In cases where these measures are not effective enough, warnings and instructions are required.

3. PERSONAL PROTECTION

The steps to protect the corporation have been taken as part of the product development process, but what about individual participants in the process? Are they protected under the corporate form and under corporate policies? In theory, the individuals responsible for the design and control of the product are somewhat vulnerable to lawsuits; however, the corporation, not individuals, has the sufficient funds to satisfy personal injury lawsuits. Moreover, a group or team under the umbrella of company operations was responsible for the design and control of the product, and the company is a much more likely target of potential litigation.

There are insurance policies that protect both the organization and its employees from losses in the event of a lawsuit. These insurance products can be all-inclusive or only catastrophic in scope. The officers of the company remain liable; however, they can be covered under some type of professional liability insurance.

The issue of personal protection should fall within the company's scope of normal operations insurance. Individual team members should have some degree of protection under this policy.

4. TESTING FOR PRODUCT LIABILITY

Testing the product from the perspective of product liability defense is an important step in the actual defense against a potential lawsuit. This testing can be thought of as an extension of the test program for design verification and validation. Its objective is to uncover potential problems in the course of the normal engagement and use of the product. It also extends to the potential misuse of the product due to either lack of knowledge or flagrant misuse.

Figure 7-19 shows a pathway for the analysis of product misuse. The testing program starts with the design use evaluation. In this analysis, the scope of use for the product is defined and agreed to. This analysis is performed on paper before the product is developed, as the product is still being conceptualized.

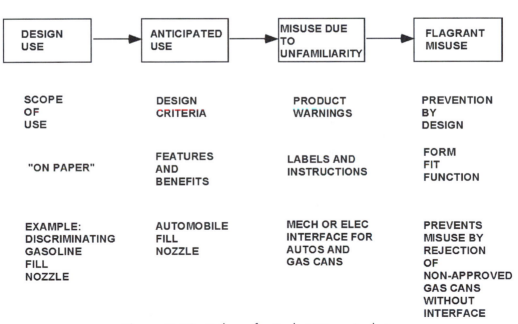

Figure 7-19. Pathway for Product Misuse Analysis

The second step in the evolution of the liability-testing model involves the anticipated use of the product. This step defines the practical and anticipated use boundaries of the product. The features, benefits and the specifications of the product define these boundaries, which are identified upon completion of product development.

Next, the testing moves toward an analysis based on unintended use and misuse of the product due to user unfamiliarity. This is a projection of common user misinterpretations of product use and tries to accommodate for them by warnings and instructions. At this stage, the company can also design criteria to prevent accidental misuse (i.e., parts not intended to fit together will not fit together; connections that are keyed with mating connections to prevent electrical shorts, etc.).

The final phase of testing is for the flagrant misuse of the product. The progression to each succeeding stage of testing shows an increasing degree of failure of user's rather than manufacturer's judgment. To the degree to which it is practical, the design of the product should try to accommodate potential misuse by prevention through form, fit, and function.

An example highlighting the various stages of product misuse analysis would be the design of a fictitious gasoline fill nozzle. The design use is conceptualized as a simple fill nozzle. It has specifications for safety and performance and can be used by individuals not normally trained to fill their automobile gas tanks. The safety issue in the example is that noncertified fuel cans are not to be filled with gasoline. An unintentional misuse of the product might be filling a noncertified can with fuel.

To prevent this from happening, a manufacturer might create a discriminating interface between the nozzle and the fuel can so that only a certified can could be used to store gasoline. This is the same interface used in the automobiles. Accidental or flagrant misuse is discouraged by this rejection scheme, as the valve in the nozzle must have "confirmation" of a certified interface before it allows flow of fuel.

5. PROJECT PHASE DOCUMENTATION

The need for project documentation does not have to be oppressive to the program. It should be a normal part of the filing and categorizing of issues related to the phases of the project. Issues should be resolved, closed out, and readily accessible. This task is made easier by the use of mechanization in the form of computers. For example, as issues are "discussed" via electronic means, they can be answered, with the decision communicated to all parties involved and filed in an appropriate manner. Retrieval of issues then becomes easier and communications are referenced to the program stage in which they occurred. Document as if you must defend your actions in court; some day you might need to.

6. LOSS MITIGATION

As the decisions concerning a product line are made and the compromises are discussed, a strategy for the defense of possible lawsuits should also be considered. In the event of an actual lawsuit, the organization can take steps to minimize the financial impact. The following steps can serve as recommendations to take in the event of a suit:

A. Notify key personnel.

B. Assign an internal liaison team to facilitate counsel's requests for information and assistance.

The defense occurs over three phases: investigation, discovery, and trial. At each phase and at each decisive point, the company must be well fortified with information and documentation to mitigate the damages. Unfortunately, failure to be prepared results in a loss that will invite similar claims.

REQUIREMENTS FOR TESTING AND QUALIFICATION

This section discusses the suite of tests required as part of the development process. This testing excludes the misuse testing referred to in the previous section, but elements might be similar as far as testing the envelope of operation. In cases in which the tests are identical, the results may be used for both analyses.

There are several tests that need to be done as the product is being developed to certify that the product meets both the original specification and the customers' needs. The suite of tests can be broken down into five major groupings: prototype testing, beta testing, pilot run testing, production run testing, and field return testing.

A. PROTOTYPE TESTING PROGRAM

The prototype testing program consists of the design verification testing. This testing compares the operational characteristics of the alpha prototype model (designed and "fabricated" by engineering) to the original design specification. In other words: Here is the list of items we want the product to do; does the initial prototype meet this functionality? It is a starting-point test to ensure that the prototype is in compliance with the design intention.

Also included in this test regimen are the initial safety and reliability tests. These can only be initial because the design is not finalized, nor is the form or fit.

B. BETA TESTING PROGRAM

The beta unit stage test program consists of the reconfirmation of design verification and functionality testing. The details of such testing can vary, depending on the degree of change from the form, fit, and functionality changes from the alpha unit.

Design validation testing also should be performed at this stage. The answer to the question: To what degree does this product meet the customer expectation? is required to proceed further. Beta testing also encompasses product testing at the customer's location in an actual application. It verifies design, requirements, manufacturing processes, and the effectiveness of literature and instruction. Extended life and reliability testing should also be included.

In this stage of testing, there also should be misuse testing. The misuse testing is needed for potential future liability defense and also to obtain feedback for improving the ease of use of the product.

The last segment in this suite of tests includes the standards and third-party certification testing for sale of the product. Third-party testing gives non-prejudicial assessment of the product and also agency approvals of governing bodies prior to sale. Confirming tests of the initial manufacturer-initiated tests, such as safety-related tests, might need to be rerun at this time under witness of the agency governing bodies. The third-party tests ensure industry validation and go a long way toward making a product liability defense.

C. PILOT RUN TESTING PROGRAM

The pilot run stage of testing consists of several tests. The pilot run is a test of the manufacturability of the product. Cost data and timed labor data will have to be verified.

The process capability of the manufacturer will be validated to ensure that there are sufficient process experience and control to manufacture the product. At this time, the database used for managing the product will be defined and put in place. Degrees of variability for the product and the processes will also be established.

D. FULL PRODUCTION-RUN TESTING

Suites of tests are required for manufacture of the product also. This suite of production tests must verify that the units are built according to the design and must test all critical components comprising the product. As experience has dictated, determining how many components represents "all components," without creating too much non–value-added production testing, is an acquired skill.

E. FIELD RETURN TESTING

This suite of tests is important for the long-term viability of the product line. When a unit is returned from the field, it contains a wealth of information about the application of the product in an actual environment. These units should be "mined" for potential weaknesses in the product. In addition, the incidences of failure should be investigated to determine root cause of the problem. With accurate root cause information, trends can be established and corrective action can be taken to reduce the likelihood of further field failures. It is at this point that the database is loaded with fresh field information.

The field returns should also be analyzed for any like issues that might arise. Often, nature uncovers issues not even considered in the laboratory or "protected" test environment. The analysis of the field returns can also be helpful in determining misuse or misapplication, and this data can then be used to alter marketing plans.

QUALITY MANAGEMENT SYSTEMS

1. WHY ISO AS A MODEL FOR QUALITY MANAGEMENT SYSTEM?

The International Standards Organization represents a homogeneous set of standards for ensuring that an organization produces quality products and services. Although this standard does not represent equivalency of every agency, approval, or quality requirement, a study of the standard provides a solid basis for constructing a quality system.

The ISO standard provides for several levels of certification depending on the type of business. We will consider version **(ISO 9001)**, which encompasses sales, engineering, development, manufacturing, and aftermarket support and service. Our consideration of ISO is not from the standpoint of certification, but rather as a framework for a quality management system. Recent update have added specifics. However, this is presented to outline the "spirit" of the system.

There are several distinct components to consider in the ISO certification. They span the transaction from initial customer engagement and order negotiation through aftermarket service.

The following represents a basic framework of the ISO standards components. They are presented in the operational order in which a contract would be executed in the company. They are as follows:

1. Quality system requirements

2. Management responsibility

   a. Quality policy

   The quality policy clearly states the organization's goals and the needs and expectations of its customers. It is to be published throughout the organization. There should be demonstrable management and employee commitment to this policy.

   b. Organization

   This is a chart showing how the company is organized; it identifies the departments and the key managers.

   c. Responsibility and authority

   Key managers with executive authority to make decisions and control all elements of the quality policy should be identified. Setting policy, taking corrective action, setting objectives, and monitoring achievement are all part of this activity.

   d. Resources

   The manufacturer needs to provide adequate resources to implement the quality policy, satisfy customer expectations within the framework of the policy, and achieve its objectives.

e. Management representative

The company needs to provide a management representative with delegated authority for arranging, coordinating, and overseeing the quality system in progress.

f. Management review

The company also needs to provide for a periodic review of effectiveness of the quality system. It must ensure adequate staffing, conformity to standards, compliance with quality policy, and feedback and resolution both internal and external to the organization.

3. Quality system

a. General

The quality system is the dynamic mechanism that implements the quality policy. Often embodied in a quality manual, the quality system is free functioning and operations oriented.

b. Quality system procedures

These procedures are required for the applicable requirements of the standard and should be consistent with company policy. The procedure should specify, who, what, when, where, and how the activities should be executed.

c. Quality planning

The manufacturer must demonstrate how the quality system elements are implemented and maintained.

4. Contract review

a. General

Contract review is the primary interface with customers. It should include what the customer requirements are and how they will be reviewed and communicated within the organization. This occurs prior to accepting an order.

b. Review

This process includes review of the requirement itself, agreement with the organization, and resolution of differences.

c. Amendment to contract

When changes occur in the contract, there should be subsequent review of the contract to confirm reasonableness and agreement. Acknowledgments both internal to the affected parties in the organization and external to the customer must be part of this process also.

d. Records

As with all activities where a potential for misunderstanding exists, there should be recordkeeping.

5. Design control

a. General

The design control is essential to ensure all of the quality aspects, such as safety, performance, and reliability of the product. This encompasses all phases of design and design process.

b. Design and development planning

The manufacturer should have established procedures, which include scope and objectives, work schedules, timing and frequency of design verification and validation, safety, reliability and performance, methods of measurement, test and acceptance criteria, and assignment of responsibilities.

c. Organizational and technical interfaces

The organization should define, document, coordinate, and control all aspects and interrelationships and interfaces within the company as they relate to development, manufacturing, and aftermarket activities.

d. Design input

The design input is the translation of customer requirements to a definition of the performance, function, environmental, and safety and regulatory agency requirements. Typically in the form of a product requirements specification, the design input is the formal starting point for development.

e. Design output

Design outputs are the requirements for purchasing, production, installation, inspection, testing, and service. A release procedure should be in place to control the design output to the balance of the organization. Design output needs to be documented to verify and validate against design input and customer requirements.

f. Design review

The design review is a formal peer review of the development results. Degree of competence of the participants should be commensurate with the technology and process under development. A review must be objective and detailed to be of value.

g. Design verification

Design verification is necessary to ensure that the design output matches the design input. Test data and demonstrations are the principal and objective means to effect the verification. Any verification should be conducted in accordance with relevant standards, practices, and predetermined acceptance criteria.

h. Design validation

The design validation checks the completed effort of the development process and compares it with the customer's original requirements. The results of all of these tests should be part of the design records.

i. Design changes

Changes to the product are required under a formal design change system. The following can prompt these changes:

1. Omissions or errors during design

2. Manufacturing, installation, or servicing difficulties discovered after design

3. Customer- or contractor-requested changes

4. Functional or performance improvements

5. Changes due to safety or regulatory requirements

6. Change due to design review, verification, and validation

7. Corrective or preventative actions

6. Document and data control

a. General

Document and data control consists of the information pertinent to design, procurement, process, quality standards, inspection of materials, and the quality system itself. It should reflect the present documentation, present data, and the historical pathway to these.

b. Document and data approval and issue

The manufacturer should have clear and precise control of procedures, responsibility for approval, distribution, and administration of internal and external data and documentation. This includes removal of obsolete information to prevent misuse or misunderstanding.

c. Document and data changes

The same level of change control that applies to the product must also apply to documentation. Procedures should be established and implemented for controlling all changes.

7. Purchasing

   a. General

   The control of procurement must ensure that purchased products that become part of or that affect the quality of the products, as well as the statutory or regulatory requirements, are adequate. Standards cover evaluation and selection, procurement specifications, performance and verification, and receipt of inspection procedures.

   b. Evaluation of subcontractors

   The manufacturer should have a means for evaluating the capability of subcontractors and ensure conformity of procured product.

   c. Purchasing data

   The manufacturer should have objective data to measure the subcontractor's product fabrication and shipment performance. This is to be kept up to date as with other documentation.

   d. Verification of purchased product

   When specified, the manufacturer may engage in supplier verification at the subcontractor's premises. In addition, there may be customer verification of subcontracted product.

8. Control of customer-supplied product

   Depending on how the manufacturer and end customer constructs the contract, the customer may supply product that will be incorporated into his or her own end product. This product supplied to the manufacturer must have controls similar to those required for any other supplier.

9. Product identification and trace-ability

   Where required, the manufacturer needs to provide a means for product identification. Similar-looking parts with differing functions might need to be marked differently. Trace-ability is required when parts must be traced back to their origin or source in the event of nonconformity and is often used to determine the affected batch.

10. Process control

   The bases of process control are to invoke procedures and methods to ensure conformity in the completed product rather than to inspect for nonconformity. Process control should include procedural control, maintenance, and essential material control.

11. Inspection and testing

   a. General

   The issue of inspection spans the organization from receiving through delivery and service. All aspects of this verification system must be under control, as the material formulates the product.

   b. Receiving inspection and testing

   Inspection and testing procedures in the receiving department ensures conformance of supplied product. It also is the first point for verification of a conforming product. Failure of this verification could result in a nonconforming product.

   c. In-process inspection and testing

   In a similar manner, the in-process inspection allows early recognition of nonconformities and the disposition of them.

   d. Final inspection and testing

   The final inspection involves examination, inspection, measurement, or testing to ensure conformance prior to shipment of product.

   e. Inspection and test records

   As always, records of activities involving the product's manufacturer and control points should be maintained for future use in solving or mitigating problems.

12. Control of inspection, measuring, and test equipment

   a. General

   Simply stated, the effectiveness of the product configuration and conformance to specification rely on the accuracy of inspection, measurement, and test equipment. Although this aspect of internal controls is often overlooked, it is mandatory for any type of certification.

   b. Control procedure

   Consistency is also required in maintaining control of the equipment.

13. Inspection and test status

   The manufacturer should have means for identifying the status of tests and inspections performed on the product as it progresses through the manufacturing system. It should denote conformance, acceptance, rejection, or a hold, pending disposition.

14. Control of nonconforming product

    a. General

    The purpose of this standard is to prevent accidental misuse of a nonconforming product. It applies to nonconforming product in the manufacturer's facility as well as to nonconforming product received from a supplier.

    b. Review and disposition of nonconforming product

    There should be a handling process and designated disposition for the nonconforming product. Repaired or reworked equipment should be reinspected according to the quality plan.

15. Corrective and preventative action

    a. General

    Corrective and preventative action represents the most powerful aspect of the entire control process, as it provides for improvement of operations. It covers assessment and action to change the process from its present form to an improved form.

    b. Corrective action

    Procedures should be in place to determine corrective actions and how to execute them. Initiation and conclusion are hallmarks of an effective system.

    c. Preventative action

    Preventative action is the proactive approach to problem solving. It takes into account the weaknesses of a system and takes steps to prevent incident(s) from occurring.

16. Handling, storage, packaging, preservation, and delivery

    a. General

    The organization must have an established set of procedures for the movement of material from receiving to product shipment. This procedure should address the uniform handling of material as it relates to the following:

    - Handling
    - Storage
    - Packaging
    - Preservation
    - Delivery

17. Control of quality records

    The manufacturer's records should show evidence of the quality system elements and components that fall under jurisdiction of ISO requirements. They should be identified, prepared, and controlled by authorized personnel, and steps should be taken to prevent unauthorized tampering of records. Some examples of these types of records are management review records, contract review records, inspection and test records, and internal quality audit records.

18. Internal quality audits

    To ensure that an organization has a vital effective quality system, it must conduct introspection on operations periodically. This means self-audits on the elements of the organization covered by the ISO system. Follow-up activities on audits ensures results that show an effective system.

19. Training

    The needs of the marketplace dictate not only change but also requirements for training of personnel. The organization must take steps to ensure adequate training and documentation and ensure that these pervade all levels of the organization. To achieve the quality objective, each and every employee should be trained in procedures and processes that affect them and that they in turn affect.

20. Servicing

    If servicing is required to maintain the products sold in normal operation and use, then the servicing organization and activities fall under ISO control. This control extends beyond repair procedures. It involves identifying responsibilities; tools, processes, equipment, documentation scope, and technical backup.

21. Statistical techniques

    a. Identification of need

       Statistical methods can be beneficial to the manufacturer to demonstrate conformance, variation, process control, and the materiality of each variance. The manufacturer should select statistical techniques appropriate to the item desired under control.

    b. Procedure

       To ensure uniformity of methods, the procedures for the treatment of statistical data should be under document control. Conducted properly, the data obtained and analyzed under these methods can be effective in demonstrating conformity.

2. IMPLEMENTING A QUALITY MANAGEMENT SYSTEM

The implementation of a quality management system should happen in stages. Installing all of the recommended procedures in an enterprise at once is simply too overwhelming. In addition, an oppressive level of procedures will much more likely be questioned for validity if not implemented gradually. In a similar vein, the implementation of such a system is more difficult if it is happening concurrently with a development project. It could take the run of several projects to implement an entire system.

If a phased approach is used, it is important to implement a limited portion of every clause and add detail in subsequent phases, rather than implement one clause at a time in its entirety. It is easier to see how the clauses interrelate this way.

3. HOW NEW PRODUCT DEVELOPMENT IS AFFECTED

As can be seen with the design control element of the quality management system, development activities are greatly affected. Each step of the development program must be executed and documented according to procedure. This process is often considered oppressive but actually will assist in the long run.

The design control process matures the development activities from a " let's see whether it can be developed" proof of concept, to development being an integral part of the customer-manufacturer interface and contract.

4. QUALITY MANAGEMENT SYSTEM AS A STRUCTURED MEANS OF ESTABLISHING OWNERSHIP

The quality management system can also be thought of as a means for establishing ownership of each element of the development program in the team members. The ownership value is diffused throughout the organization among personnel who are not necessarily identified as participating in the project. This is an important point, as the quality management system can establish a unifying function across several programs and departments.

5. IT'S INTEGRAL, NOT AN ADD-ON

The final thought on the quality management system is that it is integral to the organization and the project, not an add-on. It must be part of the core development function and be a contributing factor in all elements and all phases of the project.

INTELLECTUAL PROPERTY PROTECTION

1. BACKGROUND OF PATENT PROTECTION

The issue of patent protection is important to the overall success of the development program. There are several advantages to obtaining a patent for the company. Generally, if a product is not patented, it is, as a practical matter, available for copying by the competition.

In addition, any agreement or venture entered into by two or more companies is fertile ground for disagreement over origin and split of profits. The use of assigned patents clears up any potential misunderstanding.

The intellectual property field encompasses more than patents, as we will discuss shortly. It includes trademarks, copyrights, and trade secrets.

The information presented in the following pages is designed to serve as background information only, not as legal advice. Readers are encouraged to seek counsel for a considered opinion and direction for their own circumstances.

A. HISTORICAL PERSPECTIVES

Historically, the speed of information transfer has been slower than in recent times. Increased connectivity among product developers, manufacturers, suppliers and employees has accelerated the pace of information sharing.

Because of the accelerated activity of product development and the rapid turnover of technology, the importance of obtaining patents is even more critical today than in the past. More firms are developing technologies and applying them in products, therefore requiring protection. Their competitors are redeveloping improvements and patenting these improvements, thereby generating additional intellectual property activity.

B. IMPORTANCE OF PATENT PROTECTION

The patent protects a company's investment in research and development. It protects the invention by granting the applicant certain rights, thereby slowing or stopping the progress of competitors toward those same rights.

A patent demonstrates expertise in the technological aspect of the invention. It has commercial advantage with federal government agencies, when contracts are negotiated and let. Finally, a patent is a demonstrable asset, which can aid in securing capital financing from investors and commercial institutions. With all of these advantages, it makes good business sense to secure developers' rights by obtaining a patent.

C. IMPACT ON THE DEVELOPMENT PROGRAM

The intellectual property issue is affecting product development in an ever increasing manner. Trade secret information and technological know-how, applications, and processes are being developed for competitive advantage and are being protected. Planning for the protection of certain aspects of new products must begin at the conceptual stage, and security is ever more difficult to maintain. The level of activity of intellectual property development and protection is growing, with the result that it is becoming a vital component of new product development.

2. THE BASIC PROTECTION MECHANISMS

A. PATENT PROTECTION

The patent is basically an agreement between the inventor and the government. The inventor "teaches" the public how to use the invention in enough detail to enable a man to make and use the invention without undue experimentation. The patent is designed for the sole purpose of encouraging the development and use of new technologies and methods.

In return for this national benefit, the government then grants the inventor the sole rights to make, use, or sell the invention for a period of time. Historically, the patent grant extends 17 years, but it recently has been amended to 20 years from the date of application.

There are two basic types of patents: the utility patent and the design patent. The utility patent protects inventions that are functional and have utilitarian value. The design patent protects ornamental designs.

The utility patent is available for inventions such as new and useful processes, machines, manufacturing methods, or composition of matter. Recently, software that is also nonobvious or anticipatory can be patented. The following fall out of the scope of patentability: mere printed matter, methods of doing business, things that are unaltered from their natural state, or abstract scientific principles.

The execution of a patent is adversarial, in that there are statutory bars to obtaining a patent, and the applicant must document timing, novelty, and potential usage and benefit. These statutory bars include items as follows:

- The applicant is not entitled to a patent on an invention that is publicly known before the inventor conceives it. It sounds like an obvious fact, but it does come into play in cases in which timing and contention might be problematic.
- Patents that have previously been issued are demonstrable proof of public knowledge.
- The applicant for a patent can lose rights by premature disclosure, sale, or use in public view.
- Another way to lose rights to a patent is through abandonment. The idea behind the patent system is to move ideas for the betterment of mankind into the public use. If someone "sits" on an invention for some period of time and shows no demonstrable intent to use it, he or she might effectively lose rights to it.
- Simultaneity of patent applications both domestically and in concert with foreign application might prevent the inventor from certain rights.

The inventors must submit the application under their own names and must be the ones to have first "made" the invention in the United States. The invention must also be nonobvious to a person of ordinary skill in the art. This restriction is designed to ensure actual novelty of the invention. The merit of the patent application is judged based on the following factors:

- The scope and content of prior patents and publications
- The level of ordinary skill in the art
- The differences between the invention as claimed and the prior art
- Whether the invention provides unexpected results, fulfills a long-felt need, and/or is commercially significant

B. COPYRIGHT PROTECTION

Copyright protection is primarily intended for protecting an author's works. These can include booklets, brochures, artistic designs, maps, and architectural blueprints. In the argument of form versus substance, a copyright is oriented toward protecting the form of ideas expressed, not necessarily the ideas themselves. Publication is not necessarily required for copyright protection, but it is still an important concept.

The term of a copyright is generally considered the author's life, plus an additional 50 years beyond the author's death. If the work is a work for hire, the term is generally considered 75 years from date of publication or 100 years from date of creation, whichever is shorter. A copyright is obtained by filling out the appropriate application and submitting it with a fee for registration.

C. TRADEMARK PROTECTION

A trademark is product protection oriented toward the source or origin of a product. It is designed to distinguish goods and services of one company over another. The trademark is the vehicle that is used to capitalize the goodwill and reputation of a company through its products. Therefore, it is illegal for one company to capitalize on another by use of its trademark. Prior to considering a trademark, the company must perform a trademark search through the Patent and Trademark Office.

The protection of the trademark is directly related to its distinctiveness and non-generic nature as it relates to the product. For example, *Skater*™ would not be considered a trademark for roller skates necessarily because the term *Skater*™ does not uniquely identify the origin. At issue is the relationship between the mark and the product/market segment it is serving. The term *Skater*™ might be quite acceptable for an unrelated market.

There are four basic types of marks with varying degrees of protection, as follows:

1. Generic mark: This is a trademark that uses a generic term related to the product family. An example of this is the earlier example of *Skater*™ applied to roller skates. It simply is not unique and therefore has little value.

2. Descriptive mark: This is a mark that uses a description of the technology or part of the product to try to identify its origin. In the previous example, *Ball Bearing*™ would be the trademark—not very unique as to identity.

3. Suggestive mark: This mark requires the imagination of the consumer to conclude the nature of the product from perception and thought. An example of this would be *Lightening*™ brand roller skates. This mark invokes thoughts of extremely fast roller skates.

4. Fanciful mark: This mark is the most valued, as it is a word created expressly for the product and the trademark. Another option is the use of an arbitrary word for the product trademark. An example of a fanciful mark would be *Whoosh!*™. An example of an arbitrary word applied could be *Warp 10*™.

The basic requirement for trademarking a product is its use. Barring any other previous use in the industry or product segment, the commercial use of the trademark is generally sufficient. Trademarks are placed in use by registering them with the Patent and Trademark Office. The term of the trademark is potentially infinite; however, the registration must be renewed periodically.

D. TRADE SECRET PROTECTION

This type of protection is generally considered for proprietary information not in the public domain. A trade secret is technology or methodology or process which, by being kept secret from the competition, prevent competitors from being able to copy it, thereby inuring benefit to the owner(s) of the trade secret.

A trade secret is a fragile entity, as its value lies in secrecy. Once a trade secret becomes known, the protection is lost. This applies no matter how the secret is disclosed.

The primary disadvantage of this type of protection is that it offers no protection against someone else developing the same technology independently. Moreover, the second party could seek patent protection on improvements, thereby foreclosing on further development and use of the base technology. Certain technologies embodied in a product that can be reverse engineered are not suitable for trade secret status, as copying is relatively straightforward.

A trade secret is maintained by secure disclosure, limited access, and clear definitions of its proprietary nature on all detail drawings, printouts, flowcharts, schematics, assembly drawings, technical data, and test results. Any disclosure requires confidentiality agreements duly executed prior to disclosure.

Trade secrets can be exploited through explicit licensing agreements; however, an agreement will govern only to the extent that the second party honors it. If the party to which the trade secret has been disclosed does not honor the agreement, the trade secret is essentially lost.

E. COMPARISON OF PROTECTION METHODS

Figure 7-20 is a chart summarizing the different methods of intellectual property protection.

3. INTERNATIONAL PERSPECTIVES OF PATENTS

The issue of patent protection does not end at the boundaries of the United States. The entire world is developing and introducing technology to the international marketplace. Consequently, there needs to be a broader perspective in patent protection both domestically and internationally.

For example, there is a protocol for protection of a patent filed in the U.S. and a certain amount of time required to file in other countries. If a patent is granted in the U.S. but is not filed and executed in foreign countries, there could be a possibility that the technology is embodied overseas in a product, imported into the U.S., and sold here in competition with

COMPARISON OF PROTECTION METHODS

PROTECTION MECHANISM	LONGEVITY OF PROTECTION	DEGREE OF PRODUCTION	ADJUNCTIVE FORMS OF PROTECTION
UTILITY PATENT	20 YEARS FROM DATE OF FILING	PROTECTION OF THE INVENTION	TRADEMARK, DESIGN PATENT, COPYRIGHT
DESIGN PATENT	14 YEARS FROM DATE OF ISSUE	PROTECTS ORNAMENTAL FEATURES ONLY	TRADEMARK, UTILITY PATENT, COPYRIGHT
TRADE SECRET	POTENTIALLY INFINITE	NO PROTECTION AGAINST INDEPENDENT DEVELOPMENT	TRADEMARK
TRADEMARK	POTENTIALLY INFINITE	PROTECTS AGAINST OTHERS TRADING ON TRADEMARK OWNER'S REPUTATION	TRADE SECRET, DESIGN PATENT, UTILITY PATENT, COPYRIGHT
COPYRIGHT	LIFE OF LAST SURVIVING AUTHOR OR IN THE CASE OF WORK FOR HIRE: SHORTER OF 75 YEARS FROM PUBLICATION OR 100 YEARS FROM CREATION	PROTECTS ONLY THE FORM OF EXPRESSION, NOT CONTENT	TRADEMARK, DESIGN PATENT, COPYRIGHT, UTILITY PATENT

Figure 7-20. Summary of Intellectual Property Protection

your patented product. There is a strategy for patent applications and use domestically and internationally designed to provide maximum benefit to the inventor. Consequently, protection of an invention should be arranged in light of various international issues.

The U.S. philosophy on patents is that the first inventor (as defined by U.S. law) is entitled to obtain a U.S. patent on the invention. Subsequent inventors may patent improvements but not the basic invention. In other countries, the criterion for holding a patent is the first party to file, not necessarily the first to invent. This approach is in conflict with the U.S. system. The first party to invent is proved by the existence of lab notebooks, records, and related chronology and witnesses' documentation corroborating the testimony of the inventor.

4. PROCEDURES FOR EXECUTING A PATENT

There is a protocol for obtaining a patent in the U.S. It is procedurally legal and volleys between the patent attorney and the patent examiner. The basic deliverables for the application of a U.S. patent are as follows:

A. Title of the invention

The title is the formal statement of what the product basically is. It is a generic description of the product so that it is recognizable and comparable to other products. The title should be as brief as possible.

B. Cross references to related applications

This is a listing of U.S. applications that are prior to the one in question. They are for reference only and are selected as those related to the current application. This listing is also used in the definition of scope and further defines specifics of the invention as related to other applications.

C. Background of the invention

The background is a brief description of the field of art of the invention. It is a definition and description of the problems solved by the invention and references applicable prior art.

D. Summary of invention

The summary describes the essential elements of the invention that are particular to the claims of the applicant.

E. Brief description of drawings

A patent application generally includes drawings used to describe the invention. This section of the application describes the details, drawings, and figures of the invention.

F. Detailed description of the preferred embodiment

This is a summary showing the public how to use the invention. Every element of the invention is described, along with the drawing referencing it. Numerically ordered and referenced, the explanation is complete enough so the average person "skilled in the art" can understand the application of the invention. A number of embodiments of the invention may be presented here; however, the best one must be included at filing.

G. The claims

The claims are a definition of the specific rights inured to the inventor by the patent. They define the boundaries of intellectual property. The claims are the value of the patent in that they define the scope of protection. Drafting them properly is a job for an experienced intellectual property attorney.

The claims of the application vary in scope, starting with the most general and moving on to the most specific. This progression gives advantage to the applicant if the patent examiner challenges the initial claims. The general claims may be invalidated by relevant prior art, but the further definition of claims may not be invalidated. This progression therefore allows for highly specific protection on a particular product. The other point to remember is that claims that are too specific allow the patent to be designed around more easily by competitors.

H. Abstract of the disclosure

The abstract is a short, albeit complete, description of the invention itself.

I. Declaration and power of attorney

The declaration identifies and documents the name, address, and citizen status of the applicant(s). It also states that the applicant(s) have reviewed the application and understand the contents and that the applicant(s) believe they were the first and original inventors of the invention claimed. The law is strict with respect to the declaration, in that no changes may be made after the inventor has signed it.

J. Letter of transmittal

This is the formal communication to the Patent and Trademark Office in submission of the application.

K. Check for application fee

Also accompanying the application is the application fee.

L. Small or large entity statement

Fees may vary depending on whether the company or individual is a small concern or a large concern. Accordingly, a statement to this effect should accompany the application.

The basic procedure is as follows: The patent attorney submits the completed application with the Patent and Trademark Office. The application then gets assigned to a patent examiner who is technically capable in the field of the invention. The examiner then does an investigation to determine prior patents and the prior art of the submitted invention. Upon completion of this process, the patent examiner issues an Office Action. This is a formal response to the application and deals with the specifics of the application claims in form and language. Equally important, the examiner also cites references upon which the claims will be evaluated. The Office Action is valuable to the patent attorney in terms of working out a strategy for executing the application to a successful completion.

The Office Action then gets mailed to the patent attorney, who in turn gets it back in the hands of the inventor for response. The inventor provides detail to the attorney, and together they draft a response to the Office Action by arguing for the stated claims. This is done by specifically addressing each and every point raised by the examiner. This stage in the process is a negotiation between the examiner and the patent attorney until the exact scope of the claims is agreed to. Once there is agreement, the application issues as a U.S. Patent.

5. SECURITY INVOLVED IN OBTAINING A PATENT

The key element to patents as well as to product development is the combination of diligence and secrecy. As a normal course of development, certain aspects that might become patentable should be treated with trade secret status.

Once the team feels that it has a product concept that might be patentable, it needs to conduct a thorough state-of-the-art search of the technology and products. This can be done by canvassing the patent files already in existence and conducting an extensive literature search. The idea is to determine whether there is infringement on some previous patent. It is beneficial to conduct this search at this time, as an alteration in the product concept at this stage can avert an infringement and save considerable time and money. In this way, the patent system can be used to research available technologies.

This activity can also make research and development more efficient. By knowing what is currently available, precious time and money need not be spent on something that has already been developed.

Next is the determination of patentability. The product should be examined for items that have specific novelty by themselves or in combination with other items. The patent attorney can assist in the patentability assessment. Finally, a full disclosure should be made to the Patent Attorney in preparation of the patent.

6. INTEGRATING THE PATENT APPLICATION WITH THE PRODUCT STRATEGY

To the extent possible, a new product development is best served by integrating the patent strategy with the technical and commercial aspects of the product. Merely seeking a patent on a product after it is developed, with no plan on how to capitalize on it or how to use the timing and commercial aspects to advantage, is a waste of development funds.

An additional consideration is that a new product development and patent program must gain momentum as products are developed. The idea is to work to constantly raise the barrier to entry for competition.

Your company can raise that barrier two ways: by increasing the level of embodied technology and by patent protection on the embodiment. With a well-considered program, the industry and your competitors will find great difficulty in displacing your position as a premium and entrenched supplier of the product. Securing a patent will enhance the technological edge over the competition.

SUMMARY

This chapter represents the core development phase of a development program, and it discussed a wide spectrum of material. A section on the mechanics of development started the chapter off. This section discussed the control theory surrounding the project and the people involved. The basic execution of the plan was reviewed, along with tasks and project management systems. Each of the product development phases was discussed, and the specific deliverables were outlined for each stage to ensure completeness. Perspectives on tracking performance, removing obstacles, and backup plans were also reviewed. Dealing with the program objectives within the larger framework of corporate objectives was discussed as well.

Tools were presented for various methods for problem solving. Included were tools for issues review, cause assessment, decisionmaking, and management and planning.

A quality management system was presented, and the ISO standard was discussed as a model for an effective system.

Decision management and recovery skills in the wake of problems or failure were presented.

A section on documentation was also included to stress its importance and the many deliverables involved.

The issue of product liability was presented in light of the need for a quality management system and the need for documentation.

Finally, a section on intellectual property protection was presented for consideration as an integral part of the development process.

With the execution of the development well underway, we are ready to focus on the manufacturing development process, which is an integral extension to the development process.

CHAPTER 8

MANUFACTURING DEVELOPMENT

ABSTRACT: The development of the product is not only limited to the design of the product; it must also include the development of the manufacturing system. The efficiency of the manufacturing system is a measure of how well the organization can capitalize on the new product development. It is, in fact, the means by which the company harvests the investment it makes. This chapter outlines several perspectives on manufacturing development, ranging from design engineering to supplier involvement. These items must be integrated into the new product development process.

CONCURRENCY OF DEVELOPMENT PHASES

1. MANUFACTURING—THE FORGOTTEN DEVELOPMENT?

Manufacturing development is an integral part of the new product development process. Often, a development occurs in a manufacturing vacuum, with the result that the new product might suffer lower than expected gross margin, or worse yet, might fail to capitalize on the market opportunity entirely.

An enterprise cannot add on, factor in at a later date, or ignore the manufacturing process and still reap the rewards of the product investment. Manufacturing must be an integral process developed as part of the product. When factoring design to cost, the manufacturing element represents a significant portion of the direct and indirect costs associated with the product. Failure to address the issue properly results in a nonoptimized situation.

2. MANUFACTURING AS PART OF THE PRODUCT DEVELOPMENT PROCESS

The importance of integrating the manufacturing process into the product development process cannot be overstated. There are tradeoffs between design and manufacturing that must be reconciled as part of the development project, rather than in production, where cost usually increases. The manufacturing process must be designed while the product is being designed, and both design phases must be concurrent.

Manufacturing does not drive the product design solely, nor does development drive the design in a vacuum of other processes. All processes must work concurrently to optimize the solution offered to the marketplace.

3. OUTLINE THE DELIVERABLES

Design engineering has deliverables to mark the completion of certain steps in the design process; the manufacturing system should be developed in a like manner. One can think of the correlation between these two elements as illustrated in Figure 8-1.

Each subelement of the product design should be a factor in the manufacturing system setup. In this way, a manufacturing system can be designed and laid out to manufacture effectively, with synchronization and gathering appropriate supporting data for management reporting. As we will see later in the chapter, the manufacturing system involves more than putting assemblies together. It is a critical feedback loop in the process of bringing the product to the customer. Along with the feedback loop is the need for objective data to evaluate performance and effect corrective action.

4. LAYING OUT THE MANUFACTURING LINE

As the product is being developed, the layout of the product line manufacturing should be completed. If the product is comprised of subassemblies, then each subassembly feeder line should be laid out and optimized. The manufacturing line should function in a synchronized fashion, with all elements operating on a common time base.

5. INTEGRATING THE SUPPLIER ELEMENT

The supplier element is a key factor in facilitating the manufacturing process. The supplier is a partner in the process. As the product and the manufacturing process are designed, the third step is to bring in the suppliers as an integral part of manufacturing.

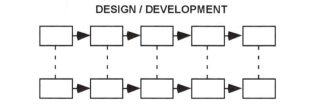

DESIGN / DEVELOPMENT

MANUFACTURING DEVELOPMENT

Figure 8-1. Correlating Design Development with Manufacturing Development

The organization can leverage off of the work the suppliers do in being at the forefront of their technologies. The suppliers can assist in the development effort by introducing new manufacturing concepts to the organization. They have the benefit of involvement with other manufacturers and can be a key linkage to state-of-the-art ideas.

INTEGRATION OF MULTIPLE DISCIPLINES INTO THE DEVELOPMENT PROCESS

1. DIAGRAM FLOW OF MULTIPLE DISCIPLINES

The development of a new product is an interdisciplinary process. All aspects of the organization should be represented in the program. These include development, quality, purchasing, sales and marketing, finance, and manufacturing. Each group contributes its portion vocationally and serves as a check against other areas where vocational conflicts naturally occur. For example, sales and marketing keep development in check against cost overruns. Manufacturing keeps development changes in check. Accounting and finance hold the purse strings tight on expenditures. Quality keeps manufacturing processes in balance. These interrelationships are represented by Figure 8-2.

2. GIVE AND TAKE BETWEEN MANUFACTURING AND DEVELOPMENT

Part of the integration process between the design-engineering group and the manufacturing group lies in the give and take between them. Neither group can secure 100 percent of the desired elements of a program to satisfy its parochial goals. Rather, the success of the new product lies in the cooperation that must take place between the two groups.

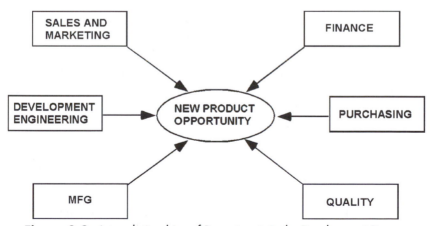

Figure 8-2. Interrelationships of Departments in the Development Process

Significant progress toward cost reduction and improved quality can be made by design engineering "assuming" responsibility for the reduction of manufacturing problems by virtue of design. If, however, the design is conducted without reference to manufacturing and the rest of the company, little is gained, as design is often unaware of the potential problems of manufacturing.

Some companies resolve this issue of cooperation by periodically assigning development engineers to work in manufacturing and requiring them to produce the same results expected of manufacturing personnel. This practice can be a healthy one in terms of highlighting manufacturing problems to the development people directly.

3. THE PARTNERSHIP

The interface between the departments therefore should be one of partnership in overcoming the dynamics of the competitive marketplace for the financial benefit of the company. Often, the relationship can degenerate into one of a competitive nature between departments; however, it is the program manager's responsibility to galvanize the departments to focus on accomplishing the greater goal. If compromise is required, there should be an accompanying action plan to optimize it at a later date.

4. DESIGN FOR MANUFACTURABILITY: IT STARTS AND ENDS WITH DESIGN

Up until this point, the focus of new product development has been from the perspectives of design, functionality, and cost. The focus must now be broadened to include manufacturability. The ease of manufacturing a product does not originate in manufacturing, nor can it be accomplished by the addition of capital equipment after the design process has been completed. Ease of manufacturing must be integral to the program. Furthermore, there are several basic tenets that comprise a manufacturable product. These will be reviewed in more detail in the next section.

The issue of manufacturability goes beyond margin improvement and the ability of the organization to harvest the new product development investment effectively. It goes directly to the issue of producibility. What if your sales channel brings in significant amounts of business—can your manufacturing organization produce? If it is ill equipped to produce, the entire effort suffers at the last stage. Keep in mind the lessons of lost opportunities, bloated inventories, incorrect inventory parts forecasts, poor throughput, and problems due to routing issues or manufacturing gridlock.

5. MANUFACTURING CAN'T FIX WHAT DEVELOPMENT DOESN'T PROVIDE FOR

In many cases, the organization has high expectations of manufacturing; however, manufacturing cannot repair or create what development has not provided for in the new product design. Development ultimately must have the vision of how the product is configured and assembled. This requires the development people to have a working knowledge of manufacturing to be effective.

The team concept is helpful in accomplishing this to some degree. In a team atmosphere, however, development more often than not has designed the product by the time the first meeting with manufacturing occurs. The manufacturing development personnel merely react as best as they can to the product concept. A better approach is to require several product concepts to be produced by engineering and manufacturing together, and then select the most appropriate platform. It is even healthy to summarily reject product concept solely from development, and to require interdisciplinary involvement in producing the concepts.

6. CONSIDERING THE DIVERSE AND FAST-CHANGING WORKFORCE

"I am not so sure about these new kids..." How often have you heard this statement? A statement like it has probably been used for each successive generation. Fundamentally, however, it underscores a basic fact—that the workforce is ever changing and is comprised of personnel with ever-divergent viewpoints and value systems. Most likely, gone are the days of the uniform "nose to the grindstone" work ethic among an entire workforce. Individualism will reign, and the manager's responsibility is to accommodate and make use of the strengths of individual personnel to accomplish the corporate goals.

For manufacturing, this means continual training issues, tardiness issues, attitudinal issues, and work quality issues. In addition, economics and shifts in focus will cause periodic spot shortages and surpluses in trained personnel. To a certain extent, design can accommodate the work quality issues by designing for one-way fits, elimination of errors, and product configuration. To leverage profits from a workforce, however, uniform and consistent effort must be directed to the manufacturing tasks at hand in spite of the diversity in the human resources applied.

7. MANUFACTURING AS A COMPETITIVE WEAPON

With a product diligently integrated into manufacturing and an effective sales channel, manufacturing can be a formidable competitive weapon. Consider the power an organization has when it has the ability to produce a product desired by the market, an organization that has the potential to generate significant amounts of profit and the ability to deliver in significant quantity relative to the market segment. This can be a prescription for immediate success and the building of long-term momentum for the corporation.

Certain segments can be defended to the point of locking out competition forever. The strength your company gains causes your competitors to become weaker not only in a relative but also in an absolute sense, as they are losing sales with nothing else to fill in the shortfall.

DESIGN FOR MANUFACTURING

At this point in the discussion, it is a good idea to review the basic tenets of design for manufacturability. The following text outlines a list of desirable attributes of a product design to facilitate manufacturing. They are mostly common-sense items when considered in the abstract sense; however, these are the items that are soon compromised as the development team begins to get into the project. In fact, a great deal of the compromise is made in the mind of the development engineer even before the first team meeting. Often, development engineers "fixate" on their concept and defend it to the detriment of the manufacturability. The challenge for management is to refocus on these basic design tenets throughout the development process—especially at the beginning, when compromise is irreversible.

1. PHILOSOPHY

The basic philosophy behind design for manufacture is to design, arrange, and position elements to go together in a preferentially compatible manner. This means that parts go together with minimal effort, force, fitting, and fastening. It takes advantage of natural forces such as gravity and accommodates the use of fixturing and tooling to speed the process while ensuring an acceptable level of quality.

2. MINIMIZATION OF PARTS

One of the easiest ways to accomplish a design for manufacture is to reduce the raw number of parts. This is accomplished integrating functionality within a more highly engineered and tooled part. By combining functionality and features in these parts, the sheer number of them can be reduced. A reduction in parts count helps reduce inventory, space, work in progress, and potential scrap. Also, it reduces the per-unit consumption of materials and supplies.

3. USE MODULAR DESIGN

A modular design reduces manufacturing costs by allowing the use of subassemblies. It allows the organization of feeder lines in manufacturing and adds quality checks and controllability earlier in the assembly process. The use of known goods and certified subassemblies in final production results in a faster final assembly as well as improved first-pass yield.

4. TAKE ADVANTAGE OF GRAVITY

As the assembly process is laid out, parts and subassemblies should "drop" into place and be secured by integral fasteners wherever possible. Taking advantage of gravity in this manner reduces the time required to manufacture the product and reduces the amount of human intervention and repositioning. It also can pave the way toward automated assembly.

5. REDUCE VARIABILITY

The design of the product should have reduced variability in parts. This means that if two parts or subassemblies are alike or nearly alike in producing different versions of the product, the designer should consider reducing the variability and designing one part or subassembly that will accomplish both functions. This tactic reduces inventory levels and also improves throughput by preventing the presence of the "wrong part only" in inventory.

A reduction in highly similar parts also improves quality and eventually contributes to reduced cost, as quantities of identical parts are increased.

6. PROVIDE EASY ACCESS

Easy access to critical or important components is both a serviceability issue and an assembly issue. Access to such components increases the availability of equipment by reducing the mean time for repair. It also reduces manufacturing time by allowing free movement within the subassembly.

Providing for easy access does not necessarily cause an increased amount of product real estate if the design is handled correctly. Creative and diligent use of product real estate can be a contributing factor to these access issues.

7. ELIMINATE FASTENERS

Fasteners are an age-old means for assembling parts and assemblies. Although they are convenient and relatively low in cost and allow a certain amount of flexibility, they add labor cost, contribute to repetitive motion illness, and increase the likelihood of assembly mistakes. Their use generally requires additional procedures and tooling to ensure proper torque of components. A part or assembly that can have a built-in fastener results in a reduction of mistakes, improvement in quality, and reduced hazard among the workforce.

8. INCREASE PART SYMMETRY

Not necessarily a requirement for manufacturability, the use of symmetry can reduce certain costs when secondary operations are required. For example, machining a certain assembly can be made easier by the use of symmetry, reducing setup time and fixture placement on subsequent operations. Symmetry also has certain advantages where material handling is concerned. Fixtures for handling can be consolidated, and additional flexibility can be created.

9. EASE OF PART HANDLING

The issues of material handling and part placement do not need to be limited to the consideration of symmetry alone. They can extend to all parts and assemblies in a product. Ease of part handling contributes to lower cost of material movement tooling, better quality due to reduced mishaps, and a more laminar process flow. The parts should be designed for ease of alignment and ease in maintaining location. They should fit into their locations easily without the use of alignment, locating, and tightening operations.

10. DESIGN PARTS WITH VENDORS

Become proactive in seeking the assistance of vendors in designing product parts. They have the knowledge of the state of the art in part design, materials, processes, and trade secret knowledge. The proprietary nature of the part can be preserved while still taking advantage of vendors' expertise.

11. CREATE PROCESSES TO ELIMINATE ERRORS

Reduce dependence on the workforce to operate as "craftsmen" by designing the parts and the processes of assembling those parts to eliminate errors in assembly. It is futile to attempt to inspect quality into a product after the manufacturing operation is completed. Splitting up the inspection process and placing it at the level of individual operations only mitigates the amount of waste; it does not prevent errors completely. The prevention of errors and subsequent waste must be systemic, not subject to human diligence.

12. REFINE PROCESSES TO REDUCE PER-UNIT RESOURCE CONSUMPTION

The processes of design and manufacturing must strive to reduce material consumption for each part consumed. If glue is used to join parts, how can the parts be designed to "snap" together and hold rather than relying on glue, the temperature at application, the cleanliness of the parts, the receipt of the correct glue product in the receiving department, and the human craftsmanship required to spread the glue and join the parts? Clearly, there are many savings in cost and improvements in quality that can be gained from this philosophy of design and manufacturing.

13. CREATE A PROCESS TO BUILD AND SHIP RAPIDLY RATHER THAN STORING INVENTORY

Executive to materials manager: "Make sure we always have enough parts and materials to build the products." Although the goal of meeting shipments and customer commitments is a noble one, the focus should be not on laying inventory in reserve but rather on designing manufacturing process that are rapid in terms of throughput. By focusing on throughput, inventory is reduced, work in process can be reduced, and the labor dollars "sitting" in inventory waiting to be consumed are eliminated. The clear directive here: Focus on rapid manufacturing throughput, not exclusively on inventory.

14. CREATE A MINDSET OF MINIMIZATION THROUGH THE MANUFACTURING PIPELINE

The product team should develop a philosophy of having manufacturing endeavor to pull the product through the process in a "stingy" manner from the perspective of materials and labor. This is to be accomplished not by degrading quality but rather by striving for the minimum material requirements while still achieving the quality measurable and customer satisfaction. Often, the product progresses through manufacturing with labor being thrown at it, accumulating costs and variability.

MANUFACTURING, PROCESS, LAYOUT

1. THE SOLUTION DEPENDS ON THE PRODUCT

There is no single optimized solution to an effective manufacturing system for all product lines. Each product line within the organization might require its own optimized solution. The general layout for a product comprised of discrete parts, however, can be represented as in Figure 8-3.

The example in Figure 8-3 consists of five stations of different operations. These are denoted as Operation #1 through Operation #5. There are parts bins that feed each of the stations and the personnel at each station to assemble the parts into assemblies. The personnel are also responsible for inputting manufacturing data into the networked computers feeding the information system. There are separate feeder lines, staffed by personnel who feed parts to the major line. All of these items must be coordinated and synchronized.

With proper synchronization and layout, this short line can produce product and keep track of manufacturing data.

2. REVIEW INDUSTRY PRACTICES

When setting up a manufacturing line for a product, it is often helpful to gather data about other products that might be similar in terms of both configuration and assembly methods. Do the necessary detective work to get detailed information about the following:

A. Processes

Determine the type of process capability the competitors have. How do they control their processes, and what are the measurements for each process used? Then design a manufacturing system that can exceed these numbers if possible.

B. Yields

What are the expected and controllable yields used in industry today? Are these numbers achievable in your environment? Determine why or why not, and effect corrective action.

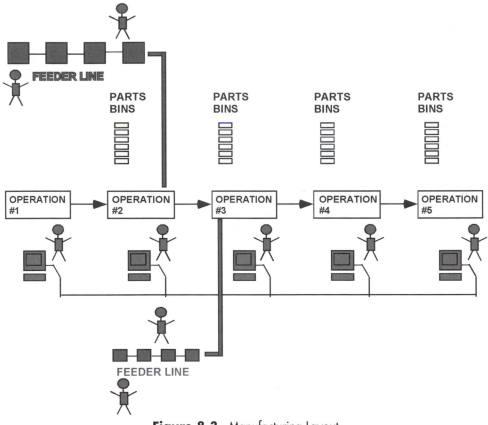

Figure 8-3. Manufacturing Layout

C. Cost structures

Find out what the typical cost structures are for the type of assembly and construction techniques used or anticipated. What is a typical labor percentage for this type of construction? Can a similar percentage be achieved at your company? If the assembly technology limits the margin by virtue of the cost structure, there could be minimal payoff in trying to optimize the existing approach. It might be preferable to focus on a newer or more radical approach to optimize profit.

D. Manufacturing throughput

Determine the manufacturing throughput available for similar processes and products in industry today. Is this the best that can be hoped and planned for in your situation? Are you in a position to increase capacity easily? Is there a plan to increase incrementally?

E. Indirect support issues

What is the indirect support required to produce the volumes anticipated? Is engineering going to be spending all of its time supporting manufacturing because of engineering circumstances? How much of the engineering talent pool will be absorbed by the need to provide indirect support?

F. Product life cycle

How rapidly will the product life cycle turn from version to version? Will an initial product platform offering have a two-year run before an enhancement becomes necessary? When will market needs demand a complete new platform? What is the window of opportunity to secure planned profit? Will there ever be a pure harvest mode of operation?

G. Redesign frequency

Do products of this type require constant redesign attention as they are manufactured? Do they have a high sensitivity to purchasing, supplier, and variability issues that require constant product maintenance, or are they stable in nature?

Each of these elements requires careful consideration and planning to understand the expectations of the market, the objectives of the organization, and the need to exceed the competitors' ability to produce product for delivery to the marketplace.

3. REALISTIC OBJECTIVES AND NUMBERS

As stated previously, there is a give and take in setting up manufacturing. The best design that calls out the best tooling and capital equipment might not be warranted if the projected lifetime product volume is too low; the return on investment would be poor in this case. Rather, the investment and tooling expenses must be commensurate with the anticipated volume, profit potential, and return on the product.

4. HOW TO SET UP FOR COMPETITIVE BATTLE

Is your manufacturing organization set up for competitive battle? Do you have the ability to fulfill large volumes of orders? Can you respond to opportunities and deliver goods and services to meet customers' expectations? These and other questions require careful consideration prior to organizing the manufacturing aspect of the new product.

If you are engaged in a high-volume market opportunity, secure arrangements with suppliers with volume commitments to protect margins and allow response to competitive price pressure. If the product is highly engineered, secure and retain engineering talent to pursue opportunities in a timely manner. If the product is purchased as a commodity, ensure that you are set up to be a low-cost producer while minimizing sensitivity to cost changes.

5. PERSPECTIVES ON INVENTORY

Inventory by its very nature serves a temporary holding function for materials and assemblies; its primary purpose is to match imbalances in and nonsynchronization of manufacturing processes. If the output of process 1 cannot go directly in a balanced fashion to process 2, there needs to be some provision for inventory so that one process in the chain does not run dry as the line is run.

In a similar manner, the use of inventory is popular as a means for delivering orders in a rapid fashion. This goal is accomplished by having subassemblies on hand for quick final assembly. The rapidity with which a product proliferates different versions through changes in hardware determines the degree to which an organization will most likely have the "wrong" inventory on hand to fill the order.

In practice, inventory seems to have an opposite effect to the one desired. It seems to slow down the ability of the organization to react nimbly to the changing marketplace. There is a cycle time associated with completing assemblies, which carries a cost in materials and labor added to it. "Stock jobs," in which a run of assemblies is completed to lay into inventory for future use, seem to absorb labor, rarely get completed, and linger within the manufacturing organization. These are signs of an organization burdened with inventory.

The goal is to synchronize manufacturing to reduce the level of inventory required to execute a given level of business. It is desirable to run the business with the minimum amount of inventory by organizing suppliers, controlling product configuration, and establishing an effective mechanism for forecasting. With these strategies in place, the correct inventory is delivered for secondary operation and completion of the product at the proper time for shipment to the customer. Stock room space is reduced, and material handling (which becomes overbearing with bloated inventories) is controlled.

6. CHANGING PARTS IN A FAST-PACED ENVIRONMENT

The impact of changing a part in an assembly has farther-reaching consequences than most give credit for. As will be discussed in the next section, the configuration of a product is almost a sacred trust between the design function, procurement, and the manufacturing function. When a part changes, it affects the product integrity, its performance, its certification, and its liability.

Therefore, changing a part must have a strict protocol, whether the impetus for change is internal or vendor driven. To understand the significance of changing a part, consider the simple model of manufacturing and the introduction of a replacement part and its effect on the overall system. Figure 8-4 illustrates the issue.

The issue of parts substitution relates to form, fit, and function. If any of these elements of a part changes, there will be an impact on the final product. Consider, for example, the manufacturing line shown in the figure. Its five operations are fed by a feeder line where subassembly is done and by parts bins located at the line.

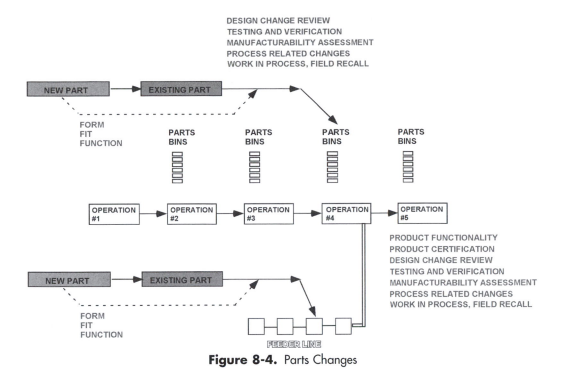

Figure 8-4. Parts Changes

If a part changes in form, fit, or function in the parts bins, several items must be considered. A decision must be made whether to substitute the new part immediately or to consume the existing parts first. An immediate substitution is denoted by the dashed line going around the existing part. In either case, the following elements are affected and must be verified:

- Design change review
- Testing and verification
- Assessment of manufacturability
- Process changes
- Work in process and/or field recall

These issues affect the final product and must be investigated. If the substitution is immediate, there is little time to investigate. If the substitution does not take place immediately, there might be more time available while the existing parts are being consumed. Scrapping existing parts can be expensive, so there might be a tendency to consume existing parts first; however, a safety-related parts change could mandate an immediate substitution regardless of the impact on cost.

In a similar manner, a parts substitution at the feeder line might require investigation of additional elements. Because the feeder line has its own manufacturability issues that dovetail with those of the main product manufacturing, the following issues must also be considered:

- Product functionality and performance
- Product certification

If the parts change, the product might have to be recertified, depending on the degree of difference of the existing part to the new part. Consequently, these changes need to be managed and grouped together or the product will have little stability in the way of design, cost, or manufacturing.

7. DESIGNING THE CHECKS UP FRONT

To minimize the disruption in parts substitution when the product is in manufacturing, there are steps that can be taken at the selection and specification stages of the parts. There are several issues that contribute to a part profile and need to be considered. These are as follows:

A. Price and price stability

Price and price stability are measures of the cost contribution to the product. They have several components that range from initial cost, long-term price stability with reference to outside driving forces, and total installed cost, including handling and inventory arrangements.

B. Availability

Availability is a measure of the organization's ability and energy required to secure the part for use in the product. Is the part readily available for general use? Does it satisfy export compliance directives in all international manufacturing venues? The best-specified part is of little value in generating profit for the company if it is not readily available.

C. Performance

The specifications of the part can play an important role in its overall use. If the part parameters are right on the tolerance edge, there could be additional qualification requirements in production to ensure functionality. If the part has parameters specified that are in excess of the minimum requirements, the tolerance/performance gap might be favorable enough to eliminate the need for additional production measures.

D. Quality

Quality is the measure of how repeatably the organization can produce parts or product that meet the requirements specified in the drawings and specifications. Clearly, the degree to which these match will alleviate long-term problems in using a part. If the production quality varies widely, there could be problems.

E. Reliability

Reliability is a measure of how long the part will meet it specifications. This must be correlated to the company's claims for product reliability and life.

F. Commonality

This is a measure of how specialized the part is. If it is a highly engineered part, you will be subject to the limitations of the suppliers in quality and delivery. If common parts are used, there is more flexibility of the product line over the longer term and of the ability of the organization to support it.

G. Ease of use

Does the part require special handling or procedures that the organization must put in place? Can nonskilled or semiskilled personnel handle the part? These types of issues determine the part's ease-of-use factor. Try to specify parts that don't require special handling, as these affect manufacturing, service, repair, especially when these latter functions are offsite.

H. Integration with manufacturing processes

How does the part fit into the present manufacturing system? Can the present or planned manufacturing process and process control handle the part? Be sure to specify parts that are correlated with capital equipment plans and capabilities.

I. Longevity of run

Determine how long the vendor plans to "run" the part. This will determine the redesign cycle for the product. Common parts seem to have longer "runs" than highly specialized parts. In addition, the highly specialized parts generally increase in cost as time progresses, as the supplier has no other customers to sell them to and learning curve cost reduction doesn't take place.

J. Last-time buy arrangements

Determine the supplier's policy for last time buys of a part. Also determine the advance warning time and the general time allowed for last-time buys. This will assist the company in obsolescence plans and redesign/replacement efforts.

K. Replaceability factor in the product design

It is important to understand clearly and document the product's dependence on the part from a performance perspective. Is the part easily replaceable, or is it designed into the product in such a way as to make it very difficult to replace? Knowing the answer will save time and development energy in the future if you don't design yourself into a corner at the onset.

L. Industry shortages and sensitivity to volume

Finally, what dynamics of the industry can affect part pricing and availability? Project the usage forward and try to determine the actions of the parts market (possibly your company's competitor) in the future; factor these into the plans accordingly.

A sample and a worksheet are available for your use in the *Toolbox*.

PRODUCT CONFIGURATION

Careful consideration must be given to how the product line is conceptualized and structured. In this section, we review the configuration of the product. The product configuration is the protocol for the various versions to be offered as well as how the versions are organized from a design and manufacturing point of view. The product configuration can contribute to the success or failure of the product. Good design practice and experience at the design level can alleviate future problems in manufacturing.

1. DESIGN STABILITY

Progress in manufacturing efficiency cannot be achieved without design stability. In the quest for a new product release to the marketplace, companies often release product to manufacturing before the design is complete. In these cases, further development changes to the product cause multiple problems in design and manufacturing because the design is still in transition. Ensure that the design is fundamentally frozen prior to a full-scale manufacturing run.

2. MANUFACTURING STABILITY

With the design frozen, the team can concentrate on manufacturing process capability. This is a measure of the performance of manufacturing in terms of quality, repeatability, cost effectiveness, and throughput. Process capability assumes that the design is stable and considers only manufacturing variability. Variability can be controlled by the use and implementation of manufacturing documentation of procedures and work instructions. These are medium independent but necessary, along with training and refresher courses to ensure consistency in the processes.

3. EXECUTING A CHANGE

Executing a change in manufacturing can have a serious impact on the integrity of the product. Previously we discussed the coordination requirements involved in substitution of a part. A change to the product is even more significant in that the entire battery of product certification testing might have to be redone. Given the seriousness of a non–part substitution change, companies should establish a procedure that must be followed to ensure preservation of the product configuration and integrity of the product as a result of the change.

Product changes can have effects in the future depending on the compatibility of the change with future product. There are two basic terms that describe the issue:

- Forward compatibility
- Backward compatibility

Forward compatibility ensures that the product line can support change in hte future by designing a product that allows a future improvement, enhancement, or change can be compatible with the host product shipped now. Today's product is designed to be repaired with future parts.

Backward compatibility ensures that previously designed and manufactured units are compatible with new parts. Yesterday's units can accept today's parts. The current product is backward compatible to the old version (parts). Both are similar. It is more of a time reference whether you are looking to accommodate the future version or past version.

The arena of compatibility is one of the most important aspects of a product change to handle correctly. Failure to do so can cause huge headaches in repair, service, and spares. One can imagine a compatibility problem when a host unit in the field needs repair and the new replacement part is incompatible. This circumstance would force replacement with a totally new unit at a significant cost penalty to either the customer or the company itself. An alternative would be to establish a new part number for the replacement noncompatible unit and to keep a stock of the old style. The problem is that many times, a part is obsolete due to inability to buy subcomponent parts that comprise it. It is worth the extra design effort to design for compatibility wherever possible.

4. FORMAL PROCESS FOR REQUESTS AND CHANGES

In order to execute a change correctly within the organization and on the product line, an effective and formal change system should be in place. Informal, "by-the-seat-of-the-pants," nondocumented changes are not acceptable in today's manufacturing environment. Because the process of development is interdisciplinary, the process of change should be so too. In addition, the execution of a change is more difficult than the original development in terms of managing actual cutover and tracking for future reference. Contrary to the beliefs of some, the process of formalized change involves more than the addition or substitution on a bill of material or a change in tooling.

As shown in Figure 8-5, the formal process must be marked by the entry of a change request that is considered by a cognizant party, an investigation and impact assessment, planning of the actual change, and only then the actual implementation. Implementation is followed by an audit of the change, a process that is frequently overlooked or forgotten.

Figure 8-5. Formal Change Process

Important elements of the change process are the initial request and the evaluation, whereby the seriousness and timing of the change are determined. These steps determine how the change will be engaged and executed by the organization.

5. WHERE THE PROCESS LEAVES OFF

It cannot be overstated that the process of changing the product does not stop at the implementation. It stops only after the audit of the change. The audit of the change considers the dynamics of the change, the specifics of manufacturing implementation, and equally important, the response from the field in terms of effectiveness and the introduction of any new issues.

6. SPEED OF EXECUTION VERSUS DEGREE OF DOCUMENTATION AND APPROVALS

What is the price to be paid for all of the diligence and approvals involved with a formal change system? The undisciplined profess that diligence will cause significant delays in any improvements. In fact, the diligence required might prevent mistakes in the "improvements." Certainly, the organization can execute the change in an informed manner. Figure 8-6 can be used to consider the time frame for changes.

The straight line illustrates the difficulty of change with the addition of diligence in executing a change. Neither graph attempts to depict the time required for a change; if there is a mistake that will have to be corrected later, it will extend the x-axis of the characteristic graphs. Referring to the curved line, the more checks and balances and approvals

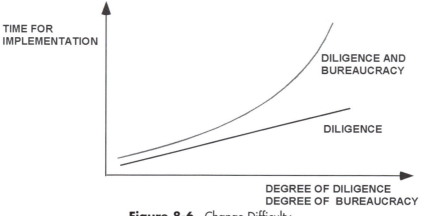

Figure 8-6. Change Difficulty

required, the longer the time required to change the product. If the organization is lethargic and laden with a certain amount of bureaucracy, the combination of approvals and bureaucracy will cause gridlock. This is unhealthy and can cause failure of the product line. The solution is not to eliminate diligence and approvals, but rather to reduce the bureaucracy.

Fundamentally, a balance needs to be struck between the need for diligence and the need to execute a change to a product in timely manner.

7. SUPPLIERS

What about supplier-driven changes? The product development team must be careful when coping with supplier-driven obsolescence. The obsolescence of parts causes disruption in the product flow, consumes development resources for product maintenance, and creates unpredictability. The organization must be fully aware of the obsolescence plans of suppliers, and last-time buys and phasing of replacement parts must be coordinated in manageable "block" changes to the product. Otherwise, the product is completely at the mercy of the vendors' plans, not yours.

8. COST INCREMENTALISM VERSUS COST REDUCTION

It happens all too frequently: A parts substitution for a soon-to-be-obsolete part is available, but it costs slightly more. It is a fact that last-time buys on parts generally carry a heavier price tag. The dynamics of the change cause the product team to weigh convenience against increased product cost. In and of itself, a simple increase in a single part will not cause product failure; however, this trend over a range of parts eventually will affect profitability.

CHANGES TO THE PRODUCT LINE: DEVELOPMENT VERSUS PRODUCTION

1. FORMAT FOR A CHANGE SYSTEM

An engineering change system consists of two basic procedures:

- Engineering change request
- Engineering change

The engineering change request is a formal request for consideration to the personnel who are cognizant of the product configuration. It is merely a request that does not initiate change but rather initiates investigative action. It results in an acceptance or a rejection or is placed in a pending state to be addressed later. A good engineering change system also provides a means for ensuring that each request is addressed and followed up on. In addition, to be effective, the system should provide for communication back to the initiator regarding status and disposition.

An engineering change is the pathway and the record of the initiation, execution, and resolution of the change to the product. It also includes provisions to audit the change at a later date. The engineering change initiates action to change the configuration of the product. A protocol is required to execute the change so that the appropriate approvals and investigative work and documentation are completed. Failure to enforce these procedures results in unauthorized, incomplete changes and loss of product configuration control.

The change in the configuration of the product should involve all affected departments and secure their approval. Those departments not directly affected should be notified at minimum, as part of the process.

An engineering change system is not a substitute for a product development system. It governs change and the recertification of the product as it relates to the change only. It is not a means for generating and certifying a new product. Major changes and new product programs should be organized under a new product development process.

A sample change procedure is included in the *Toolbox* for your review and use.

2. DEVELOPMENT VERSUS PRODUCTION

Depending on the size of the organization, the scope of the change, and the complexity of the product, it could be necessary to track changes in the product before production occurs. Tracking these changes assists in securing the correct parts off the bills of material should they change. Tracking also can yield a measure of the stability of the development effort and can be used to forecast whether development will complete its activities on time. It may be a good idea to institute this system for the development group if it has been newly assembled within the organization. For the development group, a tracking system serves as a check against the development effort constantly redirecting the organization with product changes, which would make manufacturing development all the more difficult. A tracking system *is* mandatory for production, however.

3. DOCUMENTATION

The documentation for a product change should be a small-scale set of documentation for the initial product offering. It needs to track the change from initiation through to audit. If a request is entered and rejected, the documentation should reflect the dates, the circumstances, the assumptions, and the decision-makers. If the change is carried out to completion, the documentation should chart the progress, the results, and the cutover date. The need for procedure and documentation might seem overbearing and bureaucratic; however, it could be required to trace down a field problem or defend against litigation.

4. COMMUNICATION

The most helpful partner to engineering change is a company-wide network of communication. Effective and accurate communication can overcome a lack of a system in executing a change, and the lack of communications cannot be made up for with a change system. It is crucial to keep communications open during the various phases of the change process, from request through audit. Here, too, there should be a protocol for communication between the various departments of the company and for communication to external parties as well.

5. APPROVALS

An approval protocol for any changes to the product should be in place. This protocol should be a strict protocol, as anything less will eventually degrade the change system altogether. This protocol should have backup in terms of personnel should the primary personnel be unavailable. Danger generally surfaces when the primary personnel are unavailable for approval and the backup personnel must fill in. It is important that they follow the same checklist for approval as the primary personnel. When protocol is broken, even once, it starts the decay of the system, and eventually the configuration is compromised.

Disregarding the protocol also exposes the organization to liability if there is an incident with a product. Therefore, it is essential to establish the protocol and adhere to it.

6. COORDINATION

The coordination involved with a change is generally outlined by the procedures of the change system. Care should be taken to clearly spell out the procedures, the functional areas, and the levels of approvals required. Do not leave the coordination to chance or to communication. Establish a procedure and adhere to it.

To reinforce the importance of coordination in a component change, one can project the collisions and voids caused by two parts arriving for assembly, one being the old version and the other being the new version. In a like manner, one can envision the interruption in production caused by the new part not being available and the old part depleted out of inventory. Furthermore, one can imagine the rework associated by a production run started in the absence of knowledge of an engineering change that requires immediate substitution of the part.

Fundamentally, effective coordination in product changes should be enforced, or there will be wasted time, effort, labor, and materials, and the integrity of the product is again compromised.

MANUFACTURING PROCESS CONTROL

1. MANUFACTURING PROCESS DOCUMENTATION COMMMUNICATES WHAT'S NOT ON THE DRAWINGS TO MANUFACTURING

There is always a void of information between the development-generated drawings and bills of material and the complete set of documentation required to do an effective job of manufacturing. This void is filled by the manufacturing process documentation—a "big-picture" methods and procedures document showing how the product goes together. Such documentation also sets the gauge by which manufacturing will be measured and provides the continuity needed in manufacturing. Keep in mind that manufacturing deals with people-related variables, and to establish a uniform skill level in the personnel requires an effective process documentation set.

2. METHODS AND PROCESS DOCUMENTATION

The process documentation is generally comprised of two types of documentation, namely:

- Methods
- Process

The methods documentation covers specific procedural approaches to be used by the manufacturing personnel. It describes how assemblies are to be put together. This documentation can take the form of drawings, illustrations, and exploded views that serve as procedural reminders to the personnel. They include both pictorial elements and written instructions.

The process documentation describes how the process-oriented elements of manufacturing are to be conducted and controlled. Its purpose is to ensure that a repeatable result is achieved for a given set of conditions with given equipment. These are conditional data points for machinery involved in the processes.

3. PERSPECTIVES ON DOCUMENTATION SYSTEMS

The documentation systems tend to be medium independent. It might be wise, however, to consider a computer-based set of documentation whereby a networked system can be accessed from server stations, and most important, whereby the content is controlled via the computer not by paper. Such a system renders all paper copies uncontrolled, and the computer system has the official content. The electronic means allows updating and control of the information in real time.

4. OPERATING IN A FAST-PACED, HIGH-VOLUME ARENA

Making process changes in a fast-paced environment can be a challenging situation for the manufacturer. Fundamentally, the process changes must be understood, controlled, and predictable before cutover to the change is made. Failure to follow these principles can introduce new unknowns into the process and complicate resolution to an acceptable yield. Consider the Figure 8-7, which illustrates this issue.

In this example, Operations #1 through #4 represent the standard manufacturing processes. If a process change occurs resulting in a nonconformance, the parts from

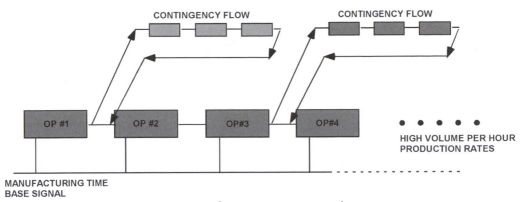

Figure 8-7. Manufacturing in a Fast-Paced Environment

Operation #1 must go somewhere; otherwise, the production line must be stopped. This example applies to the process situation, not to the discrete parts manufacturing situation. The pathway entitled "contingency flow" is simply the pathway for the misprocessed parts to take. This pathway must return to Operation #2 somehow. This situation is especially likely to arise with very-high-volume manufacturing. One can see that the inconvenience of this route can disrupt operations and play havoc with the quality of the end product. In addition, it is only a temporary "holding area" that will soon fill up and eventually stop the line anyway.

A similar contingency flow pathway is also shown occurring between Operations #3 and #4. The manufacturing time base signal depicted at the base of the illustration sets the pace for production. Detouring to the contingency pathway disrupts the timing. Therefore, it is best to certify the process change "offline" or while the line is shut down before taking a chance on phasing in the process change and risking scrap and schedule problems.

5. PROCESS CONTROL SYSTEMS FOR THE SKILLED AND UNSKILLED WORKFORCE

Differences in skill levels among the workforce dictate differences in the way the process and manufacturing control systems are handled. Training is, therefore, a crucial part of the new product manufacturing development. Figure 8-8 illustrates the differences in the treatment of training between the skilled and nonskilled members of the workforce.

As the figure shows, for skilled manufacturing personnel the training is oriented toward product-specific process training. The personnel are given process instruction and are certified to a specific level. Subsequently, they are given method sheets, which serve as reminders of the training and certification previously given.

In the case of nonskilled workers, initial training must be given for basic instruction in manufacturing technology, general procedures and methods, and safety in the workplace.

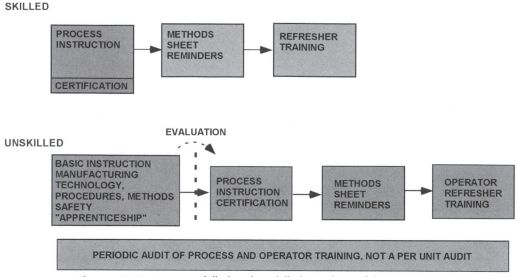

Figure 8-8. Training Skilled and Unskilled Members of the Workforce

Workers are then evaluated to determine whether they come up to a minimum level of proficiency. Assuming this is the case, they progress to the actual product-specific training and process instruction and certification. Subsequent methods sheets serve as reminders of the training, and operator refresher training is also used. Throughout the entire process of orientation to manufacturing technology and the product-specific training, there is an audit function to double-check proficiency and effectiveness of training. This type of audit is not associated with the audit steps used as part of the manufacturing process.

6. SYSTEMS OF MEASUREMENT FOR PROCESS CONTROL

As part of the manufacturing development, there should be a systemic method for performance data entry and analysis. This method is used to determine yield data and trend analysis. As the products are being produced, the results of the various operations are catalogued and posted to a report. The report is analyzed periodically to determine problems slated for correction and trends indicating a degradation of the manufacturing process.

Figure 8-9 is an example of yield data in a manufacturing analysis. It represents a four-week window of data. This chart of data can be expanded to include any pertinent number of interval data. The important factor in scaling the chart is to gather and process data at meaningful intervals to facilitate determining trends and taking corrective action.

Figure 8-9a is organized to record data on six operations labeled OP#1 through OP#6. In the first week, a total of 12,250 pieces of product were manufactured. The number of reported errors occurring at each operation is recorded and scaled to reflect the parts per million (PPM). In the second week, 15,000 pieces were manufactured, with the associated

MANUFACTURING REPORTING SYSTEM

OPERATION	VOLUME FOR WEEK 1	NUMBER OF REPORTED ERRORS	YIELD DATA (PPM)	VOLUME FOR WEEK 2	NUMBER OF REPORTED ERRORS	YIELD DATA	VOLUME FOR WEEK 3	NUMBER OF REPORTED ERRORS	YIELD DATA	VOLUME FOR WEEK 4	NUMBER OF REPORTED ERRORS	YIELD DATA	TOTAL MONTHLY VOLUME	NUMBER OF REPORTED ERRORS PER MONTH	YIELD DATA PER MONTH
OP#1	12250	3	245	15000	2	133	17500	1	57	20000	4	200	64750	10	154
OP#2	12250	1	82	15000	0	0	17500	2	114	20000	2	100	64750	5	77
OP#3	12250	5	408	15000	4	267	17500	3	171	20000	1	50	64750	13	201
OP#4	12250	2	163	15000	1	67	17500	1	57	20000	1	50	64750	5	77
OP#5	12250	1	82	15000	1	67	17500	1	57	20000	2	100	64750	5	77
OP#6	12250	1	82	15000	1	67	17500	1	57	20000	1	50	64750	4	62
CUMMULATIVE YIELD DATA	73500	13	177	90000	9	100	105000	9	86	120000	11	92	388500	42	108

A

OPERATION	WEEK 1	WEEK 2	WEEK 3	WEEK 4	CUMMULATIVE AVE. BY OP#S
OP#1	245	133	57	200	154
OP#2	82	0	114	100	77
OP#3	408	267	171	50	201
OP#4	163	67	57	50	77
OP#5	82	67	57	100	77
OP#6	82	67	57	50	62
CUMMULATIVE AVE. BY WEEK	177	100	86	92	108

B

C

MANUFACTURING YIELD DATA

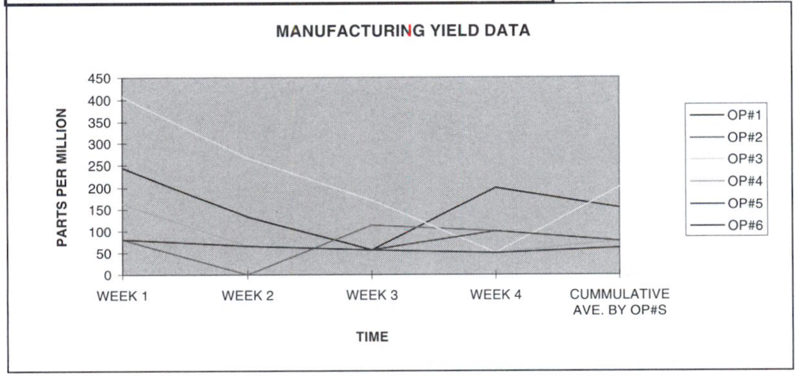

Figure 8-9. Manufacturing Yield Data

errors posted to that week's production and calculated PPM. At the bottom of the chart is the total number of operations (OP#1 through OP#6 at 12,250 each, or 73,500 total). This figure represents the total number of opportunities for errors. The number of errors are then totaled, and the PPM is calculated. This represents the cumulative yield information on the entire assembly operation by week.

The figure also shows the total production unit volume, the total number of errors, and the calculated PPM for each operation. At the bottom of the chart are the cumulative data for all parts produced, with total errors and calculated total PPM for the month. Figure 8-9c gives the summary for the month, showing totals of the data.

To deliver a sense of the trends occurring in this scenario, Figure 8-9b illustrates the trends by operation for the four-week period. Each operation in Figure 8-9c is graphed and shows the trends of the operations in terms of parts per million of errors. To the right of the chart, the intersection of the data line and the y-axis represents the cumulative average of the errors by operation number.

Through the use of manufacturing yield data, actual yield performance can be measured and recorded for the sake of manufacturing process improvement. A sample chart is available for your use in the *Toolbox*.

7. VERIFYING THE PROCESS AND PROCESS CHANGES

The importance of verification and certification of the manufacturing process must be underscored. Product quality can be repeatable only if the process by which the product is made is predictable. This means that control of the inputs and control of the environment will yield predetermined results given that the subassembly itself is stable. Figure 8-10 illustrates this point.

As shown in Figure 8-10, the discrete parts and/or subassemblies are input to the process, which is under control of the environment and input process variables. These input process variables can be temperature, pressure, humidity, and time, for example. With the process under control, the parts exiting are predetermined. Furthermore, the process can be repeated by using the same "recipe" at a later date or in a different location.

In a similar manner, any change to that process must also be verified and certified before it is integrated into manufacturing.

CERTIFICATION OF MANUFACTURING PERSONNEL

1. ALL SYSTEMS REQUIRE TRAINING OF MANUFACTURING PERSONNEL

The issue of certification transcends mere training of the team. It requires an interactive exchange of data, information, procedures, and talent development. The manufacturing personnel must be in a position to execute the procedures without direct supervision. This

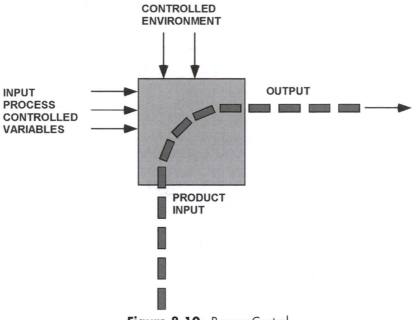

Figure 8-10. Process Control

statement might sound nontraditional at first, but in actual practice, a supervisor cannot always prevent mistakes, nor can every mistake be caught by an expensive inspection process. The goal is to have the people doing the work be knowledgeable enough to perform the tasks without mistakes in the first place. This requires training beyond the basics of assembly. Such training must address the issues of yield data, yield pathways to improvement, and data reporting.

2. CERTIFICATION OF THE OPERATION STATIONS

The goal of a manufacturing manager is to have a uniform homogeneous workforce that is trained in all operations at the same level. This scenario allows maximum flexibility when rotating personnel and dealing with interruption of the workforce. To accomplish this goal, the manufacturing manager must train and certify each member of the manufacturing team on all of the operations. In addition, manufacturing must keep the training fresh and updated.

A record of the training and proficiency of each member should be developed and kept up to date. Figure 8-11 shows an example of the data that needs to be kept on each employee. The simple record keeps track of the employee, basic training requirements such as safety, and the specific training for manufacturing operations. It confirms that the training was complete and notes the date of the training. A blank chart is included in the *Toolbox* for your use.

PERSONNEL RECORD OF CERTIFICATION									
EMPLOYEE	DATE OF HIRE	SAFETY	OP#1	OP#2	OP#3	OP#4	OP#5	OP#6
1	4/2/1990	Y	Y / 4/19/98	Y / 4/19/98	Y / 4/19/98	Y / 4/20/98	Y / 4/20/98	Y / 4/20/98
2	5/5/1991	Y	Y / 4/19/98	Y / 4/19/98	Y / 4/19/98	Y / 4/20/98	Y / 4/20/98	Y / 4/20/98
3	3/2/1979	Y	Y / 4/19/98	Y / 4/19/98	Y / 4/19/98	Y / 4/20/98	Y / 4/20/98	Y / 4/20/98
4	5/6/1989	Y	Y / 4/19/98	Y / 4/19/98	Y / 4/19/98	Y / 4/20/98	Y / 4/20/98	Y / 4/20/98
5	6/7/1990	Y	Y / 4/19/98	Y / 4/19/98	Y / 4/19/98	Y / 4/20/98	Y / 4/20/98	Y / 4/20/98
6	12/3/1992	Y	Y / 5/20/98	Y / 5/20/98	Y / 5/20/98	Y / 5/21/98	Y / 5/21/98	Y / 5/21/98
7	1/4/1993	Y	Y / 5/20/98	Y / 5/20/98	Y / 5/20/98	Y / 5/21/98	Y / 5/21/98	Y / 5/21/98
8	2/5/1994	Y	Y / 5/20/98	Y / 5/20/98	Y / 5/20/98	Y / 5/21/98	Y / 5/21/98	Y / 5/21/98
9	3/6/1995	Y	Y / 5/20/98	Y / 5/20/98	Y / 5/20/98	Y / 5/21/98	Y / 5/21/98	Y / 5/21/98
10	5/3/1996	Y	Y / 5/20/98	Y / 5/20/98	Y / 5/20/98	Y / 5/21/98	Y / 5/21/98	Y / 5/21/98

Figure 8-11. Manufacturing Employee Certification

3. REFRESHER TRAINING OF PERSONNEL

The initial training must be followed by refresher training periodically to ensure that the manufacturing process remains consistent over time. As a natural course of events, the importance of certain methods and procedures will degrade. This degradation is due to the fact that human beings are involved in the process; they possess biases, experience boredom, and have selective memories. Operator refresher training can go a long way to ensure a degree of consistency in the process in spite of degradation.

4. INSPECTION NOT A SUBSTITUTE FOR CERTIFIED TRAINING

Contrary to an older school of thought, inspection is not a substitute for training, nor can it ensure any level of consistent quality. The quality must simply be instilled at the beginning of the process. In addition, training teaches and encourages good habits and procedures. At best, inspection can only point out errors. To be effective, the errors must be prevented, even to the extent of taking extra time to train the personnel.

5. HANDLING OF CRITICAL COMPONENTS—MORE THAN A GUIDELINE

There is one additional element to be considered in the area of training and procedures: the issue of handling critical components. Every product has critical components, from electronic circuits to mechanical assembles to sensitive materials. Training in handling critical components can reduce warranty expenses and scrap significantly.

The scrap occurs in the factory, where mishandling causes the part, assembly, or material to no longer be acceptable in terms of quality or function. The warranty expense occurs when mishandling in the factory causes latent failures in the field, causing customer complaints and lost reputation. Both scrap and warranty expenses cause problems, so the issue of handling critical components must not be addressed lightly.

PROCUREMENT AND PARTS CONFIGURATION

Having addressed the issues of manufacturing process control and certification of personnel, we now focus on the procurement side of the equation. In order for the process to be in control, the input to the process must also be in control. Here, vendors play a key role in contributing to manufacturing quality. In these next sections, we focus on procurement, parts configuration, and certification of the vendor.

1. SETTING UP BILLS OF MATERIAL

The financial success of a product is related to the product configuration. The product configuration is the blueprint for how the product line will be offered to the market place, factoring in the different versions, sizes, ratings, and features. How design and manufacturing package the offering can either simplify the inventory management or complicate it to the point of ineffectiveness. The issues extend beyond inventory management and the related expense for finished goods inventory. They include the general confusion factor in manufacturing when there are many versions, offshoots, and "forks" in the evolution of the product design. The management of these "offshoots" can soon become unwieldy and begin to consume resources for maintenance and support, as shown in Figure 8-12a.

The diagram in Figure 8-12a depicts a product line that consists of a power electronics section, a logic section, associated software, and some mechanical packaging configuration. In this example, there is one version of a power module, two versions of a logic module, four versions of software, and three versions of packaging. This generates a total of 24 versions of a product offered to the marketplace, as shown in Figure 8-12b.

If the market demands a finished goods inventory in place and on a shelf waiting for disbursement, the product configuration can represent a significant value of finished goods inventory. In addition, there are two versions of logic boards and four versions of software to support. If an improvement or correction is required in the core of a logic board in hardware, for instance, then two boards will require modification, and two locations of inventory must be reconciled. If the issue is serious enough, scrap and rework will apply to the inventory already built.

Consider the product line depicted in Figure 8-12c, in which the number of versions of logic module is reduced to one and the software is conditionally compiled to allow selection of the four different versions in the factory or the field. In this example (see Figure 8-12d), the amount of overhead support required is drastically reduced to three versions. The end product can still be available in 24 different and distinct versions; however, inventory and overhead for product support are reduced.

Other points to consider when setting up the bills of material are the design and management of parts cost and vendor negotiations. Figure 8-13 shows an example of a bill of material showing the part, vendor, and other pertinent information that can be used as a management tool.

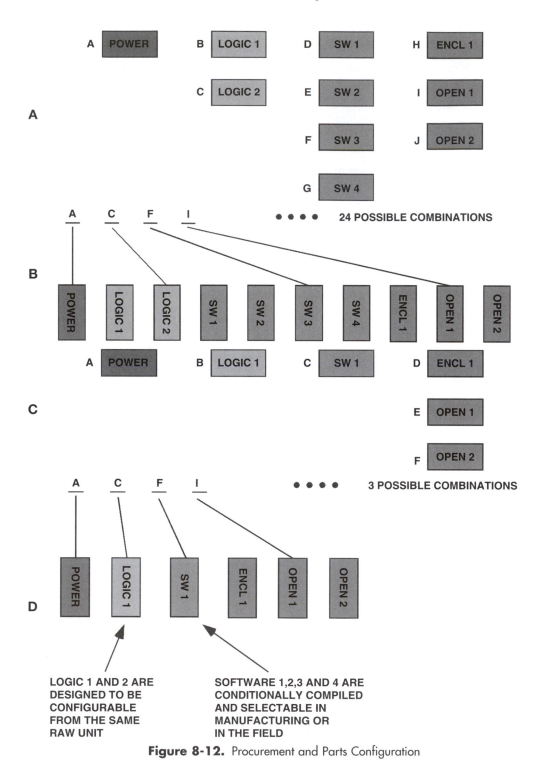

Figure 8-12. Procurement and Parts Configuration

BILL OF MATERIAL

ITEM	PART NO.	DESCRIPTION	VENDOR	VOLUME PER UNIT	INITIAL BOM PX-1 EACH	ECONOMIC ORDER QTY	FINAL BOM PX-2 EACH
ASSY #1							
1	100123	PART A	ABC CO.	5	34	1000	29
2	100124	PART B	BCD CO.	6	35	200	30.35
3	100125	PART C	CDE CP.	7	36.75	350	43.25
4	100126	PART D	DEF CO.	8	37	750	35.6
5	100127	PART E	EFG CO.	9	38.5	125	32

	DIFFERENCE EACH PART	COST IMPACT PER UNIT	TOTAL ANNUAL QTY	DOLLAR DIFFERENCE	LABOR MINUTES	LABOR COST	
ASSY #1						85	
1	5	25	4500	112500	10	14.17	
2	4.65	27.9		125550	11	15.58	
3	−6.5	−45.5		−204750	12	17.00	
4	1.4	11.2		50400	13	18.42	
5	6.5	58.5		263250	14.5	20.54	

	BURDEN LABOR	BURDEN MATERIAL	TOT FAC COST
ASSY #1	2.33	1.25	
1	33.01	181.25	373.43
2	36.31	227.625	461.62
3	39.61	378.4375	737.80
4	42.91	356	702.13
5	47.86	360	716.40
			2991.37

Figure 8-13. Bill of Material

In this example, Assembly #1 has 5 parts, each having a part number, description, and vendor. Depending on the number of identical parts used per unit, the volume number adjusts accordingly, and a total price is placed on the bill of material. Based on the economic order quantity and the product forecast, a negotiated price may be reached. This is the final price that will be used in the calculation of manufacturing cost. The difference between the initial and final cost is calculated, and the impact per unit is determined. Factoring in the anticipated volume, a dollar amount is presented for determination of materiality. Noteworthy in this example is the case in which a part (#100124) increased in price beyond its initial price; this has a negative effect on the dollars saved.

Next, the labor is factored in terms of time and expense and burdened expense (labor cost shown in white at 85 dollars/hour). Both material and labor are burdened, also shown in white in this example. Depending on the type of accounting system used, material might or might not be burdened. Finally, a total factory cost is calculated.

Although this is not a strictly formatted bill of material, it does illustrate the issues that must be addressed in defining one properly and completely.

A sample bill of material and a worksheet for defining a more complete one are included in the *Toolbox* for your use.

2. LINKING TO VENDORS

With a favorable product configuration and bill of material, the focus now shifts to establishing the part's documentation so it may be procured properly. This can be accomplished by the use of a part card.

The part card identifies all aspects of the part in order to procure it. It must be complete enough so that the wrong part is not procured. It provides the manufacturing linkage from the Bill of Material and manufacturing process documentation to the vendor and source of manufacture. Figure 8-14 illustrates a typical part card. Each item is outlined with an explanation of its contents. This is only an illustration of a possible part card format; a blank card is included in the *Toolbox* for your use.

3. SUBSTITUTES/SPECIFICATIONS

The issue of substitutes relates directly to the issue of product configuration. The procurement function is responsible to a large extent for the maintenance of the original product configuration. Substitutions made with unauthorized and unsatisfactory parameters as listed in the part card can eventually degrade the product or increase its cost.

There is one basic fact that cannot be compromised: The development engineering function is responsible for the selection of a part, not for procurement. Development is the entity responsible for the functionality and performance of the product. Development is also responsible for a product's reliability and quality. This function includes controlling the parts input to the makeup of the end product.

The corporation and the manufacturing system are responsible for creating a system that enforces the selection made by development, but the initial selection is development's, and any substitutions should be approved by the development function.

4. DOCUMENTATION OF SUBSTITUTIONS

In actual situations, the selection of an alternate component rarely has the exact parameter listing as the original. In many cases, the new parameters are examined and evaluated, and a decision is made to substitute. This is acceptable; however, the rationale and reasoning behind the substitution should be documented for future reference because there could be interrelationships between substituted parts that affect functionality and performance.

For example, substituted Part A from the bill of material shows minor differences in the parameters that by themselves may not affect the performance or the product configuration. When Part B from the bill of material is up for substitution, however, it must be evaluated within the framework of the Part A substitution. In order to keep the record effective, the original Part A substitution should be documented. Such documentation allows the development engineering personnel to evaluate the absolute effect on the product line.

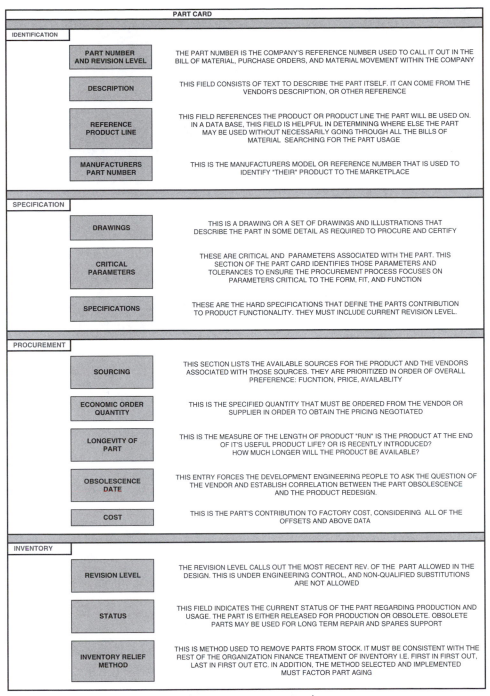

Figure 8-14. Part Card

5. BRINGING DESIGN-TO-COST TO THE VENDOR

The philosophy of design-to-cost should not stop with the company proper; it also needs to extend to a company's suppliers and vendors. The product development team can act as a conduit from the marketplace to the supplier. As the market places pressure on the manufacturer, the manufacturer should allow the pressure to be felt by the vendor. Indeed, this can be a healthy practice, as the vendor can assist with alternatives to meet the factory cost.

The consequences of no action result in the vendor as well as the manufacturer missing out on the market opportunity. So it is in the best interests of both parties to resolve the cost issue in cooperation rather than in a vacuum.

6. TRACKING VENDOR CHANGES

Up until now, we have centered the discussion of parts procurement and configuration on internally dictated changes. There are also changes forced upon the organization by its suppliers. These changes might in turn be forced on them by their own suppliers or internally generated.

In either case, vendor changes can have a significant impact on product maintenance, as development-engineering talent is consumed by qualifying vendor-initiated changes, obsolescence, or pricing actions.

Given the inconvenience and the internal cost, the changes inevitably must be tracked, documented and referenced. One way to accomplish this is to have the supplier or vendor issue a certificate of compliance on the goods shipped. A certificate of compliance is a formal document that ensures that the goods shipped against a purchase order have been certified by the supplier or vendor to be consistent with the parts specified on the part card and the purchase order. Certification gives the manufacturer some degree of assurance that the parts supplied are in fact equivalent to the specifications on the part card. Furthermore, the certificate of conformance creates a contractual issue with little margin tolerated for error or oversight.

7. FAILED PARTS DISPOSITION

The basic theme of our discussion is to encourage the supplier and vendor to become part of the business process, from specification through development to manufacturing and including field experiences. When a part fails either in house or in the field, the vendor will need the failed part as part of its customer feedback mechanism. This requires a process to facilitate the vendor's needs; Figure 8-15 illustrates the general format of that process.

8. TRACEABILITY SYSTEMS

The basic concept behind traceability is to generate a data logging system and establish access to information about the component parts and materials comprising the end product

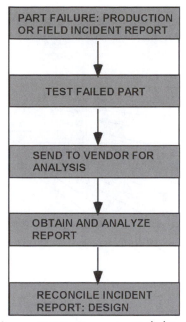

Figure 8-15. Processing Failed Parts

shipped to a customer. As the product progresses through the procurement and manufacturing processes and gets shipped to a customer through a sales channel, the ability to backtrack information on component parts and materials might be required as part of a future investigation of a mishap or for other reasons.

Consider Figure 8-16. In this simplified case, several parts comprise a product. The product is shipped through a sales channel to a customer. For whatever reason, the product fails, and an investigation into the failure ensues. The manufacturer needs to be in a position to be able to trace the parts backward through the chain, back to the source supplier of the part. This information can assist in assessing exposure for other products shipped and in identifying the scope of a recall effort or field upgrade effort.

Traceability provides answers to questions like the following:

- What batch is affected by the root cause failure of the failed part?
- What date codes are affected, and how do we correlate the product we shipped some time ago to the suspect lot or batch, to the product received from the supplier?
- Are the parts or subassemblies serialized from the vendor and posted to the product serialization?
- Have there been any other recorded incidences that are similar?
- How do we contain the problem?
- How does the supplier batch process for parts feed our inventory system?

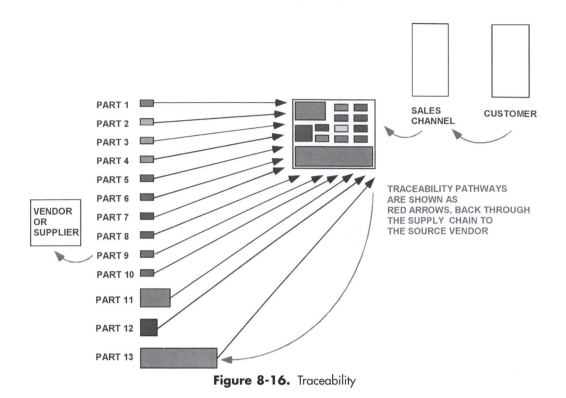

Figure 8-16. Traceability

- How were the parts pulled from inventory, and does this correlate to the batch shipped and the volume of product manufactured during the time frame?
- Has this happened before?

The best way to establish the correct amount of data and access required is to project situations and scenarios and determine what questions you will need answers to, to determine exposure. Start by gathering this data, and fill in as appropriate.

Some will say that there can never be enough data; however, one must design these overhead-oriented manufacturing systems to balance the need for data with the need to run the business.

Traceability systems are often used where these subsequent actions cannot be taken lightly. They are used when critical action to contain the problem must be taken and must be effective.

CERTIFICATION OF VENDORS

1. THE BENEFITS OF CERTIFICATION

The amount of activity required to double check each and every one of the items we have discussed would absorb human energy to the point of noncompetitiveness. How, then, can a manufacturer ensure that the parts and materials supplied meet the manufacturer's requirements each and every time? The answer lies in vendor certification. Vendor certification is an activity whereby a supplier is brought into the procurement process with understanding of its impact on product quality. It entails a series of procedures and internal controls agreed to by the manufacturer and the vendors, whereby the materials and parts manufactured at the supplier's location have been carefully specified, documented, manufactured, and shipped as part of an overall arrangement with the manufacturer.

2. MODEL OF VENDOR CERTIFICATION

Figure 8-17 is a conceptual flowchart illustrating a vendor certification transaction. There is a protocol for the vendor/manufacturer interface that results in a higher degree of assurance that materials and parts shipped will conform to development's specifications.

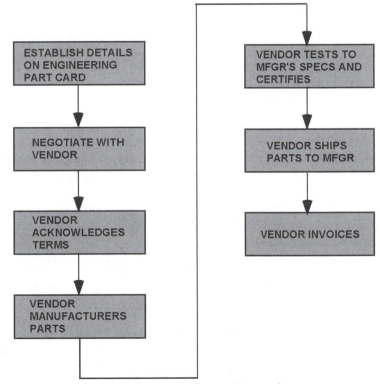

Figure 8-17. Model for Vendor Certification

3. VENDOR CLASSIFICATION PROGRAM

It simply is not practical for every part to be certified and every shipment qualified. Vendor certification should be used in cases in which critical parts to the design affect product performance parameters. When the form, fit, and function of the end product is affected, these parts are certain candidates.

Parts that affect product liability also require some sort of vendor certification. To make this process work, a vendor classification program may be used. The vendor classification program ties the importance of the part's product configuration to the degree of certification needed to ensure consistent quality levels. Figure 8-18 illustrates the correlation between types of suppliers and certification needs.

At the top of the figure is a gradient of vendor types, ranging from a fully integrated vendor to a commodity-oriented supplier. Along the left is a gradient describing the degree of contractual certification of goods shipped.

The following describes the gradient of four types of suppliers:

Integrated supplier: The integrated supplier is one who has an intimate relationship to the manufacturer. What one party does will directly affect the other. The dynamics of the relationship are symbiotic in nature.

Specified supplier: The specified supplier is characterized by a long-standing and stable business relationship between vendor and manufacturer. The "company" is specified along with the part.

Generic supplier with value added: This designation applies when there are many vendors who have a certain capability but only a select few are willing to modify or add value to the parts to suit the manufacturer's needs.

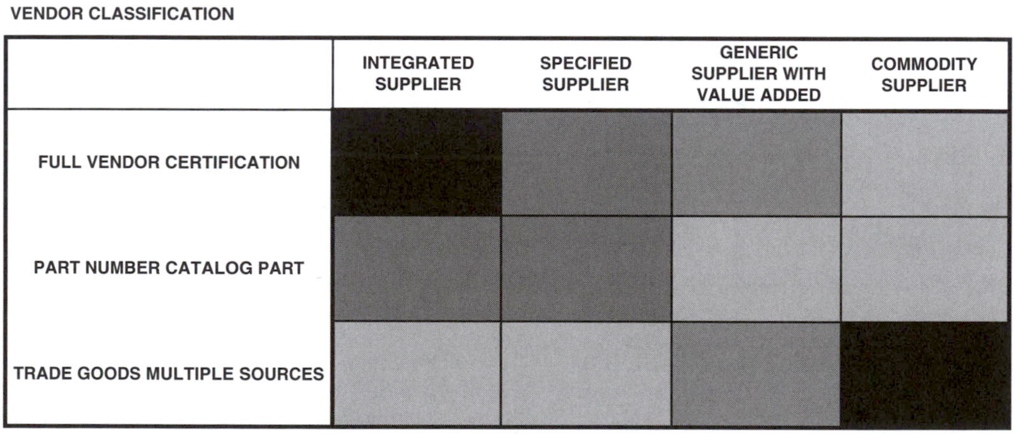

VENDOR CLASSIFICATION

	INTEGRATED SUPPLIER	SPECIFIED SUPPLIER	GENERIC SUPPLIER WITH VALUE ADDED	COMMODITY SUPPLIER
FULL VENDOR CERTIFICATION				
PART NUMBER CATALOG PART				
TRADE GOODS MULTIPLE SOURCES				

Figure 8-18. Vendor Classification

Commodity supplier: The commodity supplier supplies materials and parts that are readily available from many sources. Contracts are awarded based on price and delivery. Performance and quality are assumed and are generally industry standards, either formal or *de facto*.

The following describes the gradient of certification:

Full vendor certification: This designation applies when there is a full, contractual, interactive exchange of data between the supplier and the manufacturer. There is little left to assumption on either side. The transaction is documented completely.

Part number, catalog part: This applies when the supplier has a listing of parts in its catalog and the manufacturer selects a standard part and uses it in its own design. Depending on clarity and configuration controls issues, there is some documentation accompanying the transaction.

Trade goods, multiple sources: This applies to cases in which there are multiple sources of the same part or material. The properties are readily known, and individual part certification is not required.

The shading of the boxes indicates a correlation between the criteria: the darker the shading, the stronger the correlation. As can be seen, there is a strong correlation between the integrated "intimate" partner relationship between the vendor and the manufacturer and the degree of certification required. The closer to the commodity status the vendor's relationship, the more likely that the need for rigorous certification diminishes. However, each vendor should be classified based on your specific requirements as illustrated in the tradegoods-commodity square.

4. DELIVERY AS A PREREQUISITE TO QUALIFICATION

Part of the vendor qualification process involves determining the track record of a vendor with respect to delivery. The right product at the right price must be accompanied by the on-time delivery performance, or the vendor can cause significant disruption and wasted effort within your organization. Choose the vendor and supplier of the components parts carefully. Their performance contributes to your own!

5. MULTIPLE SOURCES

To what extent can the company rely on a single vendor? Most companies desire this type of relationship. With care and some relationship engineering, it is actually possible to achieve it for a time. As time progresses, however, conditions change, and a puristic reliance on a single source might not be practical in every sense.

As the part is being specified on the part card, alternate vendors are usually called out. A first-tier vendor is listed along with a secondary source. In actuality, it might be better to source from multiple vendors initially, to keep both channels open in procurement. This decision depends on volume, price breaks, and product cost sensitivity, all other things being equal.

Every supplier would prefer to single-source its customers, but from the manufacturer's perspective, this arrangement might not be advisable. A balance must be struck between the needs of procurement and the needs of manufacturing. Set up multiple sources when single-source failure would cause product-line failure for your company. In addition, for highly engineered parts, contractually reserve the right of self-fabrication or third-party fabrication in case of primary vendor failure. This should protect the company in steady state so that only the transition must be managed.

6. SOURCE DATA—PURCHASING BOOKS

Traditionally, as a part is specified and qualified for use on a product, it generally goes into some medium often referred to as a purchasing book. This is a collection of prequalified and available parts. This medium should be constructed from two perspectives as follows:

- Parts listing
- Vendor profile

The parts listing can be organized by description, part family number with a database link to description, or by some other classification. It should be interactive so that if a development engineer is seeking a part to accomplish a function, it should be possible to scan the available parts currently in use in the company. This would be accomplished by use of the part card discussed earlier in this chapter. The part card is the source data for the database system.

The vendor profile is the "reverse view" of parts supply, whereby the profile of the vendor is kept on company record. Vendor profiles are helpful for negotiations, future part design, and general business transactional analysis. An example of a vendor profile is shown in Figure 8-19.

As shown in Figure 8-19, the vendor profile can yield a great amount of information about the potential supplier. Be sure to also include the sales route to market and the service mechanism if applicable, to assist in resolving field problems at a later date.

In addition, a vendor profile format is presented for your use in the *Toolbox*.

By setting up all of the part cards on an interactive database, this search capability should be possible to accomplish. In addition, interactivity could be linked to the vendor for updated information in real time. As pricing and status changes of announcements come from the vendor, the information could automatically be loaded into the database for retrieval.

The vendor profile is also part of the interactive database, whereby information about the vendor can be loaded in real time and newly announced changes in the vendor's product line and organization can be updated automatically. An example of the workings of an interactive database is represented in Figure 8-20.

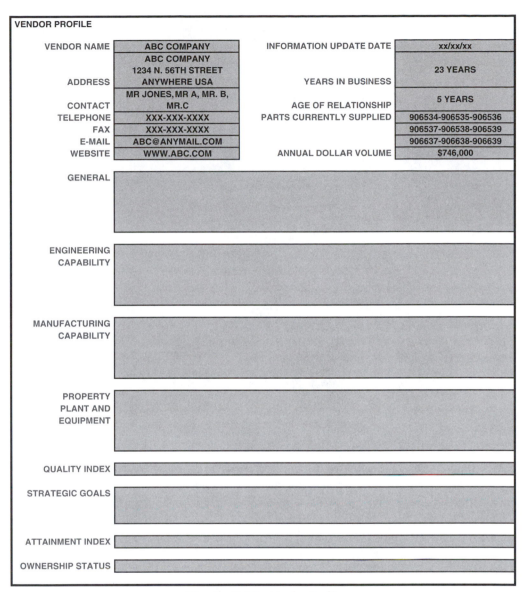

Figure 8-19. Vendor Profile

In this model, the vendor is linked to the company's management information system. The company has established records for procuring parts for the end product. The procurement record consists of two elements—the part card and the vendor profile. The part card can be the real-time record of updated cost information and lead times. In this way, both technical issues and logistical issues can be updated on a real-time basis.

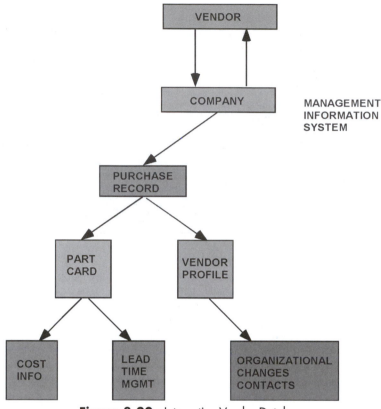

Figure 8-20. Interactive Vendor Database

The vendor profile record can also be linked to the interactive database in real time to update the manufacturer on personnel changes, obsolescence plans, new product lines, and any organizational changes that might affect the vendor-manufacturer relationship.

The system should be able to interlink all of this information for all of the parts and the vendors involved with the manufacturer.

INFORMATION RETENTION AND RECORDKEEPING

1. TYPES OF MANUFACTURING INFORMATION

Given the retention of certain vendor information, what is the appropriate amount of information to keep for general manufacturing data? How much data is enough, and how much is too much? These questions must be deliberated to engineer a data logging system in manufacturing that can assist in the resolution of field problems later.

Ideally, data should be captured at the point of labor content. Each operation should log the labor content and the materials involved. The materials and parts should be serialized where appropriate, and the database should capture this serialization. Taking it one step further, if the part is used in an assembly, it should be serialized and fit into a hierarchy of serialization for the end product, as illustrated in Figure 8-21.

As shown in Figure 8-21, each critical part of an assembly should be serialized. The database should then post the serial numbers of these parts to the subassemblies. The completed subassemblies should then be assigned a serial number. Each of the subassemblies should have its own serial number, and the list of these serial numbers should be posted to the main product serial number. In this way, a field problem can be traced back to determine critical information required for remediation.

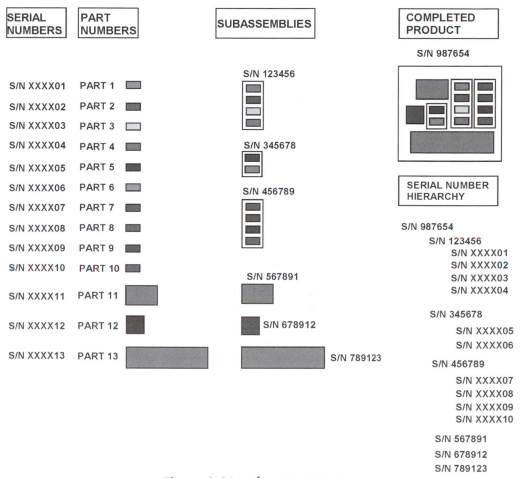

Figure 8-21. Information Retention

2. REASONS FOR RETENTION

As can be seen, retaining the information just discussed can be very helpful in managing a product liability incident. During one of these incidents, critical parts information is required to determine the company's exposure both financially and legally. By keeping track of the vendor's serial number and the product hierarchy of serial numbers, the scope of damage can be assessed readily using a database. This information is also useful in determining billing in a warranty situation. The serialization numbering system should be tied to manufacturing or receipt dates. The uncertainty in backtracking depends on inventory withdrawal practices. The parts are to be consumed in the order in which they enter the company.

The result of determining exposure to liability issues is generally a band of product serial numbers not in compliance rather than a specific list of parts. This is simply because of the size of the database that would have to be supported if every part were serialized and logged to the database. These high traceability requirements are part of the reason for the expense of military and nuclear products. Moreover, the better solution is to determine what scenarios might occur and design a practical system to minimize the amount of liability exposure.

3. DATA INTEGRITY AND ACCURACY REQUIREMENTS

The degree, to which the database system can be helpful is the degree to which accurate records are kept and posted. If the database becomes compromised, its usefulness in a product crisis becomes negligible. Be sure to make the database for manufacturing recordkeeping an integral part of the quality requirement for manufacturing. Getting the correct serial numbers posted to the parts, subassemblies, and products is as important as correct assembly. In effect, it is an integral part of the manufacturing process itself.

4. MODEL FOR DATA, ROOT CAUSE ANALYSIS, AND CORRECTIVE ACTION

In the next section, a model for the data set, root cause analysis, and corrective action will be presented. Although this model will not necessarily fit all circumstances, it does represent and illustrate the philosophy of product control and disposition of field feedback.

The model consists of two basic reports driven by a database. The first report is the record of product return and failure. It represents the source data for the incident. The second report is a database report that accumulates information from the incident reports for compilation and analysis. The second report makes use of the sorting capability of the various fields to determine trends and the status of specific issues currently affecting the product, as well as trends of historical experiences.

This model is intended to demonstrate the value of a database that can be sorted for information. Depending on the specific information desired, the database could be sorted differently. Sorting capability is where the real value of this type of database lies.

Depending on your own information system and data retention methodologies, the implementation might vary somewhat; however, the basic structure remains the same. Figure 8-22a shows a sample format for an incident review.

As shown, the incident report is divided into five major sections. They are as follows:

A. Incident report information

B. Repair analysis

C. Repair information

D. Failure analysis

E. Remedial action

The incident report information cites basic information about the failure or return of the product. It contains logistical information and cites appropriate serial numbers and product descriptions to be used in the database.

As shown in Figure 8-22b, after the unit is received and logged into the repair department, the repair of the unit commences. The repair analysis record cites detailed information about specifics such as part numbers, quantities, repair orders, personnel involved, production dates, and revision levels. It also records the exiting revision level of the product after the repair.

The repair information then dovetails into logistical repair information regarding billing and product disposition, as shown in Figure 8-22c. The disposition might be to repair and return, to scrap the unit, or to repair and place it in rotating repair stock.

Next, as shown in Figure 8-22d, the failure analysis is performed as part of the repair process. This process translates the repair of the product into specific categories, which might require subsequent action. In a rigorous system, every failure must result in some action to prevent future failures. The failure analysis is designed to determine the root cause of the incident and force it into a specific category so that the failure can have an associated corrective action. In this example, the vendor was chosen as the root cause of the failure. It could have been any of the other causes listed, such as design, customer abuse, or even no fault found. The information cited in these records is used to generate the summary report, which will be reviewed later.

Finally, the analysis must result in some action to remedy the incident or failure. In the sample case shown in Figure 8-22e, the vendor failure was serious enough to warrant a design change that designs around the part or the vendor. A cognizant party reviews the repair and failure analysis, and a list of remedial projects is generated for engineering change. The possible action types in this section are as follows:

A. Engineering change

B. Manufacturing operator refresher

C. Alternate vendor selection

INCIDENCE REPORT INFORMATION

A

REPORT DATE	XX/XX/XX
REPORTED BY *(INTERNAL)*	EMP #2
REPORTED BY COMPANY	ABC CO.
REPAIR ORDER REFERENCE	R 123456

PRODUCT FAMILY	1000
USER COMPANY NAME	BCD CO.
CONTACT	MARY
RECEIPT DATE	XX/XX/YX

REPORTED ITEM DESCRIPTION	PRODUCT CHASSIS	REPORTED ITEM NUMBER	312-345	MFGR'S SERIAL NUMBER	3025678.01

DESCRIPTION OF FAILURE OR SYMPTOMS	UNIT FAILED CATASTROPHICALLY

REPAIR ANALYSIS

B

ANALYZED SYMPTOM	UNIT INOPERATIVE, FAILS FUNCTIONAL TEST

FAILED ITEM PART NUMBER	123456
FAILED ITEM DESCRIPTION	COMPONENT
REPAIR ORDER	R 123456
REPAIRED BY	1668
TECHNICAL ADDITIONAL INFORMATION	UPGRADED TO REV 4

SERIAL NUMBER . ECL	9945.03
PRODUCTION DATE	XX/XX/XY
REPAIR DATE	XX/12/XY
UPDATED SERIAL NUMBER . ECL	9945.04

PART NUMBER	DESIGNATOR	CODE	DEF QTY	DESCRIPTION
315-569	R-12	UPGRADE	2	PART XX12

REPAIR INFORMATION

C

PRODUCT UPGRADED FOR IMPROVED PERFORMANCE AT NO CHARGE

REPAIR DISPOSITION		RETURN	SCRAP	ROTATE

FAILURE ANALYSIS

D

ROOT CAUSE PART NUMBER	315-569	DESCRIPTION	PART XX12

COMMENTS
NONE

DEFECT TYPE	VENDOR	DESIGN PROCESS VENDOR LIFE ABUSE SETUP NONE UPGRADE

REMEDIAL ACTION

E

ACTION BY:	1668	DATE:	XX/13/XY
REVIEWED BY:	1234		

ACTION TYPE:	DESIGN CHANGE	ECN OP. REFR NO ACTION ALT. VEND MFG REV.

ACTION COMMENTS:
COMPLETED DESIGN CHANGE FOR PRODUCTION, FIELD RETURNS NOT YET COMPLETED

Figure 8-22. Incident Information

DETAILED ANALYSIS / EXAMPLE SORT				
PRODUCTION DATE	PART DESCRIPTION	DEFECT TYPE	SERIAL NUMBER	REPAIR ORDER
XX/XX/XY	PRODUCT CHASSIS	VENDOR	3025678.01	R 123456

RECEIPT DATE	REPORTED SYMPTOM	ANALYZED SYMPTOM	REPAIR INFO	ROOT CAUSE COMMENT
XX/XX/YX	UNIT FAILED CATASTROPHICALLY	UNIT INOPERATIVE, FAILS FUNCTIONAL TEST	UPGRADED TO REV 4	PRODUCT UPGRADED FOR IMPROVED PERFORMANCE AT NO CHARGE

ROOT CAUSE P/N	ACTION TYPE	ACTION COMMENT	ITEM ECL
315-569	ACTION TYPE:	COMPLETED DESIGN CHANGE FOR PRODUCTION, FIELD RETURNS NOT YET COMPLETED	UPDATED SERIAL NUMBER . ECL

F

Figure 8-22. Incident Information

D. Manufacturing process revision

E. No action taken

The chronology of events and data involved in the incident is included in the incident review record. This information will be used to construct the summary discussed next. The value of the summary lies in the ability to determine trends and chart the incidence and resolution of field problems with the product by sorting the data in various ways to expose the information. Figure 8-22f is a general format for the summary. It contains only the information shown in the example record. By itself, the one entry of data does not indicate any trend; however, the configuration of the data and the flexibility to sort demonstrates its value.

One of the most useful ways to organize data is by sorting all of the incident reports in a multilevel format to expose information desired. Figure 8-23 illustrates the order of the multilevel sort.

For example, one can sort by product description, then by root cause, and then by production date. This type of multilevel sort arranges the data (and carries all collateral data) by the product description first. Then it goes through each of the product descriptions in the database and groups all like descriptions together, along with the collateral data associated with them. It will further sort within the product description categories according

SORT ANALYSIS

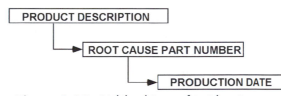

Figure 8-23. Multilevel Sorts of Incident Data

to the root cause of failure causing the incident. Next, all of the like root cause failures within a product description are grouped. A third-level sort sorts this group of root causes by production date.

A sort like this can show the incidence of a field problem and the resolution. The resolution is shown by a corrective action being put in place and the absence of incidences after the production dates with the "fix" in place.

A simpler sort of product description and root cause yields the raw data for a Pareto analysis. A sort of product description and then defect type can show development performance by charting all of the design-related defects. In short, the multilevel sort can be an invaluable tool in minimizing field problems by early recognition and resolution.

A sample incident report form and summary are available in the *Toolbox* to model a system suited to your specific needs.

5. IDENTIFYING POTENTIAL FIELD PROBLEMS

The cognizant manager needs to constantly review and monitor field reports. This need is greatest immediately following introduction of a product; however, periodic review is required throughout the life of the product. What are the criteria that constitute an "event"? The answer to this question lies in the acceptable level of field returns and cost of warranty. Ultimately, the answer lies in the customer's aggravation level experienced in the usage of the product.

Product quality events typically take on a characteristic profile. The intensity can vary; however, the shape of the profile seems to be consistent. Figure 8-24 illustrates the expected profile for a product "event."

As shown in Figure 8-24, the incident starts with a response from the field through repair or direct returns. As the incidents occur, the data is logged in. In this example, all like incidences are charted via the sorting mechanism, and like root causes are listed with the production date. Assuming the organization is responsive, a design change is qualified and implemented to root cause Part Number 312315 on 1/10/98.

The database then tracks failures due to root cause part numbers after this date to determine effectiveness of the "fix." The database is then monitored to see whether the field failure occurs after the "fix" implemented on 1/10/98. The database shows one entry involving a different root cause Part Number on 12/1/98. This would indicate that the product "event" has been resolved effectively.

The absolute dates are important here, as design changes needed to "fix" the design are implemented in an absolute manner. If the fix is real, there will be no more repairs or field incidences due to that particular part.

Repairs or incidences with production dates before 1/10/98 at any point in the future (depending on the type of failure mode) might continue to trickle in. The numbers will rise to a peak level and then die off as time passes on, as illustrated in Figure 8-24.

EVENT MONITORING

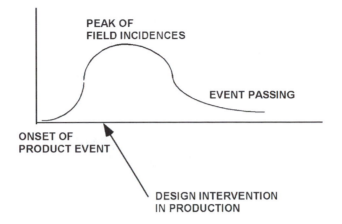

PRODUCT DESC.   ROOT CAUSE   PROD. DATE

PRODUCT DESC.	ROOT CAUSE	PROD. DATE
MODULE A	P/N 312315	12/1/97
MODULE A	P/N 312315	12/5/97
MODULE A	P/N 312315	12/8/97
MODULE A	P/N 312315	12/10/97

**DESIGN INTERVENTION ECN TO CORRECT
IMPLEMENTED 1/10/98**

MODULE A	P/N 312315	1/5/98
MODULE A	P/N 312315	1/6/98
MODULE A	P/N 312315	1/7/98
MODULE A	P/N 312315	1/9/98

NO FURTHER RECORDS ON 312315 ROOT CAUSE

MODULE A	P/N 312300	12/1/98

Figure 8-24. Event Monitoring

To manage the process of a product correction effectively, both types of reports need to be monitored. The profile indicates the financial and company exposure, and the sorted chart of data indicated dates and "fix" effectiveness.

6. EVENT TRIGGERING

If the manufacturing process tracks manufacturing performance in terms of parts per million (PPM), then the field incidence should trigger on an appropriate level also. Depending on the stability of the design and the stability of the manufacturing process, the trigger points may be adjusted. The operative point is to set the trigger at a point to get an early warning of a product problem so the company can effect a solution and implement it in a timely manner.

FIELD PROBLEMS AND EVENT STATUS MONITORING

1. THE IMPORTANCE OF FIELD MONITORING

Field monitoring of product failures or "events" helps the company from three basic perspectives:

A. Financial exposure

B. Product reputation

C. Company reputation

From the financial perspective, the product problem can be quite a financial drain on the company. Throughout history, many companies have been able to point to a product problem or field recall that financially stressed the company. Such problems also cause cash redeployment that negatively affects future success. Consider the financial model in Figure 8-25 (it shows the initial development investment and the return).

As Figure 8-25 shows, the original investment is offset by the original planned return. The sales revenue increases disproportionately early in the life of the product. The field recall issue then causes cash to be spent on remedial activity rather than on funding the investment required in inventory and accounts receivable during the normal growth pattern. This pattern of resource allocation then causes the return of the product to be significantly lower than expected. In addition, the energy and intellect that should be applied to reading the marketplace and generating the next opportunity are spent on the remedial work at hand. This example shows why it is imperative to confront field problems at onset and act decisively to mitigate their impact on the organization. By acting early, the installed universe of product is smaller and more manageable.

It is also important to preserve the product's reputation as much as possible. This is effectively accomplished by acting to correct any problems as early as possible. A poor reputation gathered by a product early in the introduction can decimate the sales effort and initial

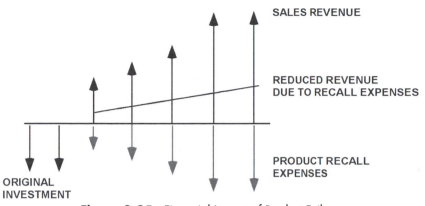

Figure 8-25. Financial Impact of Product Failure

launch. In some cases, damage to the product is done through initial feedback through the sales channels rather than from direct customer feedback. The company must act to reduce the impact of bad press through its own sales channel to pave the way for increasing sales after the product problem is resolved.

Left unresolved, a product problem can eventually reflect poorly on the entire company's reputation. Given the rhetoric of quality and continuous improvement, failure to act will indict the company for poor performance. The customer base is generally forgiving for mistakes, nonlethal product issues, and issues of inconvenience; however, they have little patience with companies that choose to ignore problems or that fail to effectively address them in a timely manner.

2. THE MYTH OF THE THROWAWAY PRODUCT

"We don't have to worry about the product after it is shipped, it's a throwaway product." This familiar quotation reflects a manufacturing attitude that provides no feedback loop to determine and improve quality. The useful life of a product might have nothing to do with its required quality level or its acceptance in the marketplace.

For example, a syringe for infusion of drugs into a human vein has a limited useful life. It is, in effect, a "throwaway" product after use; however, the quality requirements for tolerances, hygiene, and protection are at very high standards. Just because the product is disposable is not an excuse for poor quality.

Furthermore, the manufacturer must not ignore field feedback simply because the product is disposable. The initial feedback is invaluable to the manufacturer in terms of improving processes and design criteria; it closes the loop on the entire product offering. For high-volume product lines, failure to heed the warnings of initial field feedback can result in massive recalls, product liability issues, and expense in replacement. The future success of a product lies in the details of product failure.

3. PARETO ANALYSIS

Pareto analysis was discussed in Chapter 7 but will be reviewed here for completeness as part of the ongoing field monitoring required to ensure delivery of a high-quality product. The product failure and resolution summary should be consistently mined for trend data and to recognize the onset of product failure problems. The criteria should be set up for a trigger analysis and the reports monitored for the characteristic profile of a problem. Development personnel should then be deployed to rectify issues immediately as they occur. These practices formulate the product maintenance portion of the development effort.

4. A MODEL FOR QUALITY MANAGEMENT

To tie together the presented concepts into a framework that closes the loop on the product development offering, the product quality system example in Figure 8-26 can be used to illustrate the interrelationships and information flow among reports, summaries, and databases.

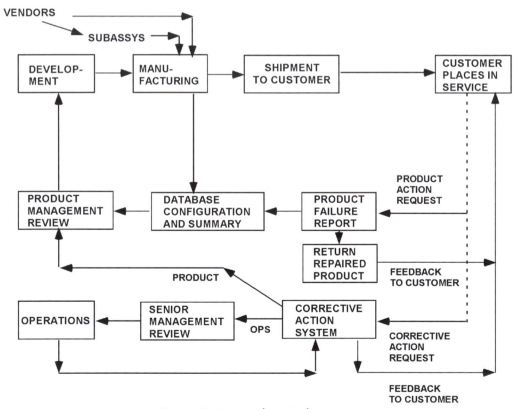

Figure 8-26. Product Quality System

As shown in Figure 8-26, the process starts with the development of the product. (In real-world practice, the complete process is much larger and more involved than this, but for purposes of illustration within the framework of the chapter's subject of manufacturing development, this example should suffice.)

Development feeds a fresh design to manufacturing. Along with vendor participation, manufacturing then builds the product and ships to the customer. The customer then places the product in service. Manufacturing contributes the appropriate records to the database configuration and summary system. Assuming the product does not fail, this example represents a typical transaction.

If there is a failure, the product will be returned to the company for analysis via a product failure report. The database is then updated with the product failure information and the root cause posted for product management review. The output of this information might be a design change, a vendor change, or a part replacement that must be executed by development engineering. Feedback should then be given to the customer about the failure and any product actions initiated.

If the customer issues a formal complaint with the company, the complaint is entered into the corrective action system for senior management review and disposition. Product-oriented requests and complaints are directed to product review. Operational requests are directed to operations. Any changes made are implemented through operations, and the customer receives communication regarding the results of the complaint.

This example shows a general outline of how information should flow for the purpose of acknowledging and resolving a product issue. Depending on the individual circumstances, the flowchart might be more complex or simplified. The operative point in the analysis, however, is to confront product failure issues directly and to resolve them accurately and quickly.

FORECASTING

1. THE LINK BETWEEN THE CUSTOMER SALES/MARKETING AND MANUFACTURING DISCIPLINES

"Our crew can build the products and get them out on time; just give us a forecast!" So goes the manufacturing lament. The forecasting process is a highly important one for balancing the expenses and the performance of manufacturing. Incorrectly forecasted demand causes inventory errors and contributes to starts and stops in manufacturing cycle, which can affect overall product quality. Consequently, forecast data does not begin and end with the marketing commitment to get the product development initiated.

The forecast process is, in effect the "clock rate" at which the procurement and manufacturing machines must run. A high forecast drives more materials and demands more labor quickly during the ramp-up process. Specificity in the forecasted product versions brings manufacturing to act on materials. If the materials are not the versions needed in the marketplace, inventories will swell needlessly.

The operative point of discussing forecasts is that neither product development nor manufacturing can be totally effective without an accurate sales forecast. The same care and diligence used in the development of the product should be applied to cementing the real-time link between the sales/marketing forecast and the manufacturing machine.

2. THE IMPORTANCE OF MATERIALS, LABOR, AND THROUGHPUT

Manufacturing has three basic elements to contribute to the new product's financial success. They are the procurement of materials, the application of appropriate skilled labor, and the organization to translate the materials and labor into shipments. Throughput allows the organization to reap the financial rewards of the development. The forecast is used to direct these elements so that throughput satisfies the customers.

Consequently, the throughput is only as beneficial to the overall program as the accuracy of the input. It matters neither what the quality level of a product is, nor the specification on performance, nor the time required to deliver it—if the product is not the one requested by the customer or marketplace, there is no customer satisfaction.

Furthermore, the materials, labor, and overhead of the manufacturing machine all have costs associated with them. If the cost is not offset by a disproportionate level of revenue resulting in a profit, the manufacturing machine becomes inefficient.

3. THE REQUIREMENT FOR ACCURACY

The requirement for accuracy of forecasting has already been demonstrated; there must, then, be some optimum method for generating an accurate forecast. Unfortunately, there really is none. Rather, the "art" of forecasting is a work in progress, whereby continuous improvement is the driving force.

From the "iron salesmen" of the 1970s automobile companies (who drove finished goods inventory from Detroit into the showrooms of dealers) to the mass customization efforts of many products (whereby individual orders drive this afternoon's manufacturing), forecasts can contribute to or prevent financial success.

The process is twofold. The first step is to feed back to manufacturing what the marketplace needs and wants. It is also a planning means to place goods where the company feels the demand will be. To be successful, the process must include elements for both market intelligence and company planning. Figure 8-27 illustrates the point.

As shown in Figure 8-27, the forecast process illustrated by the face "looks" out into the marketplace to determine usage patterns. These "eyes and ears" to the customers must either transcend the sales channel or become an integral part of the sales process. The forecast process also enables introspection into the future desires of the company and makes it possible to work toward a strategic objective. The forecast function then coalesces this information into a usable forecast. The process doesn't stop there, however. See the two-way arrows? The future profile of the company must be driven. This means

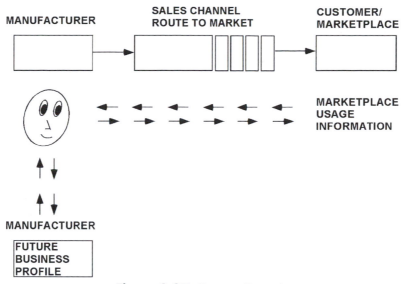

Figure 8-27. Forecast Dynamic

both determining the marketplace needs and directing those needs toward the end goals of the company.

The process is dynamic and changing. Success and accuracy are in the details of continuous improvement.

4. REFERENCE BACK TO PRODUCT CONFIGURATION: DON'T DESIGN IN FUTURE SHORTAGES

As discussed previously in Chapter 7, there are strategies for configuring a product that can minimize the problems of version forecasting. Not all problems can be eliminated, however. The product design can reduce dependence on forecasts for different versions of a product by incorporating the version selection into the product itself. The result is usually some cost penalty, but if the market can tolerate the cost/price impact, this can be a method to make the issue more manageable, given poor forecast information from the sales personnel.

Refer back to the product configuration to determine the company's critical dependence on forecast.

5. SALES PROGRAM SYSTEMS

The company should have a sales program and forecast system in place. In fact, these should be components of the new product identification vehicle. If the programs in place are not sufficient to generate accurate forecasts, the company should take time to correct them and make them useful.

To further the point, forecasting for a new product is not software. It is market intelligence. Be sure to have the sales force stay close to the customers to determine usage patterns and forecast levels. Software cannot do this unaided; it is only part of the information system used by management, and it is a medium for communication. The information must be assimilated and analyzed by the sales and marketing personnel.

CYCLE TIME MANAGEMENT

In this section, we review the dynamics of the cycle of manufacturing. This cycle deals with the processes of order entry, material procurement lead times, and labor scheduling. The cycle time of a product involves multiple aspects, many of which are beyond the direct control of the manufacturer.

The challenge is to manage the cycle time in such a way as to make it appear from the perspective of the customer as though the manufacturer has total control of it. The customers place their orders and trust in the manufacturer for completion in a timely manner.

1. PLANNING THE LEAD TIMES

"So, what is the lead time on this product, anyway?" This frequently asked question, rooted in the dynamics of intercompany cooperation, has multilevel complexity. The process by which materials are ordered and parts are manufactured, shipped to the primary manufacturer, and assembled into the final product for shipment involves many schedules, priorities, and sequencing. To illustrate this complexity and interdependence, consider Figure 8-28, which shows a feeder supplier to a manufacturer.

As shown in Figure 8-28, the process of delivering a manufactured product to a customer is divided into three basic segments. Referring to the customer at the end of the process, the order for the product is placed to the manufacturer. The manufacturer performs the contract review, procures parts, and builds the product.

The parts, however, are probably available from a supplier. The manufacturer, then places orders to the supplier. The parts might be standard parts or customized, highly engineered parts. Each follows a slightly different pathway through the procurement process. If parts are standard, pre-engineered parts, the order from the manufacturer placed to the supplier is reviewed in contract review, materials are procured, and the part is manufactured and shipped to the end product manufacturer for reshipment to the customer. If they are customized parts, the process includes a stage in engineering. The several parts comprising a bill of material are depicted by the arrows shown entering these supply chains to the end product manufacturer.

Each of these parts then can be sourced by a subcomponent vendor or supplier. This process is shown on the top rung of the illustration. Here, a subvendor goes through the

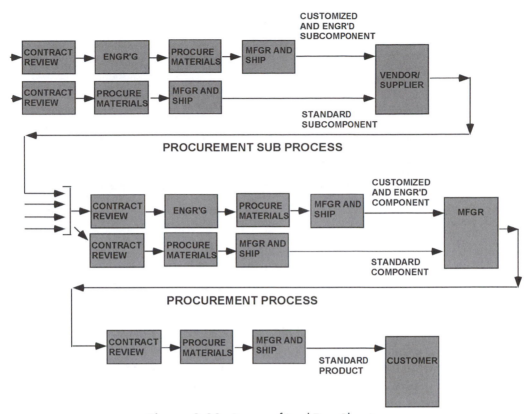

Figure 8-28. Process of Lead Time Planning

same process to deliver the part used in the bill of material. These parts are depicted as either customized or standard.

Figure 8-29a shows a more concrete example of the process that generates lead times in producing a product.

In this example, the end product (Module A) is comprised of 10 parts labeled Part A through Part J. Each of these parts has an associated lead time in procurement. In this example, Part A has the longest lead time, 24 weeks. The chart also shows the quantity of parts used in each product and the quantity on hand. The volume of end product that can be produced with quantities on hand can be calculated. This chart also indicates the quantity of product that is currently on order and the date on which it is due. Most manufacturing information systems drive these types of screens. This graphic and subsequent ones in this section are shown for illustration purposes only.

If we investigate Part A further; we see in Figure 8-29b that it is comprised of five subparts, shown as Parts A1, A2, A3, A4, and A5. Each of these parts has an associated lead time. As indicated, the third subpart, A3, is the part with the long lead time of 24 weeks. This lead time affects the delivery of Part A, which in turn affects the delivery of Module

PRODUCT SUMMARY

VENDOR LEAD TIME SUMMARY

BOM   MODULE A

ITEM	PART NUMBER	DESCRIPTION	QTY REQ'D / UNIT	QTY ON HAND	UNITS AVAIL	QTY ON ORDER	DUE DATE	LEAD TIME
								WEEKS
1	600123	PART A	5	50	10.0	200	XX/YY/ZZ	24
2	600124	PART B	4	73	18.3	220	XX/YY/ZZ	9
3	600125	PART C	3	123	41.0	220	XX/YY/ZZ	21
4	600126	PART D	2	232	116.0	220	XX/YY/ZZ	12
5	600127	PART E	5	99	19.8	220	XX/YY/ZZ	5
6	600128	PART F	4	57	14.3	220	XX/YY/ZZ	12
7	600129	PART G	3	76	25.3	220	XX/YY/ZZ	4
8	600130	PART H	2	28	14.0	220	XX/YY/ZZ	4
9	600131	PART I	5	23	4.6	220	XX/YY/ZZ	2
10	600132	PART J	4	10	2.5	220	XX/YY/ZZ	2

A

MAX UNITS 2.5   MAX LEAD TIME 24

Figure 8-29. Lead Time Summary

A, the end product. There are additional bills of material for the various Parts B, C, and D, and so on. In this example, Part A was the long lead part (due to Part A3). By investigating the bill of material further, we see the following (Figure 8-29c): Part A3 is comprised of five parts, namely: A3A, A3B, A3C, A3D, and A3E. Each of these parts has an associated lead time. In this example, Part A3C is the part with the longest lead-time (24 weeks). This penalizes Part A3, which then penalizes Part A, which in turn penalizes Module A.

Conversely, by focusing and improving the lead-time on Part A3C, the entire lead time of the end product can be improved up to the point of the lead part with the next longest lead time. With some parts, the lead time cannot be compressed easily; therefore, it could become necessary to forecast these parts more accurately or stock them. Often, however, it is difficult to forecast them, as they are "buried" in the product hierarchy and product versions. To forecast these types of parts, you need to forecast the versions of units of the end product, and manufacturing and material procurement should not necessarily be making this forecast—marketing and sales should.

For reference, a sample lead time summary is available for your use in the *Toolbox*.

2. MATERIALS MANAGEMENT

The basic definition of materials management is the timely presentation of materials and supplies at the point of labor with minimum amounts of inventory, carrying charges, and procured costs, and minimal movement between point of entry and usage. The function of materials management is to confront nonlinear order entry with a laminar flow of materials while achieving the cost requirements set forth in the original design specification.

BILL OF MATERIAL | PART A

ITEM	PART NUMBER	DESCRIPTION	VENDOR	QTY ON HAND	QTY ON ORDER	DUE DATE	LEAD TIME WEEKS
1		PART A1					10
2		PART A2					9
3		PART A3					24
4		PART A4					9
5		PART A5					3
						MAX LEAD TIME	24

BILL OF MATERIAL | PART B

ITEM	PART NUMBER	DESCRIPTION	VENDOR	QTY ON HAND	QTY ON ORDER	DUE DATE	LEAD TIME WEEKS
1		PART B1					9
2		PART B2					5
3		PART B3					8
4		PART B4					3
5		PART B5					6
						MAX LEAD TIME	9

BILL OF MATERIAL | PART C

ITEM	PART NUMBER	DESCRIPTION	VENDOR	QTY ON HAND	QTY ON ORDER	DUE DATE	LEAD TIME WEEKS
1		PART C1					21
2		PART C2					4
3		PART C3					20
4		PART C4					7
5		PART C5					3
						MAX LEAD TIME	21

B

BILL OF MATERIAL | PART D

ITEM	PART NUMER	DESCRIPTION	VENDOR	QTY ON HAND	QTY ON ORDER	DUE DATE	LEAD TIME WEEKS
1		PART D1					12
2		PART D2					3
3		PART D3					5
4		PART D4					8
5		PART D5					2
						MAX LEAD TIME	12

BILL OF MATERIAL | PART E

ITEM	PART NUMBER	DESCRIPTION	VENDOR	QTY ON HAND	QTY ON ORDER	DUE DATE	LEAD TIME WEEKS
1		PART E1					2
2		PART E2					4
3		PART E3					3
4		PART E4					3
5		PART E5					5
						MAX LEAD TIME	5

BILL OF MATERIAL | PART F

ITEM	PART NUMBER	DESCRIPTION	VENDOR	QTY ON HAND	QTY ON ORDER	DUE DATE	LEAD TIME WEEKS
1		PART F1					12
2		PART F2					2
3		PART F3					3
4		PART F4					5
5		PART F5					7
						MAX LEAD TIME	12

Figure 8-29. *Continued*

BILL OF MATERIAL		PART A1						
ITEM	PART NUMBER	DESCRIPTION	VENDOR	QTY ON HAND	QTY ON ORDER	DUE DATE	LEAD TIME WEEKS	
1		PART A1A					8	
2		PART A1B					3	
3		PART A1C					7	
4		PART A1D					10	
5		PART A1E					5	
						MAX LEAD TIME	10	

BILL OF MATERIAL		PART A2						
ITEM	PART NUMBER	DESCRIPTION	VENDOR	QTY ON HAND	QTY ON ORDER	DUE DATE	LEAD TIME WEEKS	
1		PART A2A					4	
2		PART A2B					6	
3		PART A2C					9	
4		PART A2D					4	
5		PART A2E					7	
						MAX LEAD TIME	9	

BILL OF MATERIAL		PART A3						
ITEM	PART NUMBER	DESCRIPTION	VENDOR	QTY ON HAND	QTY ON ORDER	DUE DATE	LEAD TIME WEEKS	
1		PART A3A					4	
2		PART A3B					5	
3		PART A3C					24	
4		PART A3D					8	
5		PART A3E					4	
						MAX LEAD TIME	24	

Figure 8-29. *Continued*

There are mechanized ways to procure material, but little can make up for an informed, interactive communication between the customer interface and the manufacturing element.

Frequently, a company might risk the expense of inventory to have parts available for and to anticipate market demand; however, this process is risky if the parts in question are expensive.

3. WORKLOAD MANAGEMENT

The basic definition of workload management is the timely presentation of the appropriate skilled talent, applied to the labor requirement to transform materials into finished goods. The function of this activity is to confront a varying workload that has varying requirements for talent, with human resources capable of completing the tasks at hand. The balancing of resources against workload requirements can be a difficult one, especially given the cyclic nature of business. It is difficult to retain personnel in a downturn, and their replacement is not immediate in an upswing, especially considering the demands of retraining new personnel.

4. VENDOR COOPERATION/COMPONENT SELECTION (AVAILABILITY)

To make the manufacturing cycle somewhat predictable and repeatable, the vendors and suppliers must become integral partners in the process. The development function can assist in the basic selection of parts to a large extent; however, the relationship between the manufacturer and the vendor will go a long way toward creating the laminar flow of materials and parts needed by manufacturing.

To make the relationship work, there must be value to and benefit from the relationship on both sides. The manufacturer or purchaser cannot always be the benefactor at the expense of the vendor, and the vendor won't continue obtaining business from the manufacturer at the latter's expense. Volume sales and uninterrupted schedules help both sides make money; consequently, both parties need to work toward those ends.

5. REVENUES ARE THE GOAL AND REQUIRE TIMELY SHIPMENTS

In the area of cycle time management, revenue generation—not necessarily the individual performance, productivity, and efficiency measurements of individual groups within the organization—is the primary goal. Each group needs to work toward the goal by making its specific contributions. It does not serve the cause of revenue generation to expedite only some but not all of the parts with long lead times, or to negotiate price concessions for one part of the product only to be offset by delivery problems and expedite charges on another part.

SYNCHRONIZATION

1. MODEL—THE RHYTHM OF MANUFACTURING

The model for the manufacturing process should be rhythmical in nature. Ideally, all of the individual departments and groups should function to create a rhythm in manufacturing. This synchronization is where leverage begins to occur and where the target costs for the product line begin to be achieved.

The product team needs to orchestrate the following issues. In the circumstances specific to your company, there could be additional issues; however, these examples will serve as a starting point:

- Product configuration
- Product design stability
- Quality requirements
- Parts selection
- Field experiences
- Forecast process

- Vendors
- Manufacturing process output
- Personnel consistency

All of these design and manufacturing issues affect the success of the product. Doing a diligent job on each of these will ensure that the product is delivered to the marketplace with the same expectations the company had of the product in the planning stages.

2. THE BENEFITS OF SYNCHRONIZATION

Synchronization does not occur naturally in business; it must be fostered, promoted, and instilled into willing participants in the business venture. Failure to accomplish this results in forces that contribute to disarray, overtime, errors, lower yields, increased costs, reduced quality, increased stress levels in the workforce, and higher rework and scrap costs. The benefits of efforts expended toward synchronization contribute to success in the new product venture.

The same care exercised in the development of the product should be exercised in the development of the manufacturing of the product. This ensures that the human and manufacturing energy will be applied to generating revenue for the corporation.

SUMMARY

In this chapter, several topics were reviewed with respect to the manufacturing of the new product. One of the objectives of this chapter was to establish the link between the development function and the manufacturing development function. Another objective of the chapter was to equip the reader with some rudimentary tools for setting up manufacturing process controls, given the availability of commercially available programs.

The chapter started with a discussion on the concurrency of development phases between design and manufacturing. The discussion reemphasized the importance of the interdisciplinary aspect of the new product development process and further emphasized the design for manufacturing philosophy.

Next, a discussion of the manufacturing layout was presented within the framework of the product and industry practices. The importance of maintaining product configuration and controlling changes to the product line was presented as a precursor to quality management.

Manufacturing process control was discussed, including certification of manufacturing personnel.

The focus then shifted to parts selection and procurement and the construction of the bills of material. Issues such as vendor qualification, documentation and certification were also covered.

The next section dealt with the quality database and the management of field product problems. Finally, information was presented on forecasting, cycle time management, the synchronization of manufacturing processes, and their effect on revenue.

With the basic design, qualification, and manufacturing elements in place, the fruits of everyone's labor is prepared for launch into the marketplace. This is the focus of Chapter 9, the pre-launch checklist.

THE PRELAUNCH CHECKLIST: SETTING UP THE ORGANIZATION

ABSTRACT: This phase of the product development activity deals with all of the preparation just before product launch. It discusses details generally left out in a nonstructured program. Each of the elements is discussed to ensure a completed program at launch, with the entire organization prepared and motivated to achieve success with the new product. It addresses final development-related issues, such as product certifications, the field test program, the incorporation of field feedback, and the training of personnel. It also reviews the literature preparation, the applications' support, the sales order entry system, and the final price setting. With the new product development program in its "home stretch," these are the finishing touches to ensure success!

PREFLIGHT CHECKLIST

1. SIMILARITIES BETWEEN LAUNCH AND FLIGHT

There are many similarities between the new product launch and the preflight checklist. The preflight checklist is a methodical review of all the elements associated with ensuring a successful and safe flight. The same holds true for the new product development prelaunch checklist. There is a protocol that must be executed to ensure all of the proper elements are addressed before launch. This represents good practice in the product development arena.

2. THE CHECKLIST

In Chapter 7, which deals with the actual development engineering, a list of deliverables was presented to ensure that all of the requisite steps are completed during the development process. This could also serve as a checklist at this stage of the project to double check that all deliverables have been addressed sufficiently and properly.

Often in the course of development, details may be missed, activities may remain incomplete, and specifics may be glossed over. It is a vocational hazard that can occur quite naturally because of the differences in interpretation of events, procedures, and degree of importance of certain issues.

The checklist ensures that at a single point in time (namely at this point in the project), all of the specifics of the program get addressed. The list forces the product development team to evaluate each item and assess degree of completion.

3. RISK ANALYSIS FOR MISSING ELEMENTS

If the project runs late, or increased pressure is placed on the development team, it may be tempting to cut certain activities out of the process to accelerate the program. It is not advisable to do so, and the following recommendation is not an endorsement of it; however, if it does occur, a risk analysis should be performed to determine financial and product liability exposure. The danger lies in the rationalization of selecting many items to be eliminated.

This results in a less-than-complete program. It also exponentially adds to the list of caveats and special handling issues for the program. It complicates the ability to perform these steps after the fact of introduction and maintain any resemblance to a product configuration control system. The best practice is to execute the planned steps before introduction, without compromise.

4. ACCOUNTING FOR DEFICITS

Assuming certain deficits are identified and the management team decides to proceed anyway, the team must account for the missing items. This means planning and executing in the future, or establishing procedures and documentation instruction to work around the deficit.

The establishment of "special" or "conditional" procedures complicates the work content of the personnel that must embrace the new product and participate in the launch. For example, let's assume the product is incompletely tested in one aspect of application. For the sales personnel and the applications support personnel, "special" procedures must be invoked when dealing with customers. This is especially difficult, as those incidences in which the customer is specifically inquiring about that particular aspect of the application. The sales personnel must exercise burdensome conservatism, while the application personnel must spend additional time "qualifying" the product for use in this area of the application. With a complete development and testing program, the question could be definitively answered in a timely manner without having to place the burden of uncertainty on the customer.

5. THE DANGER OF INCOMPLETENESS

There is also a more parochial aspect to the incomplete development program. This issue is the inherent product liability the company is exposed to. The nature of product liability exposure is such that one issue can be somewhat controlled or mitigated. Two issues could be possibly identified and controlled; however, three or more issues begin to statistically approach the point where the risk simply cannot be controlled if there is more than one individual involved in the product.

The company then acts to enforce the strictest control over these issues, and can run the risk of disenchanting the sales channel with all of the "caveats."

CONFIRMING AGENCY CERTIFICATIONS

1. AGENCY CERTIFICATIONS ARE AN INTEGRAL PART OF THE DEVELOPMENT PROCESS

Modern product development and marketing demands the use of agency certifications to complete the process of development. As a result of government requirements, to determine adherence to specification, and to provide a degree of public safety, agency certifications affecting the industry are becoming more of a factor in the product development process. In fact, certain products cannot be marketed without the agency's approval. In some cases the absence of it is also a huge marketing hurdle to overcome.

The securing of these certifications is part of the development; it is not added onto after the development is completed. It is virtually impractical for a product to undergo agency testing and approval after the product is developed and expect things to remain unchanged. The process is a give and take between the agency and the manufacturer where certain elements are subject to interpretation. Where the requirements are clear, the manufacturer can often anticipate the agency's needs and design them into the product. When the certification is initiated after the development is completed, there may be cost adders to the product to ensure compliance. These cost adders can be significant. The agency certification exercise should be a confirmation at this point of the program and *not* an initiative.

2. MODIFICATIONS AFTER BETA SITE TEST PROGRAM

Given that the agency approvals are not an initiative at this point, the involvement here is more for the benefit of any changes made as a result of the Beta field-testing. Certain changes demanded by the customer base may have to be cleared through these agencies. Consequently, the agency approval issues dealt with here should be on the specific Beta test requirements.

3. PLUG THE AGENCIES INTO THE ORGANIZATION TO ESTABLISH RELATIONSHIPS

One of the most effective methods in prosecuting these last-minute changes is to have an ongoing relationship with the agencies involved. In effect, you will be requesting a "favor" of time to be granted to the company to process these changes. The best way to garner a favor is to have a working relationship established whereby the agency begins to move from a policing orientation to one of consultant orientation and confirmation.

This is where agency-approvals coordinator within the company can provide a valued service in time management for the project. Their role is one of technical liaison and information expediter between development and agency personnel. They can accelerate the approvals process or delay it, based on their own diligence and drive.

Another method to facilitate the integration of the agency into the product development process changes is to immerse the agency personnel into the organization itself. Have them communicate directly with development personnel on details and interpretations. This can accelerate results and create a more intimate understanding between the company and the agency.

4. DON'T LAUNCH WITHOUT CERTIFICATIONS IN PLACE

It can be quite tempting to launch the product without tying up the loose ends of these agency approvals. However, do not succumb to the temptation, because market pressures and timing could work against your company effort. Once you launch the product the sense of urgency is diminished greatly. It can be very difficult to keep the pace going once the product is launched.

It is usually best to launch with all approvals in place. It keeps the pressure on both the agency and the company personnel to complete the project correctly, and as soon as possible, given all of the requirements.

5. CERTIFICATIONS ADD LEGITIMACY TO THE NEW PRODUCT DEVELOPMENT

The agency certifications can add legitimacy to a program when new innovations are being implemented. When working with the agency, these new methodologies and implementations can be reviewed by an independent third party, and sanctioned by them for use in the applications they govern. This can make the job of introducing the technology and implementations to the marketplace easier, because the company can leverage the legitimacy of the third-party certification agency into the new product implementation.

The agency certification process is looked on by some as a bureaucratic necessary evil to the product introduction process. In reality, however, agencies can be a partner who can bring information and methodologies to the process, just as vendors bring to the development process when they are considered a partner.

PILOT RUN MANUFACTURING

1. THE PURPOSE OF PILOT RUN

The manufacturing pilot run is a verification of several aspects of the newly developed product. The verification has an orientation that is internal to the organization. Its primary function is to verify manufacturing infrastructure effectiveness, process documentation, and test manufacturing yield capability.

The pilot run is analogous to the midterm exam for education, the dress rehearsal for a theatrical performance, or the exhibition game in professional sports. It is an opportunity for the company to flesh out the manufacturing processes, to test throughput expectations, verify the functionality of the quality database system, and to determine effectiveness of personnel training. It is a smaller-scale test drive of a larger-scale production run for the purpose of verification and correction.

2. PERSPECTIVES ON THE SIZE OF THE PILOT RUN

Since this is a practice run of manufacturing, the size of the pilot run is an important factor in the verification. There is a significant difference between the manufacturing run of five pieces and 100 pieces, or 1000 pieces. It is easy to "hand build" five pieces, but it is another matter to manufacture 1000 pieces and have them integrated into the testing and quality database.

The problem of lack-of-manufacturing procedures and systems does not go away if a company chooses to go with a disproportionately small pilot run. In fact, the issue will only get more complex as time goes on. Furthermore, scaling the manufacturing organization up from a small quantity to a large quantity can be disastrous, given the lack of sufficient manufacturing systems and controls.

The size of the pilot run should be reflective of a normal condition in manufacturing and should be embraced as an opportunity to confirm manufacturing systems capability.

3. DEGREE OF VARIABILITY ALLOWABLE

One of the salient points of the Beta test program results is to determine the tolerable deviation that the company and the customer base would be willing to accept from the original plan. There needs to be a balance between the severity of the deviation and the esoteric purity in compliance.

A degree of practicality is the best course of action, in that minor deviations can be resolved quickly during the pilot run. Major deviations should be resolved outside of the pilot run proper. In either case a large number of deviations, significant or not, indicate that the process setup for manufacturing of the product is not in control. A process that is not in control negatively affects the quality level of the product.

The pilot run allows the quantification of the deviation and the development of a correction plan to bring the processes into compliance.

4. MEASURES OF PERFORMANCE

When the product is first developed, the manufacturing plan establishes certain expectations of performance. The pilot run tests the data-gathering capability of the system to document performance. The benefit is that it removes the specter of interpretation of data by a third party. The data is objective and can be sorted by categories, and corrective actions can be assigned to individuals responsible for their particular area.

Manufacturing yield data should reflect a rapid approach to the objective with each successive manufacturing run. Given a diligent layout of the entire program, the expectations for yield should be able to be met.

5. CORRECTIVE ACTION

The corrective actions required as a result of the pilot run need to be in place before any subsequent pilot runs. In addition, the corrections should be able to be evaluated individually for effectiveness. The vehicle for correction should be the engineering change as it relates to process changes (assuming that is the only type of change needed).

These engineering changes should be addressed immediately after the pilot run, and another qualifying run scheduled as soon as possible.

The degree of risk one may want to take here is one of scale. If the next run is a full-scale production run with large volumes, there is a possibility for significant rework costs to correct manufacturing problems.

The temptation is to correct the product as is it being made; however, in its purist form the pilot run should not require any "hand-holding." It should be a process that is deterministic and that can be turned on and off at any time with repeatable results.

6. TIMING AND STAGING

The timing of the pilot run is anticipatory to the product introduction. It should occur in the general sequence specified in the deliverables list included in Chapter 7. The pilot run must utilize the procured parts and the designed processes. Therefore it is considered the first production run from a corporate infrastructure.

The disposition of the product coming off the pilot run depends on the degree of product conformity. In many cases these products are used for the Beta test program with certain qualification, and are possibly even used for initial sales efforts.

During this time in the product development program, the intensity and sense of urgency increases dramatically, because of management intervention to accelerate sales demand from the sales channel and customer demand. There is a tendency to try to race through the process of the pilot run, its qualification, and the Beta test program and feedback integration. The product development team needs to be diligent in its activities so as to complete all of the details. With the sense of urgency increasing, the team needs to bring effort to bear on the details of the tasks and to not treat its importance as perfunctory.

7. PROCESS IMPROVEMENTS

The pilot run results can point out areas of proposed improvement in the processes used. These improvements should be incorporated into the product line as part of the continuous improvement effort. Barring any showstoppers in yield, quality, or manufacturability, these improvements can be made as soon as time permits.

They should be integrated and measured for overall effectiveness. Some process improvements require volume product to implement; therefore the team should identify those process improvements that can be implemented in the lower volume initial production runs from those improvements that require significantly more volume to implement properly.

BETA TESTING PROGRAM

1. THIS IS THE REPORT CARD ON THE ENTIRE EFFORT

If the pilot run is the internal measurement of performance of the manufacturing system, the Beta test program is the external measurement of performance. The Beta test program will yield the results of third-party usage of the product in a nonlaboratory environment.

This process tests the entire new-product development and manufacturing process from one end to the other. It is the report card on the entire product development effort. The customer base has the opportunity to engage the product, receive it from the manufacturer, place it in service using the manufacturer's documentation, and evaluate the performance against the product claims. It gives the manufacturer feedback on the details of the product function, ease of usage, and performance characteristics.

The Beta test program should be an organized evaluation by the selected customer, with a beginning date and an end date established for the test and feedback. Agreement should exist between the manufacturer and the customer as to what test criteria will be used, and what type of product feedback the manufacturer will need. The test article should be removed from service and examined by the manufacturer.

Often these articles are shipped to customers for testing and may not even be placed in use. Many times a less-than-diligent job done by the factory will result in no feedback at all. There are even conclusions drawn by the manufacturer that no response is good news (i.e., no complaint is good news). In some cases the customer has no response because they have not tested the product yet!

The reality is, structured evaluation done in a timely manner that results in specific feedback to the manufacturer, makes the Beta test program valuable to the manufacturer.

2. CHECKING ALL ELEMENTS OF THE PLANNED PROCESS

There are several aspects to the Beta test program. They are as follows:

A. Customer selection

   The customer selection should be made by factoring in the customer's ability to test the unit and cooperation in giving feedback, and by the company's liability exposure in case of a product failure.

B. Customer contact

   The customer contact is the primary point of communication between the manufacturer and the testing customer. Communication should be bidirectional and open.

C. Contractual issues

   The manufacturer should have some form of agreement that the customer will sign and honor for the duration of the testing process. The agreement should cover legal exposure issues and responsibilities on both parties. It should also cover test duration, timing, and terms and conditions.

D. Waiver of liability

   Due to the nature of the Beta test program, there should be some type of waiver of liability established in the agreement to protect the manufacturer.

E. Structure and timing of feedback

   Each type of product being tested has certain requirements for testing. In addition, each project schedule has different demands for timing of feedback. The test program is to confirm previously completed development work and assumptions, not create a forum for new discovery that will delay introduction. Consequently, exhaustive testing and timely feedback is key to a successful test program.

F. Number of test sites

   Depending on the uncertainty and degree of confirmation required of the testing, the number of test sites may change. The Beta test program criteria should be established early in the program, and consideration given to the potential sites and number should be done throughout the development cycle.

G. Conditions to be tested

   As part of the agreement, the specific test criteria and test conditions should be identified. The customer should then agree to perform the tests under these conditions. A customer unwilling to test completely or thoroughly is not a good candidate.

H. Demographics of selection

The demographics of customer selection should represent the demographics used in the marketing of the product. To be thorough each segment of the customer base should have the opportunity to test the product and give feedback.

I. Recall and exchange terms and conditions

As part of the agreement, the timing of product recall for the test unit should be identified. If there is a replacement product to be given to the customer after the test period, the terms and conditions should be spelled out.

J. Documentation of feedback

All customer feedback should be loaded into a database for comments. The database should be categorized into critical issues, non-critical needs, and noncritical wants. Critical issues must be effectively addressed before introduction. The database can then be sorted by need type, demographics, degree of importance, etc.

K. Disposition of feedback information

The company should have a plan to do something with the field feedback information. There must be a structured plan for incorporation into the product in the same manner as the initial market needs and wants. Often if the feedback allows for it, the features or wants are incorporated in the next revision of the product, so as not to hold up introduction.

L. Recommendations

In a similar manner, recommendations made by the customer should be factored into the analysis and acted on.

M. Plan to mitigate any discovered liability

As a result of the test program, some undiscovered product liability issues may surface. These issues must be addressed in some effective manner before introduction. This does not necessarily mean that the introduction must be delayed; however, the issues must be addressed in some manner, either through warnings or instruction.

There are two forms included in the Tool Box that are available for your specific use. The first is the Beta Test Form, which outlines the basic requirements of the Beta test program and the expectations of the manufacturer. It also allows for the listing of test conditions and for specific feedback from the customer. The second form is a spreadsheet summary of all of the Beta test customers and the current status of the program.

3. MODIFICATIONS AS A RESULT OF BETA TESTING

How should the modifications be selected and made as a result of the test program? As stated earlier the selection must be made on an evaluation that prioritizes issues, from serious safety-related issues to customer preference. This should be done in decreasing order of importance, as in the original qualification of the specification. Figure 9-1 illustrates the point.

As shown in Figure 9-1, the Beta test feedback shows several responses, some of which should be implemented before introduction because they are safety-related issues. Others are less critical in nature and are desires vocalized by the customer. These can be phased into the product line during subsequent revisions. In this example, items 1, 2, and 3 related to safety will be implemented immediately, and will preempt product introduction. The modifications have minimal cost impact and may be as simple as additional warning labels and/or warnings in the user's manual.

Still others are suggestions and comments that may or may not be implemented. There is also a cost impact for each of the items to be included. This format serves to document and provide some orderly disposition of the customer feedback and a plan for integration. *A sample format is included in the Tool Box for your specific case.*

BETA TEST SITE FEEDBACK

CUSTOMER COMMENTS

	FEEDBACK ITEM	ACTION	DEGREE OF CRITICALITY			COST IMPACT	PHASE IN DATE
			PREEMPT INTRODUCTION	REMEDIAL	ENHANCEMENT		
SAFETY RELATED ISSUES							
	ITEM 1		YES			0.50%	IMMED.
	ITEM 2		YES			0.75%	IMMED.
	ITEM 3		YES			0.30%	IMMED.
CUSTOMER DEMANDS							
	ITEM 1	NONE					
	ITEM 2			YES		2%	NEXT REV.
	ITEM 3	NONE					
CUSTOMER WANTS							
	ITEM 1				YES	1%	NEXT REV.
	ITEM 2				YES	1.50%	NEXT REV.
	ITEM 3				NO	N/A	
CUSTOMER SUGGESTIONS							
	ITEM 1	NONE					
	ITEM 2	NONE					
	ITEM 3	NONE					
GENERAL COMMENTS							
	ITEM 1	NONE					
	ITEM 2	NONE					
	ITEM 3	NONE					

Figure 9-1. Beta Test Feedback

4. SERIES A OR B AT LAUNCH, WEIGHING THE RISK

The product team must assess the liability risk of the field feedback. The safety-related comments must be addressed and documented. The risk of accommodating the issue with warning labels to keep the product introduction on schedule must be assessed and decided on. The worst path to take is to ignore the feedback and go ahead with the launch without making any provisions.

Hopefully the issues returned from the field will be small and not safety related, allowing the team time to organize the requests and integrate them. If they are serious, then a revision of the initial planned offering or Series A may have to be completed and launched with a Series B device.

5. UNITS MUST BE INSTALLED AND EXERCISED

For the Beta test program to have any tangible value, the units under testing must be exercised fully. This is not an area where the product's "hand should be held" through the application. The Beta test form referred to earlier, and included in the *Tool Box*, outlines specific test conditions in which the product undergoes normal field stress.

The customer who signs on for the Beta test is, in effect, signing on to ensure that these test conditions are met. More directly stated, the test is virtually a waste of everyone's time if these specific conditions, designed to stress the product in some specific way, are not met. This should be discussed and agreed to as part of the Beta test agreement.

6. BE SPECIFIC ABOUT WHAT YOU ARE LOOKING FOR

In the same spirit the manufacturer must be specific about prescribing the test conditions to the prospective customer. They cannot determine the stress conditions; the manufacturer needs to test to, so it is the manufacturer's responsibility to prescribe the test regimen, the timing, the customer's commitment, and the expectations.

7. SELECTING THE SUITE OF TESTS

The suite of field tests should exercise the product in normal usage, and somewhat beyond in a few sample cases. The tests should verify functionality, ease of usage, and performance of the product with reference to the customer's expectations and needs. The test also verifies the engagement of the company and the customer, in terms of these expectations, so that the marketing program designed for the product is confirmed.

If a specific instruction set is required to explain the test regimen to the customer, then textbook instructions must be used.

If the procedure required is specific to the product and needs clarification, then an application team at the factory should be in place to assist and log in the customer satisfaction or frustration data.

LITERATURE

1. HOW THE TERM *LITERATURE* IS DEFINED

Literature is a term used to describe the management and dissemination of product (line) information. Literature can take several forms and utilize different media. The advent of computer mechanization allows more options for the manufacturer, and will continue; however, the information about the product, published for its use and also for buying decisions, does include several basic elements. They are as follows:

A. Information to make a purchasing decision

The information needed to make a purchasing decision can take on several forms. Depending on the product, there can be printed matter, electronic mediums, or voiceovers in a video commercial, etc. In any case the objective is to provide the prospective customer with the appropriate information so that they can make an informed choice. In some cases this can extend to a competitive comparison of other products. This can be referred to as a brochure.

B. Information to familiarize the user about the product

The next form of product information is used to familiarize the prospective customer about the product, its uses, and any caveats associated with it. It generally is provided before the sale of the product (or immediately after) and serves as background information. In some cases this information is merged with the information to make the purchasing decision. This can be referred to as a product data sheet.

C. Information to use the product

The next form of information is specific information that accompanies the product in shipment to the customer. It is the company's formal document that instructs the customer how to use the product. It has general information, receiving inspection, storage information, theory or sequence of operation, and other logistical information. It may also include troubleshooting and repair information, but for purposes of this definition sequence we will consider them separately. This is often referred to as a users' manual.

D. Information to train about the product

Some products are rather complex and require training beyond the users' manual. This may include start-up information, procedures, and certification issues for the personnel using the product. These may be entire programs designed to facilitate the use of the product and to mitigate the liability to the manufacturer.

E. Information to apply the product

Depending on the complexity and the product configuration, additional information may be required to facilitate its use. For example, some products are components of systems and do not operate in a vacuum of a system. Consequently, these products require application training to instruct the user about the operative limits of the product. This is to ensure a product is not applied in a system that stresses the unit beyond its intended design parameters.

F. Information to effect repair on the product

For those products that require repair during their useful life, there is information given about the repair of the product. As the pathway of information and literature is developed, it can be seen that the information about the various operational aspects of the product follows its lifeline. Sometimes the product is repairable by the customer, and other times it must be repaired by an authorized service center.

G. Information about warranty

Finally, there is information given about the warranty and how to put it into effect. Included will be the rules of engagement and the terms and conditions.

These presentations of information outline the various types generated for the product line. Depending on the specific product line, the types may be combined or issued separately. For some service oriented "products," this information may be negotiated as part of a contract and as such, there is no standard per se. A summary of these presentations is presented in the form of a matrix in Figure 9-2.

As shown in Figure 9-2, there are five distinct types of product information that can take the form of six basic types of medium. Each product line will demand use of the various types of information and the medium of choice. The product-development team should

MATRIX OF PRODUCT INFORMATION

TYPE		PRINTED MATTER	NETWORK	FIXED CD	MULTIMEDIA	VIDEO	PERSONAL
A	PURCHASING DECISION						
B	PRODUCT USAGE						
C	PRODUCT TRAINING						
D	PRODUCT APPLICATION						
E	PRODUCT REPAIR						
F	WARRANTY						

Figure 9-2. Matrix

determine the best use of the medium and the arrangement of the types of information to maximize sale of the product. A sample matrix is available in the *Toolbox* for your specific case.

The following are example outlines of the five different types of information:

TYPE A: Effecting the Purchasing Decision

A. Introduction/Background
B. Features/Benefits
C. Application
D. Product Configuration
E. Pricing
F. Selection Criteria
G. Sizes and Ratings
H. Availability

TYPE B: Product Usage Information

A. Information about your safety

B. About this manual

   Who should read it

   Conventions used

C. Getting started

   Overview

   Working life

   Unpacking

   Storage

D. Mechanically-oriented information

   Environmental

   Mounting consideration

   Dimensions and weights

E. Electrically-oriented information

   Wiring standards and codes

   Connection specifications

Grounding

Power connections

Control connections

F. Power up and commissioning

Preparation

Power up

Operator information

Modes of operation

Diagnostics

G. Specifications

TYPE C: Product Training

A. Introduction to product
B. Application of the product related to operational characteristics
C. Precautions
D. Procedures
E. Certifications
F. Caveats
G. Operator training
H. Continuing training

TYPE D: Application of the Product

A. Introduction
B. Application criteria
C. Experience base
D. Limits of the product line
E. Specifications
F. Interfacing of the product line
G. Mechanical
H. Environmental electrical
 I. Human factors
 J. Safety considerations
K. System dynamics
L. System performance
M. Significant performance and applications features document

TYPE E: Effecting Repairs on the Product

A. Product configuration
B. Internal repair
C. Field repair
D. Processes
E. Procedures
F. Parts and authorization
G. Certification
H. Turnaround time
 I. Flowchart of repair logistics
J. Repair locations
K. Billing coverage
L. Upgrades
M. Referencing the quality database

TYPE F: Prosecuting the Warranty

A. Warranty statement of coverage
B. Timing and extent of coverage
C. Product upgrades
D. Explicit versus implied
E. Rules of warranty engagement
F. Third-party arbitration

For your use, each of these outlines is available in the *Tool Box*. The needs of the product will dictate variations in these outlines; however, they can be fleshed out and modified.

SETTING UP THE INFRASTRUCTURE

1. MANAGEMENT SUPPORT

The process of managing the development of a new product involves repeated checking and verification of many elements of the program. Now is the time to reassess management's commitment to the program. This is necessary as you are embarking on the initial launch of the product. This launch will consume human energy, time, and funds. Consequently, the effort must have management's complete support.

The support does not end with the investment. Often management may feel that their part is done with the investment; however, their support may be required to adjudicate priorities of collateral departments within the framework of the launch.

Management must be made to understand the importance of the continuous effort up to and including the launch. Do not let them abdicate this responsibility.

2. IDENTIFYING SCENARIOS AND DEVELOPMENT OF RESPONSES

The marketing of a new product requires that a certain infrastructure be in place and functioning at the time of product introduction. This infrastructure exists to accomplish several things, including the entry of orders, the clarification of the product's use, application information, availability information, and so on. The specific products will dictate the requirements and needs; however, for purposes of illustration, Figure 9-3 is an example that is skewed more toward the monolithic.

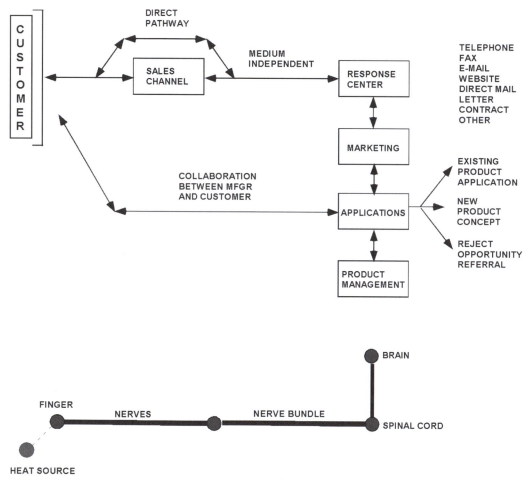

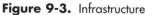

Figure 9-3. Infrastructure

As shown in Figure 9-3, the interface between the customer and the organization may contain several layers. It can be analogous to the synapses that occur within the nerves when a finger touches a heat source. When the finger comes in contact with the heat source, a signal is sent via the nerve bundles to the spinal cord. Depending on the severity of the heat source, a decision is made at the spinal column or at the brain to pull the hand away from the heat source. Often the decision is made locally at the spinal column and the hand is withdrawn immediately.

In a similar manner the customer sends in a query to the sales organization. This "stimulus," which is medium independent, is recognized by the organization and interpreted to determine the opportunity. The decision to act on the stimulus may be made locally or may be made within the organization. The query would traverse the response center through marketing to applications. The applications would then confer with the customer to see if a fit exists. Finally, product management oversees the entire transaction.

The response may be to apply the existing product to the opportunity, generate a new product concept to be evaluated, or summarily reject the opportunity and refer the query to someone more suited to the opportunity.

In any case the operative point is that there is a process and procedure and rules of engagement to field inquiries coming into the organization and proper disposition of them.

3. AMASSING THE REQUISITE WARMWARE

The structuring of a responsive organization to field requests and interrogatories does not stop at the front door of the factory. Nor does the responsibility lie solely at the sales channel. The entire organization needs to be imbued with the sense of urgency and responsiveness that allows the prospective customer to have immediate confidence in the company.

The customers make their decisions based on the product specifications, the company reputation, and the current state of doing business with the company that they perceive. This is why the best possible impression is to be made upon receipt of the customer inquiry.

The company should mirror the responsiveness exemplified at this interface.

4. SETTING UP AUTOMATIC FEEDBACK AND REPORTING OF CRITICAL INFORMATION

Another important point in the structuring of this interface is to create a systemic method for automatically logging in the customer information; the purpose being to remove the dependence of the company on the biases and interpretation of the personnel supporting the interface. The idea is to prevent applications and other personnel from summarily dismissing a lead to an inquiry based on their *own* opinion. This can cause the company to overlook significant opportunities that are real and achievable, but do not necessarily fit the *exact* form the personnel expected.

In addition, this automated system should log in critical customer comments and feedback so it can be distributed and absorbed by the rest of the organization. Comments regarding safety, ease of usage, misunderstanding, and product failures caused by misapplication derived from root company information or recommendations, are all candidates for this critical information requiring documentation and retrieval.

5. PLUGGING THE CUSTOMER INTO THE ORGANIZATION

The interface should essentially "plug" the customer into the organization. They should become an intellectual value-added appendage to the organization with free-flowing information coming in about the product, company, personnel, and logistics. In its best form, the relationship between the customer and the company should be one of cooperation and open communication. To foster these properties of the relationship there must be accessibility first. This is what is meant by plugging the customer into the organization.

Young organizations with a smaller universe of customers are very adept at this. The larger an organization gets, and the more diffuse the customer base, the tendency is that the company essentially isolates itself from the customer. The department titles may not imply this with their organization chart and title blocks; however, the reality often is that the operations and development, in effect, become isolated. This is what ultimately can contribute to the demise of market share.

Keep close to the customer, understand their problems and concerns, and act on them with your company's products and services.

6. THE INTERFACE IS A COMPONENT OF THE PRODUCT

The customer interface is an important component of the product. Without it few products can flourish in the marketplace. A product that has all of the elements covered in the development process will have a difficult time being successful if the interface is flawed. So important is this interface, that many times, mediocre product lines survive and flourish because of the effectiveness of the interface.

To ensure that the new product has every opportunity to flourish, pay attention to the interface between the customer and the company.

TRAINING FOR PERSONNEL

1. IS THE ORGANIZATION SET UP TO HANDLE THE NEW BUSINESS?

This chapter deals with the final aspects in the new product development process. This means that if the organization has not been modified to effectively handle the new product, the team must implement needed changes at this time.

Take a thorough look at the organization dynamics as they will relate to the new product, not necessarily how they relate to the existing business. The following questions need to be resolved:

A. Are there enough people in place to sell, manufacture, support, apply, and service the products?
B. Have the objectives for the product line been made clear to the organization?
C. Has the organization been oriented to the new product in balanced fashion, compared to the existing business?
D. Are the people in place motivated to make the new product a success?

Figure 9-4 is a log of training that will assist management to determine skill levels of the personnel associated with the new product. As shown in the figure, the log can keep track of the employee's certifications. It is also useful from a manpower-loading point of view, to see who has what certifications. The flexibility the manager has is based on the degree of mobility of the certified personnel. If one area needs strengthening, the personnel can be moved in from another area.

TRAINING OF PERSONNEL

CATEGORY	DESCRIPTION	EMPLOYEE 1	EMPLOYEE 2	EMPLOYEE 3	EMPLOYEE 4
SALES ORDER ENTRY	QUOTE PROCESS	X			
SALES ORDER ENTRY	PRICING	X			
SALES ORDER ENTRY	SALES ANALYSIS	X			
SALES ORDER ENTRY	NEGOTIATING SKILLS	X			
SALES ORDER ENTRY	APPLICATIONS	X			
SALES ORDER ENTRY	SALES ORDER ENTRY	X			
SALES ORDER ENTRY	ORDER ACKNOWLEDGEMENT	X			
SALES ORDER ENTRY	SCHEDULING	X			
SALES ORDER ENTRY	FOLLOW-UP COMMUNICATION	X			
SALES ORDER ENTRY	CUSTOMER SATISFACTION TRAINING	X			
SERVICE - IN HOUSE	PRODUCT REPAIR TRAINING		X		
SERVICE - IN HOUSE	TROUBLESHOOTING ANALYSIS		X		
SERVICE - FIELD	CUSTOMER INTERFACE			X	
SERVICE - FIELD	TECHNICAL DATA			X	
SERVICE - FIELD	TROUBLESHOOTING ANALYSIS			X	
PARTS	SPARE PARTS				X
PARTS	REPLACEMENT				X
ADDITIONAL SKILLS					
CAREER PATH PLANNING					
	SUPPORTING TRAINING				
	SUPPORTING TRAINING				
	SUPPORTING TRAINING				

Figure 9-4. Training

Also the chart can be used to outline the certifications not directly related to the present job description, as well as the employee's desire for the future. Supporting training, fostering employee growth, can then be planned. A training record is in the *Tool Box* for your use.

2. PRICING/QUOTES, SALES ORDER ENTRY, SERVICE, AND PARTS

Previously discussed was the external and internal interfaces to the company. We also looked at how a customer interrogatory is fielded within the organization. In this section the sales transactional analysis will be reviewed. Not every product will be quoted, nor will every product require service; however, when the company's product is sold to the consumer by a third party, the company must service that third party like a customer.

For example, they are issued a quote on the goods, they are issued a delivery date, and their stock order must be entered for prosecution. Every organization has established practices and procedures for this part of the business. The objective at this point is to ensure that the procedures in place will be sufficient to support the business generated by the new product.

Record retention and retrieval from databases must be tested and verified. This is because critical information coming in at the initial product launch must be analyzed and acted on to ensure a product success.

The procedures should specify how communication with field organization and the customer should take place. The procedures should also specify how to handle refusals, referrals, and those interrogatories that will not result in an order for the company.

3. AUDIT THE SYSTEM

For any system to continue to be effective, it must be audited. The sales order entry system is one example in which a periodic audit can have huge payoffs at the revenue line. Keep in mind that this portion of the company is the primary contact point with the customer. The customer formulates an opinion about the company based on this narrow viewpoint.

To keep this area operating correctly, periodic audits are required. They should be done in an unannounced way, and from the perspective of the customer. At the risk of sounding like a "cloak and dagger" exercise, the degree to which the organization does *not* know it is an audit, is the degree to which the audit will have value in assessing the performance. Take the necessary steps to secure the information without affecting the outcome.

4. CORRECTIVE ACTIONS

During the course of the investigation and final checkout of the organization, there may be some elements that require correction. There may be a tendency to rationalize and procrastinate with respect to correcting these elements. This should not be allowed. The product success is based on certain assumptions, which involve certain systems be in place and functional. To ignore the issue and fail to take corrective action contributes negatively to the product. This may be serious enough to cause product line failure. If there are corrections needed in the process, correct them now. They do not have to be absolutely perfect; however, they *must* be effective.

APPLICATIONS SUPPORT

1. RESPONSE CENTERS AND AFTER-HOURS SUPPORT

The applications support function should have clear, measurable objectives for supporting the new product development initiative. For example, the response center, a central location for all communication between the field and the company, should have a short throughput time that is measurable and can be evaluated for performance.

The after-hours support function shall answer incoming communiqués in some defined period. This is so the inquiring customer has some idea of when they can expect a response. Establishing a response center that falls short of customers' expectations will cause the company to lose market share rapidly. In addition, many products require a certain amount of support to make them successful in the customers' venue. To this end, the response center must be effective, accurate, and timely in their assistance.

2. CUSTOMER SUPPORT TEAM

Assemble a customer and applications support team to enhance the reception of the product in the marketplace. This is for customers who may not be knowledgeable or well versed in the product. In these cases an applications support team can enhance sales and establish immediate customer confidence, before and after the sale. The key to applications support is product and applications knowledge. To that end, do not staff this function with personnel who are not fully qualified in the specific area the product serves.

The customer support team should have three basic tenants in the operations: namely, resident expertise, motivation, and thoroughness.

The *expertise* will allow immediate feedback and customer assistance, negating the "need to check and get back to you." These delays in communication erode customer confidence. By having the expertise resident in the mind of the support personnel, the information transaction will proceed much smoother.

The *motivation* aspect underscores the requirement for the entire customer interface and sales channel. Customers do not wish to interface with people who are bored, lack enthusiasm, and are difficult to get answers from. Consequently, it is critical to the success of the product to have the support personnel motivated to grow the business through increasing sales.

The *thoroughness* aspect protects the company from misapplication and also establishes high-confidence levels among the customer base. With the personnel exhibiting diligence and doing a complete job at this level of the transaction, the customer will be more at ease in proceeding to the next step with the company.

3. TECHNICAL ASSISTANCE

In a similar manner the technical support given by the company must be accurate and documented. This is required for future possible liability in which the misapplication results in a possible litigation. The technical assistance must support the applications of the product and warn of misapplication. All this must be effective via whatever medium that may be in place between the customer and the company.

4. SIMULATIONS

Depending on the type of product and the need for direct communication with the manufacturer, there may be a need for the customer assistance function and the applications personnel to simulate the product's use in a system. Here again, it is in the best interest of a sale to have these people and their systems in top-operating form.

The simulations systems should be state-of-the-art and contain a certain amount of mystery to the customer. In this way, they have value. By establishing the value in the customer's mind, there is goodwill created, and abuse of the function is kept to a minimum.

5. EXPERT SYSTEMS

As technology evolves and products become more and more complex, the ability to use traditional tools to support the product line falls short of customer expectations. This is driving the need for expert systems to be implemented. The expert system can mechanize, to a certain extent, the tribal knowledge resident in the applications personnel. In addition, it can analyze a set of conditions and draw conclusions, based on the input data and these conditions, according to a prescribed algorithm.

This can mechanize a response to the customer and free up applications personnel to develop algorithms for the lesser-known issues. This process makes the development and implementation of the algorithms in the expert system continuously improving and able to handle more complex situations, thus adding value to the service.

6. MODEL FOR CUSTOMER SUPPORT DATABASE

One of the most effective tools for improving customer assistance and the logistics involved is to create and implement a call-in database. This database will allow the organization to field inquiries from customers in a systematic fashion. The inquiry will be addressed and prosecuted according to a procedure and a protocol. It will also force resolution of the issue, if the format is followed diligently by the personnel involved with the process.

Figure 9-5 can serve as an example model for this database.

As stated many times before, the needs of the specific businesses will drive the configuration of the database. Figure 9-5 is presented as a model only, and will require modification for your specific situation. *For your reference, a sample database format is available in the Tool Box.*

CUSTOMER SUPPORT INQUIRY LOG / DATABASE

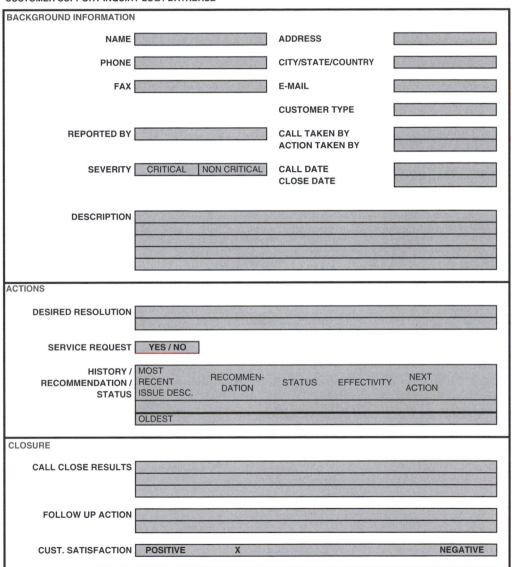

Figure 9-5. Customer Support Database

The database format shows three basic sections in prosecuting an inquiry. The first section is the background section. It contains identity information and some logistical background information. An important component of this section is the degree of severity of the incoming call. If the call is a critical situation, it must be acted on immediately with priority over other calls. If it is noncritical, it can be handled in due time.

The second section deals with actions prompted by the call. The customer articulates the desired result of the call. If the product warrants it, a service call may be in order and is noted here. The next section is one of the most valuable in the database because it contains the history of the incidence, presenting the most recent first and listing them in reverse chronological order. This history consists of several elements, which include the following:

A. Issue description

This is the complete description of the problem with the product, as given by the customer. It is the root information that the response center has to go on to assist the customer.

B. Recommendation

This is the company recommendation given to the customer, based on the description and analysis of the response center.

C. Status

This is the status as a result of the recommendation and some action effected by the customer to correct the problem.

D. Effectiveness

This is an assessment of how effective the recommendation is in terms of the action taken.

E. Next action step

This is the best guess as to the next step required in resolving the problem.

The third section deals with closure of the call-in. It is the resolution in the mind of the customer that they have corrected the problem and no further communication is required. It has elements, such as the results at the time of the close of the call, any recommended follow-up action, and an assessment of the customer's degree of satisfaction with the incident. In this example a gradient, ranging from a positive reaction to a negative reaction, is given.

FIELD ORGANIZATION SET-UP

1. SALES TEAM PROVIDES THE FOLLOW THROUGH ON LAUNCH

The best marketing program for the best product generally cannot succeed without the field sales organization providing the follow through in the customer's venue. The sales team, in whatever form it may take or whatever channel it uses, is the vehicle to execute strategies of the company. This section deals with getting the sales team positioned and motivated to launch the product and prosecute sales.

There is a certain amount of demand created within the field sales organization concerning the new product, while in gestation. This generates enthusiasm and healthy apprehension among the sales force. When the product is released for general sale, the product team can leverage this enthusiasm into initial sales volume.

In the long run, however, the product team has a significant job in maintaining this activity level. In addition, depending on the sales force's compensation structure and the sales channel's means for revenue, both of their outlooks may be shorter than the manufacturer may want, to sustain the sales campaign.

The sales force simply must be directed to carry the torch of new product introduction and light off the sales channel effectively. There should be no compromises here. The gospel of new product development must be preached and followers must be cultivated.

2. REVIEW THE TYPE OF FIELD ORGANIZATION IN PLACE

What type of sales organization is in place within the company? What are their strengths and weaknesses? These questions (first posed to the new product development team in Chapter 1), in the assessment phase of planning, must be looked at in-depth to ensure the human energy of the sales group is focused on the tasks of a launch.

As a general rule progress is marked by effort directed at the elements of a program that will generate sales. If the sales department is averse to correcting certain things, cited as "weaknesses," they must be corrected. There is no finessing of weaknesses in the area of a new product launch. Weaknesses must be transformed into strengths. The introduction and prosecution of a sales campaign is an activity that requires complete, targeted effort. If one element is missing, it cannot be compensated.

Is this the highly motivated and effective field organization that will carry the launch of the product to secure sales? Is this the organization you had envisioned and made assumptions about at the beginning of the program?

If not, then the organization must be modified to meet the requirements outlined in the original assumptions. If conditions have changed that allow for this area to be able to be compromised, then proceed cautiously. However, in reality this rarely happens. The markets and customer needs usually increase to the point where more is demanded of the sales force than originally envisioned.

3. WHAT ARE THE DYNAMICS FOR SELLING AND SERVICING?

What are the dynamics required in the sales organization to effect the forecast volume? Do they match up with the customer service team? To answer this question it may be necessary to assess the validity of trying to use the existing sales organization, without improvement, to launch the product. Consider the evaluation tool example in Figure 9-6, and fill it out for your situation.

As shown in Figure 9-6, the example chart categorizes several aspects of the sales organization's skills and evaluates them in terms of priorities, capabilities, cost-effectiveness, technical expertise, compensation, and employment issues. With each aspect, the initial assessment of the organization is compared to the product launch and sales skills required. The improvements required are noted, and an action plan generated with a completion date assigned. Status indicates the degree of progress toward the goal.

For example, reviewing the sales force prospecting capabilities, the initial assessment of the team indicated a tendency toward route salesmanship, rather than a balanced effort that

SALES TEAM EVALUATION TOOL							
CATEGORY	INITIAL ASSESSMENT OF EXISTING CUSTOMER SUPPORT TEAM	INITIAL ASSUMPTION	CURRENT PRODUCT REQUIREMENTS	IMPROVEMENT REQUIRED	ACTION PLAN	COMPLETE DATE	STATUS
INTERDIVISIONAL PRIORITIES	DIVISION FAVORITISM EVIDENT	BALANCED REQUIRED	20% FOCUS	20% FOCUS	MGMT DECREE	AT INTRODUCTION	PREPARE COMPLETE
PROSPECTING CAPABILITY	TENDENCY AWAY FROM PROSPECTING	NEEDS IMPROVEMENT	BROADEN SKILLS IN PROSPECTING	NEEDS MAJOR IMPROVEMENT	TRAIN AND PRACTICE	DURING INTRODUCTION	NOT STARTED
TRACK RECORD							
CHANNEL MGMT ABILITIES							
SALES PER EXPENSES							
CALL FREQUENCY	RIFLE SHOT METHODS CURRENTLY USED	NEEDS IMPROVEMENT	NEEDS BROAD COVERAGE FAST	NEED INCENTIVE FOR SALES	SEE ATTACHED	AT LAUNCH	40% COMPLETE
MOBILITY	INSIDE SALES ONLY	NEED OUTSIDE SALES ALSO	DEVELOP NET NEW ACCOUNTS	CHANGE PERSONNEL	SEE ATTACHED	JUST PRIOR TO LAUNCH	50% COMPLETE
EFFICIENCY							
AVE. TERRITORY SIZE							
TECHNICAL EXPERTISE							
COMPENSATION SYSTEM							
MOTIVATION							
DRIVING FORCES							
EMPLOYMENT FACTORS							

Figure 9-6. Customer Service Evaluation Tool

included prospecting new accounts. The initial assumption was that by the time of product introduction, (a long way off, at the time) the characteristics would change in the sales force. Now is the time when it should change!

The current state for the product's marketplace demands a fair amount, so improvements are required. Training and practice can help change the attitudes, and this can be accomplished during the introduction for this example product.

By using an evaluation and planning tool such as the example in Figure 9-6, the product team can ensure the sales function will be trained and ready to launch the new product.

A sample of the Evaluation Tool is available in the *Tool Box* for your specific company situation.

4. DON'T LAUNCH WITHOUT THE SALES TEAM'S SUPPORT

The frame of mind of the sales team at the time of product launch is critical to a sustained, successful effort in moving the product to market. There are many barriers that must be overcome by the sales team, and the energy needs to be devoted to overcoming these external factors. There is no time, energy, or patience to overcome problems *in* the sales team at launch. All of these problems should have been overcome before launch.

Consider Figure 9-7, which shows some of the barriers to entry the sales team must overcome. The route to market is shown as the manufacturer funding a program specified by product planning. The marketing group, on completion of the development, outlines a campaign for the sales personnel. In this example the sales personnel are company employees that work through a network of representatives indicated by the sales channel. They in turn sell the product to the marketplace.

Each stage of the process has barriers associated with it. The product planning and marketing group must compete for corporate resources. They must navigate the financial and corporate politics in the assignment of funding to get the funding for the best program. They have to keep the corporation on strategy in the selection of programs.

The marketing personnel have to reprioritize the sales force's activities to accommodate the new product. They may have to do the "missionary work" necessary to launch into a non-traditional market. They will have to cultivate new accounts and redirect energy and focus. The sales force must be aligned and balanced to support both the new product introduction and the existing business.

The sales personnel now have to carry the corporate message about the new product and drive the requirements home through the sales channel. If the sales channel is comprised of independent businessmen their priorities must be aligned by the corporate sales force, such that the new product receives the appropriate time and energy from the channel. To break down this barrier the sales personnel must demonstrate how the agent in the sales channel can make money with their new product.

Finally, the last barrier in the model is the customers' themselves. They have a host of issues, such as familiarity with the new product, technology, and application. There is the

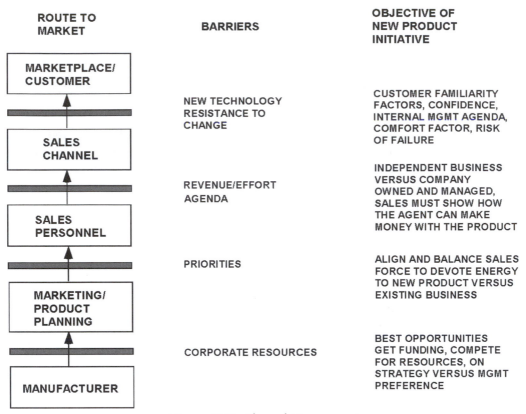

Figure 9-7. Channel Management

confidence factor with the product, their fear of failure, aversion to risk, and any internal customer agenda that may exist, preempting the use of the new product. Breaking down these barriers paves the way for the success of the new product.

5. REFINE SALES ORDER ENTRY SYSTEM AND REPORTING

The sales order entry system should mimic the product quality system in that each request for information, each quotation, and each interface with the customer should be cataloged and mined for market data to further the sales of the product and reposition the company for the future product offerings.

The system should have sales status, either positive or negative, attached to each query. In this way the database can be updated automatically.

The system should feed back data to the individual sales office to allow measurement of their effectiveness against the company's marketing plan. In addition, the results of the marketing plan can be evaluated in aggregate, for its effectiveness by company management.

6. THIS IS TIME-CONSUMING IF FIELD OFFICES ARE INVOLVED

There must be a balance struck to determine the best time to involve the sales force in participating in the new product development. The sales force can be especially helpful in being the eyes and ears of the company in the marketplace. However, there is a risk in involving them at a point where the program may be vulnerable from an intellectual property standpoint. Consequently, each program must be evaluated on its own to best determine the optimum time. Certainly, though, the sales force must be plugged into the new product development when formulating the commercial marketing strategy. There simply is not enough time and energy for the marketing staff to release the product and light up the sales force in one initiative. Sales must be brought on board early in the development of the commercial plan so that at product launch, they are ready to prosecute the program.

7. TEST MANAGEMENT COMMITMENT

Throughout the entire program there has been a need to test the management commitment to the product. In this section the idea is to verify *sales* or *regional* management commitment. To get to the point, the product group has secured the management commitment, but now must ensure the support of the sales force. If this means priorities must be adjusted, then that is what must be done to ensure success.

8. SALES MOTIVATION AND COMPENSATION

The issue of sales compensation can be a complex and difficult one to administer in a company. The system should be fair and reward those who perform. There really need not be a provision for those who do not perform, since they should be dealt with by senior management. It does no good to keep marginal sales personnel on the payroll, and attempt to legislate performance and behavior by the administration of compensation. It generally takes too long and has little effect. In addition, the new product launch does not have that luxury of time.

The sales force should be operating to an internally generated plan to move volume of product. Price cannot be the issue at this point.

Given these assumptions, the individual should be rewarded for their results. There should be several components to a compensation system, and emphasis can vary on the components based on the type of product. Consider Figure 9-8 as a format for compensation for company-employed sales professionals.

In this format (Figure 9-8), the compensation system for the sales force is comprised of several components, including a base salary and bonuses tied to performance both individually and through the achievement of company goals.

The key point of the chart lies in the percentages of the sales professional's compensation that is a result of each of the initiatives. Fundamentally, the sales effort must result in

FORMAT FOR A SALES COMPENSATION SYSTEM						
ITEM	DESCRIPTION	MAINTAIN EXISTING BUSINESS LEVELS	EXPANSION OF SALES VOLUME	MARGIN PRESER- VATION	EXPANSION OF AVAILABLE MARKET	BUDGET MAIN- TENANCE
1	BASE SALARY	60	50			
2	BONUS - VOLUME OF NEW PRODUCT	10	20			
3	BONUS - VOLUME OF EXISTING PRODUCT	10	10			
4	BONUS - NET NEW ACCOUNTS	10	10			
5	BONUS - GROSS MARGIN	5	5			
6	BONUS FOR CORPORATE GOALS	2	2			
7	PERSONAL EFFORT	2	2			
8	NET NEW IMPROVEMENT	1	1			
9	TOTAL	100	100			

Figure 9-8. Format for a Sales Compensation System

orders, and orders will be generated by sales professionals in response to incentive compensation. If you want to legislate their results, structure their compensation system accordingly.

Many companies lose sight of this simple axiom. All of the speeches by management, pleas for balanced selling, and requests for focus on product lines do not achieve the same result as a simple change in the compensation system. The chart shows different focus placed on the various categories, based on the desires of the company. If expansion of the market is the critical objective, then this category is weighted heavier toward expansion from a compensation point of view.

A sample spreadsheet is available for your use in the *Tool Box*.

FINAL PRICING

The setting of final pricing is one of the crucial elements of a product launch. All of the assumptions about the marketplace, tolerance for pricing, features versus benefits, and attaching value to each, will be verified or negated by the refined pricing at this time. This section is designed to present several perspectives on this important step.

1. FIELD FEEDBACK HELPS SET PRICING

Just as the product cannot be developed in the vacuum of the marketplace, neither can the pricing be developed in a marketplace vacuum. The information available from the field personnel, both in sales and service, is used to refine the pricing of the product. To that extent, do not ignore the feedback received from the field. Their perspective can yield clues to the reception of the product from the pricing standpoint.

When the sales force is company employed and giving honest feedback, they are giving valuable information within the context of the company's position in the marketplace and their take on how successful it may be. If the product is targeted at a premium price strategy in a commodity marketplace with no collateral value added, the sales force could see potential problems in their area that may not be evident to the corporate marketing staff.

Factor in their input on pricing, make a decision on the pricing, and execute consistent pricing at product launch. This underscores the philosophy about the pricing, which is *consistency*. Do not change the price during the launch. This confuses the field people and undermines the launch effort. Wait to determine the full effects of the price selected and then take corrective action after launch, if required.

2. REVERIFY YOUR POSITION IN THE MARKETPLACE

Throughout the development program time in the marketplace is passing. Competitors are participating with their own programs and product introductions. Now is the time to evaluate the product and company position in the marketplace and ensure the pricing that was originally set can be held.

In all of these perspectives in this section, the objective is not to rationalize a price cut just before introduction. Remember, this is what will generate profits to pay for the initial product development investment and to fund the next program.

Rather, the idea is to ensure reasonableness in the pricing to maximize the impact of the product launch. Reverify the position in the marketplace to make sure the product offering can support the pricing.

3. IS YOUR PRODUCT STRATEGY STILL VALID?

In the same manner, reverify the original product strategy. Do the features and benefits generated in the embodiment of the product still support the price? Are the target markets and the target volumes still reasonable? Has anything in the marketplace caused a change in the strategy? In short, now is the time to confirm the pricing from the product strategy perspective. Ensure that the product can still command the price, given all of the other changes occurring in the market during development.

4. RERUN RETURN-ON-INVESTMENT CALCULATION

The investment in the new product is essentially "sunk" and the costs are nonrecoverable. It is an interesting exercise to rerun the return-on-investment calculation, using the actual figures instead of the estimated figures. How does the program appear now? Has it changed for the better because of lowered costs in development and product costs, or is it worse, to the point of not being "sellable" to management if it were a fresh program?

Use this data as an operative lesson in future programs to see which assumptions work and which do not. These are the lessons in new product development that are learned over several programs, and how a product development team is grown and nurtured over several projects.

5. DO YOU HAVE A LOW-VOLUME, SPECIALIZED OR A HIGH-VOLUME, STANDARD PRODUCT?

At this point, it is wise to check out the product offering and how the customer base, the market, and the sales force perceive it. Does the product still have the initial high-volume appeal, serving a wide range of demographics with the same embodiment, or has it degenerated into a universal platform, attempting to serve all embodiments to all customers with little or no standardization?

These are the perspectives, which one must understand just before launch. Did the team lose sight of the new product goal and deliver a product that will be an uphill battle to launch and sustain in the marketplace?

Can changes in pricing offset these issues? Consider these issues carefully, and set the pricing accordingly.

6. SPEND THE APPROPRIATE AMOUNT OF TIME HERE; IT'S ONE OF THE ONLY MODIFIABLE WEAPONS AFTER LAUNCH

At this point in the program, the price is one of the last elements of the product offering that can be modified. When setting the pricing, consider all of the aforementioned factors and lock onto pricing for the launch. Given all of the diligence performed throughout the program and the final checklist evaluation, the pricing set should be workable in the marketplace. The customer base will tell you if the pricing is out of line. If you have done the homework, it won't be.

TYING IN SALES ORDER ENTRY SYSTEMS

1. REMOTE TERMINAL SALES UNITS WITH BIDIRECTIONAL COMMUNICATIONS

The remote regional offices should be tied into a network of communications with the factory, whereby information flows freely between the two. This is illustrated on Figure 9.9, which depicts a sales organization that has several regional offices that work through remote sales offices. These remote offices can be company owned or individual businesses. In either case they get their direction from the regional sales office, and then the regional sales office gets their direction from the factory.

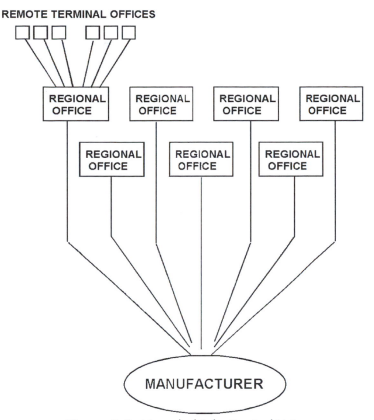

Figure 9-9. Networked Sales Terminal Units

In this way a uniform, new product story can be diffused throughout the sales organization. Policies are communicated in a consistent format and information is fed back to the manufacturer in a consistent fashion from the regional sales office.

It is very important to have accurate and consistent information transfer between the terminal units and the regional offices and the factory. This is because new product decisions may have to be made early in the introduction to facilitate success. Given this fact, accuracy of communications cannot be an issue.

Because every product is different, and the sales route to market may vary, this model is not intended to be the single solution to all situations; however, it does illustrate the need for accurate and directed communications between the manufacturer and the remote sales offices.

2. TRACKING THE SALES PROGRESS

Another basic need to launch a product is the ability to evaluate the ramp up in incoming order rate among the various offices. Figure 9-10 can be used to chart the progress of a

SALES ORDER ENTRY ANALYSIS

REGIONAL OFFICE	PRELAUNCH	MONTH 1	MONTH 2	MONTH 3	MONTH 4	MONTH 5
REGION 1	9	11	17	101	81	73
REGION 2	11	14	69	48	65	58
REGION 3	14	18	26	46	62	56
REGION 4	11	14	21	36	49	44
REGION 5	13	16	24	43	58	52
REGION 6	16	20	30	53	71	64
REGION 7	19	24	36	62	84	160
REGION 8	24	30	45	79	106	96
REGION 9	29	36	54	95	128	116
REGION 10	22	28	41	72	97	97

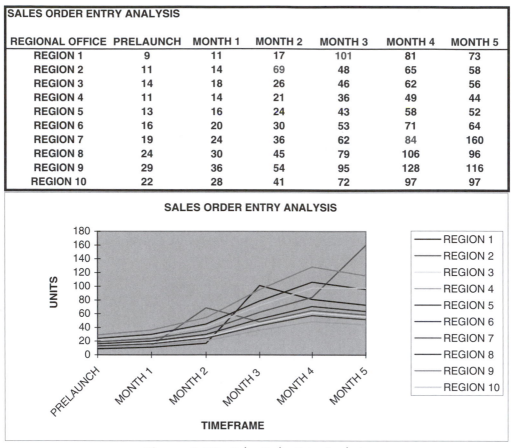

Figure 9-10. Sales Order Entry Analysis

new product launch by region. This chart is accompanied by the graph, which pictorially shows the progress.

As depicted in Figure 9-10, every region shows a nice progression in sales. Some interesting characteristics, however, are shown by region 2, region 7, and region 8. In region 2 a rapid increase in orders is shown in month 2. This is followed by a decline in month 3 and slower growth in succeeding months. This gives cause for questions in what is causing the behavior. Did one large, unexpected booking cause the rapid growth in month 2? Similarly, what caused the behavior of region 7, which shows an increase in month 3, followed by another higher increase in succeeding months, and another dramatic increase in month 4? Conversely, what causes the decline in bookings in region 8 after month 4, after such a positive progression in the previous months?

A sales analysis such as this can give the manager a wealth of information about the market, the progress of sales, the activity level of each region, and the comparative analysis of one region against another. In addition, a look at the specifics of the activities, compared to the results, will give an indication of performance of the region.

This information is extremely useful in assessing the launch progress and the office's participation in it. A sample chart and graph is included in the *Tool Box* for your specific use. *In addition, a more comprehensive sample for sales representative management is also included in the Tool Box.*

3. QUOTES

Depending on the type of product and marketing arrangement, there may be quotations used as a means for selling the products. Although this is not a model for every sales need, it does serve to establish the point about quotations and follow up of quotes to secure orders. Figure 9-11 is an example of a typical model for the quotation database.

As shown in Figure 9-11, the sample database allows management to keep track of quotation activity and to follow quotes through to a resulting booking. It contains pertinent information about the activity and "walks" the sales personnel through the next steps to secure an order. In addition, the percent probability and timing information can be used to generate materials demand in absence of a formal forecast.

A sample Quote Database is included in the *Tool Box* for your specific use.

4. SALES REPORTS

Another required item in the sales management arsenal is the use of sales reports. These reports summarize each project's details and outline the market activity surrounding each project. Consider the sales report format in Figure 9-12. There are several elements, but it generally follows three main items, namely, background, quotation information, and status.

The first section in Figure 9-12 is the sales report section. It contains all of the background information about the project being tracked. This section is used by the database for retrieval of quotations, general information, and logistics.

The second section is the quotation stage. This section summarizes the listing of firms quoted, the pricing, the calculated margin, and the date quoted. This section also contains the competitive summary for any competitors participating in the contest to secure the business.

Finally, the third section summarizes project status indicating the project status, the successful bidder, and the particulars about the commercial issues. One of the most important elements of this section is the "comments," which requires reasons be given for being successful in securing or losing the order. A sample of this spreadsheet is included in the *Tool Box* for your specific use.

QUOTATION DATABASE

BUSINESS SEGMENT A

QUOTE #	DATE	CUSTOMER	REGION	REGION MANAGER	VALUE	PROJECT STATUS	FOLLOW-UP	% PROBABILITY OF RECEIVING ORDER	TIMING	SALES STATUS	COLLATERAL OUTLOOK
1											
2											
3											
4											
5											

BUSINESS SEGMENT B

QUOTE #	DATE	CUSTOMER	REGION	REGION MANAGER	VALUE	PROJECT STATUS	FOLLOW-UP	% PROBABILITY OF RECEIVING ORDER	TIMING	SALES STATUS	COLLATERAL OUTLOOK
1											
2											
3											
4											
5											

BUSINESS SEGMENT C

QUOTE #	DATE	CUSTOMER	REGION	REGION MANAGER	VALUE	PROJECT STATUS	FOLLOW-UP	% PROBABILITY OF RECEIVING ORDER	TIMING	SALES STATUS	COLLATERAL OUTLOOK
1											
2											
3											
4											
5											

BUSINESS SEGMENT D

QUOTE #	DATE	CUSTOMER	REGION	REGION MANAGER	VALUE	PROJECT STATUS	FOLLOW-UP	% PROBABILITY OF RECEIVING ORDER	TIMING	SALES STATUS	COLLATERAL OUTLOOK
1											
2											
3											
4											
5											

Figure 9-11. Quotation Database

SALES REPORT

PROJECT   QUOTED BY

RELEASE DATE   EMPLOYEE #

QUOTE #   REGION

DESCRIPTION

QUOTE STAGE

	PRICING	MARGIN	DATE

QUOTED TO: 1
2
3
4
5
6
7
8

COMPETITION
1
2
3
4
5

STATUS

PROJECT LET

PROJECT PENDING

PROJECT CANCELLED

	PRICING	MARGIN	DATE

SUCCESSFUL BIDDER

COMMENTS/REASONS

Figure 9-12. Sales Report

5. DEMOGRAPHIC MARKETING DATA

The management tools discussed in the previous section must be used in concert with the demographic data generated by the manufacturer, in order to evaluate sales performance. There is no point managing a region that may appear to have poor performance simply because the market may not be available to the extent originally forecast. This assignment of forecast in a fair and equitable manner is the responsibility of the marketing group located at the manufacturer.

6. LEAD-TIME MANAGEMENT

The lead-time of the products should be communicated to the sales force in a real-time fashion. Given the availability of resources to generate the labor portion of the manufacturing equation, the quote database and sales management systems can be used to generate a materials requirements forecast, based on quotations and probability of securing the orders. Although this model may not work for all products, it does illustrate the point that the manufacturing side of the business should be tied very closely to the sales and marketing side of the business to effect on-time production and customer satisfaction.

7. RAPID DELIVERY NETWORKING WITH OTHER AGENTS

One of the ways to alleviate the lead-time issue is to stock the product at the remote terminal units. By having each of the terminal units containing stock, the critical orders requiring shorter lead-time, not supported at the manufacturing level, can be overcome by shipping product from one remote sales unit to the other. In this way the customer satisfaction level can be maintained without overdue burden to the manufacturing organization. Local stock at the remote terminal units also will "encourage" market development at the local level to "turn" the units' inventory over.

8. PRICING ADMINISTRATION

The pricing administration of the new product can be accomplished by the sales order entry system. This system can lock pricing in during the product launch and thereafter, in order to preserve the pricing integrity. This must be accomplished so the product team can understand the margin dynamics. By holding the pricing fixed via the sales order entry system, the variances in manufacturing and procurement can be resolved faster.

If a mechanized network is used to link the offices and manufacturing, the pricing can be communicated to the remote sales units, and can be modified uniformly by the manufacturer when the product team initiates the change.

MATERIALS PROCUREMENT

1. ORDERING MATERIALS IN ADVANCE OF THE PILOT RUN

In any new product environment there is always the management of risk. In the later stages of new product development the risk in procurement of longer lead-time material, in advance of the pilot run, is financial. This financial risk is associated with unusable inventory, if the results of the pilot run and testing program necessitates changing one of these long lead-time or specialized parts. Given the assumption that the product team is diligent in its selection of components during each stage of the program, this risk can be mitigated. In any event the fundamental requirement for the pilot run is to have the processes and material in place. The risk of a material change should be low, in comparison to the risk of not being able to initiate the pilot run and subsequent Beta testing.

2. ANALYSIS OF RISK OF CHANGING PARTS AFTER THE PILOT RUN

The issue of inventory and disposition of obsolete parts, based on a change after the pilot run, is one of percentages. Given a complete Bill of Material, if a small percentage of the parts change (necessitating scrapping of the parts), their value pales in comparison to the value of lost revenue in arriving late to the marketplace. Even if they are expensive parts, the issue is one of loss mitigation at this stage of the project; time is more valuable than material for the most part.

3. ALLOCATION OF COMMON PARTS WITHIN STANDARD INVENTORY

When introducing a new product to the manufacturing arena, there is another issue to concern the product team. This is the issue of common parts usage with the existing business. The existing business, which is comprised of products that can use some of the same parts used in the new products, can be a problem when increased volumes in existing products may "rob" parts from the pilot run or initial production runs of the new product. In these cases it is important to reserve parts with some type of new-product restriction code implemented in the management information system of the company.

4. BLANKET PURCHASE ORDERS TO VENDORS WITH RAMP UP

Another element of the materials equation is to establish agreements with the vendors to utilize blanket purchase orders for annual quantities of material. This material could then be delivered in stages, as required by a monthly forecast. This affords the procurement function to take advantage of quantity pricing breaks and allows the vendors the ability to plan for their production requirements.

For a blanket purchase order to be effective for a vendor, the same requirement for the manufacturer applies. The product on the blanket purchase order is committed to and must be shipped. Failure to do this will result in cancellation charges that must absorb material, any advanced production done for efficiency reasons, and disruption in schedules. The same holds true for the vendor supplying parts to you. The commitment is made at the time of the purchase order. If you change parts, you own the original blanket order quantity that is unused.

5. PURCHASING MANUFACTURING CUES VERSUS STANDARD PARTS

If the parts being procured are noncataloged parts, the blanket order secures a manufacturing cue at the vendor. There is a vendor material risk in a blanket order that may change also. In this case it may be a good idea to secure, through purchase order, a "manufacturing cue" at the vendor. In this way the parts you may need are secured by virtue of a reserved manufacturing time and duration. However, the down side of this arrangement may be the procurement of material at the vendor level. If it has a long lead-time, little may be gained, since the vendor's material will not arrive in time for the manufacturing cue to be used up.

SUMMARY

This chapter can best be characterized as a checklist for the new product development. The development phase and manufacturing arrangements have been made, and now the focus has shifted to the preparation work for product introduction.

The chapter starts out with the preflight checklist for the new product line. It then moves into the agency approvals and pilot run manufacturing. The product receives external validation in the Beta test program. The information system associated with the commercial aspects and usage of the product followed.

Next is the examination and confirmation of internal support systems to prosecute business generated by the product. These systems involve training of the personnel, the application support personnel, and the field sales organization. In all of these functional areas there is a change in the operations that must be introduced and cemented into the organization.

Final pricing is established at this point as the program is very near launch. To effect a successful launch, the sales order entry system must be functional and effective. Finally, the material requirements in the stages of transition, from development prototypes through initial production, are discussed.

Each stage of the development process brings closure and anticipation, simultaneously. With the completion of the elements discussed in Chapter 9, the product is poised for the transition the product team has planned for—The Launch.

THE LAUNCH

ABSTRACT: The product launch is the culmination of all the effort and work the new product development team has invested. It is the decisive point in which the market must respond to the total effort. This chapter deals with all of the issues surrounding the product launch. Now, with all of the development, planning, and execution in place, it's time for the major thrust in getting the new product business off the ground. As each phase brings feelings of apprehension, accomplishment, and then anticipation, the launch is what the group has been working toward.

THE PRODUCT ROLLOUT

During the development phase of the project, there is a constant balance struck between the maintenance of secrecy of pertinent areas of the development and the need for promotion of the program to various parties. The rollout is the time for all efforts to be directed to the introduction and promotion of the product. In this next section the various aspects of the rollout will be reviewed.

1. THE ROLLOUT FORMALLY LAUNCHES THE PRODUCT

The product rollout, or introduction, is the formal means to launch the new product. It is specifically designed to get the word out to the customers and marketplace that the new product is available for sale, and consists of the certain features and benefits. The rollout should be timely, pervasive, and far-reaching. Unless there is a compelling reason to have a phased introduction, the rollout should simultaneously announce the product in all sectors of the marketplace.

The rollout should also be a time to generate enthusiasm about the product. This enthusiasm needs to pervade the sales group, the sales channel, and the customer base. In some industries, the rollout is the media event anticipated by the marketplace because of pre-introduction activities. This anticipation helps cultivate the enthusiasm needed to carry the product.

2. MEDIA USAGE IS CRITICAL TO LAUNCH PLANS

Usage of the media is critical to the launch of a new product. This is because the media, in whatever form, creates leverage. This leverage enhances the effect of the initial launch. It brings the company and product name to the forefront in the minds of the marketplace. Without this the manufacturer could not devote enough energy, time, and funds to create the same effect. There are several different types of media to use in the initial launch of the product; the following list represents a few of these types, along with their explanations. (Each product and business will have its own needs and requirements for media usage. These examples are for reference.)

A. **Trade press.** The trade press involves those publications and other related media that participate in the industry the new product will serve.

B. **Newspaper.** The newspaper can assist in a local fashion by introducing the new product to the general population that reads the paper. It provides completeness in the introduction, and often "sparks" interest of other publications that comb the regional news for new information.

C. **Direct mail campaigns.** The direct mail campaign targets the customer directly in the hopes of soliciting interest in the new product. If the nature of the product is such that the details of the product are already known, the direct mail campaign can focus on availability.

D. **Seminars.** Seminars are an effective way to introduce a new product to a specific market segment. It is based on the assumption that the product presentation and problem/solution development can be imparted to a group of people participating in an industry within the framework of a generic presentation. The seminars can also be specific to product introduction solely.

E. **Professional awards.** The ability to win a professional award or an industry award can be an effective tool in product promotion. The award secures product legitimacy, and further legitimizes the launch.

F. **Trade shows.** The use of trade shows reaches industry and market segment individuals in an effective manner. While these people are at the show, and in the booth looking at the product, you secure their entire attention outside of their traditional decision-making venue. It gives the manufacturer an advantage in introducing and promoting the new product.

G. **Internet/Website.** Explosive growth in the usage of these media and offshoots can get information to the customer directly in immediate fashion. This assumes the customer requests the information and knows where to secure it. The use of a website still requires the customer to search for the information or log on to a known location. Certain links can be used to direct a generic request for information to the company's site also.

H. Magazine articles. Industry trade magazine articles about the problems and product solution development can assist in promoting the product. It adds legitimacy to the product, personnel, and the company when a reputable magazine publishes such an article.

I. Vocational white papers. White paper or generic subject primers are useful in training an audience in a new product area. Since the paper is generic, there cannot be overt advertisements, and still be called a white paper. However, it does position the authoring company and the personnel involved as "experts in the field" and a source for solutions.

3. TELLING THE NEW PRODUCT STORY

The new product story must be effectively told to the marketplace. At first blush this may seem like a daunting task; however, the strategy, tactics, and execution requirements are all in the original business plan. Refer back to this plan for the various aspects of the marketing program that will be needed at this time.

Traditional marketing of products generally follows a certain pattern (i.e., problem solution development). This means the marketing program outlines the problem to the prospective client and then presents its products as the best alternative to solve the problem or need. This has been an effective means for promotion of new products, and will continue to be. There are, however, different motivations for the use of products and as such, the marketing of these products will be altered accordingly.

4. ESTABLISHING MOMENTUM

The issue of momentum is as important in the product launch as it is in the development program. The product launch is analogous to lighting a candle. Energy is placed at the candle to light it. When the energy or match is removed, the candle should stay lighted. The same holds true for a product launch. The company invests energy in the form of funds and human energy to initially launch the product through a campaign to generate enthusiasm. The sales channel must maintain the "fire." In addition, the better analogy for a large launch effort might be a fire, which is lit by a match. The small amount of energy is leveraged into a larger and growing fire.

5. INCENTIVES

For a sales organization to be effective, a clearly spelled-out and equitable incentive program should be in place and functioning. This incentive program should account for a new product introduction also. If it does not, there may be a need to create a separate incentive program to launch the new product. There is a danger with this, however, in that certain windfalls may occur, or a total lack of incentive participation in certain circumstances. The incentive program ideally should factor in the new product promotion so as to drive a successful introduction.

INITIAL MONITORING OF RESULTS

1. MODELS FOR NEW PRODUCT ROLLOUT

There are several models for a new product rollout. There can be regional rollout in which the introduction is limited to a certain geographic region. There can be a national rollout in which the introduction is within the nation's boundaries; and finally, there can be an international or global rollout in which the product is introduced across the entire global marketplace. Each has a cost associated with it, and results that can be expected. There is also the company side of the rollout, which is separate from the geographic side. This focuses on the product introduction from the depth of the sales organization and route to market (Figure 10-1)

As shown in Figure 10-1, the matrix of rolling out a new product is two-dimensional, with one aspect being geographic and the other focusing on the sales channel's route to the customer. Additional effort is required to establish a more pervasive introduction geographically as well as organizationally. For example, pushing the envelope in the channel requires training on the product and how to sell it. Pushing the envelope geographically will require promotional funds and time.

2. MONTHLY MONITORING

After the initial rollout, there is a need to monitor results on a periodic basis. Depending on the specific product and business, this may vary. However, for illustrative purposes (and more often than not), the monthly monitoring model will be adequate. This can be best represented in Figure 10-2.

MATRIX OF ROLLOUT	REGIONAL ROLLOUT	NATIONAL ROLLOUT	INTERNATIONAL ROLLOUT
COMPANY SALES ORGANIZATION			
SALES CHANNEL DEPTH			
MARKETPLACE			
CUSTOMER BASE			

Figure 10-1. Models for Rollout

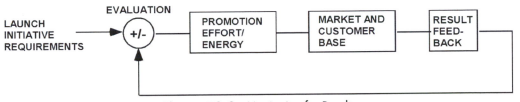

Figure 10-2. Monitoring for Results

As shown in Figure 10-2, the promotional efforts are under closed loop control. This means that the results of the effort are being monitored via the market reaction and compared with the initiative requirements. The assessment is made and the promotional effort is adjusted accordingly. If the results show lack of sales at a given effort level, then more effort may be required to equal the launch initiative requirements.

To place this in more quantifiable terms, consider Figure 10-3 as an example of a launch rollout evaluation, in which there are five basic regions of a company that need monitoring. Each region is evaluated in terms of investment and results. They can also be evaluated in terms of how they performed in comparison with each other.

Figure 10-3 shows the performance of five regions participating in the new product launch. The numbers represent the raw data and the percentages. There is diversity among the regions. Region 1 achieved 80% of the sales requirement, with only 57% of the expenses. They expended 75% of the available time during the period.

Regions 2 and 3 performed the best of all five, since they delivered in excess of the budgeted sales. Region 2 even outperformed Region 3, expending only 56% of the allocated time and only spending 61% of the expenses.

INITIAL FEEDBACK MONITORING

	ACTUAL EXPENSE ($)	BUDGETED EXPENSE ($)	ACTUAL TIME (HOURS)	TOTAL AVAIL (HOURS)	ACTUAL SALES ($) (UNITS)	BUDGETED SALES ($) OR (UNITS)
REGION#1	20000	35000	600	800	400	500
% BUDGET	57.14		75.00		80.00	
REGION #2	22000	36000	450	800	600	550
% BUDGET	61.11		56.25		109.09	
REGION#3	33000	40000	590	800	850	750
% BUDGET	82.50		73.75		113.33	
REGION#4	44000	35000	230	800	190	550
% BUDGET	125.71		28.75		34.55	
REGION#5	66000	50000	120	800	250	1250
% BUDGET	132.00		15.00		20.00	
TOTAL	185000	196000	1990	4000	2290	3600
% BUDGET	94.39		49.75		63.61	

Figure 10-3. Initial Feedback Monitoring

Conversely, Region 5 spent 132% of the funds and delivered only 20% of the sales required. In addition, they show little participation in terms of expended time. This would need immediate correction if Region 5 is to contribute effectively to the launch.

The data also show that with 94% of the funds expended and 50% of the time devoted to the launch, 63% of the result was obtained. This indicated required action and intervention from headquarters to place the program on track. Later in this chapter is a section on measurement, which outlines additional criteria for evaluation.

3. MAKING THE STRATEGY WORK

There is a difference between formulating a strategy on paper and executing tactics to actually make it work. During the launch phase of the project, it is important to focus on those tactics that will implement the overall product strategy. Refer back to the business plan for the strategy, and formulate the tactics to execute it at regional and local levels.

Each region or office knows their customer base and marketplace well enough to formulate and execute successful tactics. If the tactics employed do not effect the results, one of two things may be happening: The local people have not bought into the program, or the strategy is unachievable given all of the prevailing conditions.

The problem of buy in can be resolved from a personnel standpoint. The other issues are much more far-reaching in scope. The strategy under which the program was launched is suspect. This should be a very rare occurrence and certainly not pervasive, since a diligent job of assessment throughout the program prevents misunderstanding in the global sense.

4. CHECKING THE PRICE/VALUE MATRIX

The launch is a good time to recheck the price/value matrix, which was constructed earlier, when the product was being specified. The initial launch will indicate how the market responds to the product, its features, and the value placed on it. The idea here is not to alter the promotional campaign or to redirect it, but rather to take necessary steps to refocus on certain features that the market may find more appealing than originally understood. This is designed only to enhance the launch effectiveness, not to correct it.

5. IS THE STORY GETTING OUT?

One of the most nagging issues in managing a new product introduction is determining if the new product story is being diffused accurately into the field. Is the story, cultivated internally within the new-product-planning group, being used in the marketplace to secure sales, or are the individual salespeople generating their own market presentation? Consider Figure 10-4 where there is a feedback system to confirm this.

In Figure 10-4, the pricing of the product and the value associated with the features are presented to the marketplace. The customers establish purchasing patterns based on these

PRICING VALUE MATRIX

Figure 10-4. Price/Value Matrix

factors. Through after-sales assessment and lost-sales assessment, it can be determined if the presentation is being used in the regions, and if it is working correctly.

6. IS THE SALES FORCE PREACHING TO ALL (UNCOVERING HIDDEN STRUCTURAL FLAWS)?

The other factor to consider in the sales assessment is if there is adequate homogeneous coverage of the marketplace, or if there are voids in the coverage. This factor is best represented in Figure 10-5.

7. OVERCOMING OBSTACLES

New product development is a process of overcoming obstacles. There is uncertainty in marketplace dynamics, uncertainty in development engineering, uncertainty in manufacturing, and uncertainty of product acceptance during the promotion. These obstacles must be effectively overcome to launch the product properly. In addition, the bulk of the effort in producing a new product is behind you; do not let problems in promotion hold back the business.

If the obstacles are physical, work around them. If the problems are people, replace them. Now is not the time for a conservative approach. The product launch is intense and short-lived. Use the energy to create product momentum with it.

8. ENERGY IS THE ONLY WEAPON AT LAUNCH—USE IT!

As you progress through the product development process and execute the various phases, the scope and the commercial terms of the product offering become relatively defined. As you near the launch, there are very few changes that can be made; perhaps, only very slight reorientation of direction and goals. The only real weapon you have at launch is human energy to bring to bear on the launch itself. Focus that human energy potential

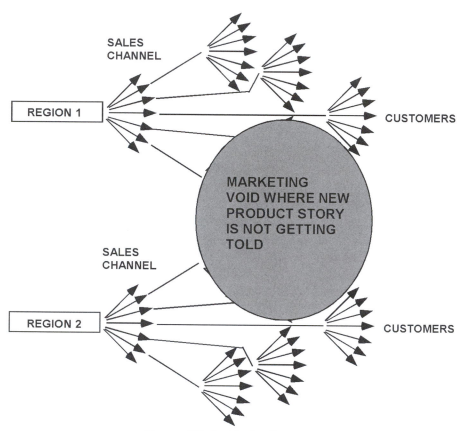

Figure 10-5. Market Coverage

directly on the tasks at hand. Establish and instill the enthusiasm in the sales channel to leverage the introduction. Create and apply the energy to show quick decisive results. This alone will help promote the enthusiasm needed for a successful launch.

EARLY MODIFICATIONS FOR SUCCESS

1. LEAST DESIRABLE THING TO DO AT THIS POINT IS CHANGE THE PRODUCT

So the greatest product in the history of the world has been launched, and early results indicate that the product is foundering in the marketplace. What do you do? An often-used excuse, by customers, not to accept a new product, is to cite certain lack of features or package of values, among other reasons. The temptation is to immediately change the product to exact approval by the customer. Unfortunately, there may be other reasons for the lack of interest, and even these added features may not sway the customer into a purchasing decision.

Considering the fact that the product group was diligent in their activities, there may be no compelling reason to even consider changing the product. Changing the offering in a significant fashion is the least-desirable action to be taken during the product launch. It sends a very mixed message to the sales channel, the marketplace, and the customers. It adds confusion to the product line. It virtually destroys the legitimacy of the "marketing story" that is used to launch the product.

Changing the product contributes to a sense of disorientation for the company and its relationship to the prospective customer base. The operative point here is "do not be too quick to change the product offering."

2. GIVE THE LAUNCH A CHANCE

It is important to take care when considering changes to the product in the wake of a launch that fell short of expectations. There are several factors that may have contributed to the lackluster results, some of them obvious and some latent. Be sure to clearly understand the issues before making changes, and also determine if the issues affect the launch in a homogeneous manner. This means to determine if the same issues produce the same result in the various demographic situations. What you may feel is harming you in a certain demographic may actually be assisting you in another.

3. REPACKAGING OF VALUES

If a decision is eventually made to alter the product, one of the easiest ways to create the alteration is to redefine the package of values offered. This adds value to the product offering without necessarily adding a lot of cost. For example, offering a package of hardware and different software sets can enhance the saleability of the product. If a product change is anticipated, reconsider it as a first step. If you missed the mark on the first try, with all the planning and execution, what makes you think you will hit the target in the middle of an introduction? Stay the course and see the production through. Another means for inducement to purchase is the use of rebates. These are simply cost reductions that are employed in a temporary fashion and allow the initial pricing to be somewhat preserved. Other promotions can be used as well. The basic idea is to leverage sales without incurring too much additional cost.

4. COMBINATION OF DIFFERENT PRODUCTS LEVERAGE THE EXISTING BASE

Another way to leverage the existing customer base is to offer the new product, along with the existing repetitively purchased product. This utilizes the familiarity the customers have with the existing product and allows the new product to "tag" along, in the hopes the customer will see the value and start using these products. Each case depends on the individual circumstances.

5. RECALL: IF NECESSARY, DO IT!

It is no secret: One of the greatest fears in new product development is the product recall. The product recall is one of the worst things to administer during a product launch. Completely halting the launch depends on the degree of confidence in the product correction and the stage of implementation of the correction. If the correction is made and is implemented in production, the launch can continue. If not, the launch most likely must be terminated and rescheduled for a later time. In either case the issues are very difficult to navigate around.

If the issue warrants a product recall, plan the steps and execute the recall decisively. Many products are successful after relaunch. The marketplace can forgive a mistake and an inconvenient correction. It is less tolerant of negligence.

The product recall has a specific protocol that should be followed. The issue of product recalls will be discussed in more detail in Chapter 11, "The Pursuit"; however, the protocol, along with the explanation of each step, is included here. The protocol is as follows:

A. Contact

The contact portion of the protocol involves the company contacting the customer in some organized fashion to advise of the recall. This can be a registered letter with return receipt indicating delivery and receipt, telephone, fax, e-mail, or some other verifiable means for contacting the customer and advising of the recall status.

B. Explanation

The explanation outlines the specific defect, the potential effects, the impact of the defect on the user, and the degree of seriousness of the defect. It must identify the scope of impact that will cause appropriate and timely action on the users' part.

C. Actions required

This is a detailed outline of the actions required to resolve the recall issues for the customer. The actions define the scope of the customer and company requirements, the specific actions to take, the timing of these actions, and the responsibility of each. It should also prescribe the terms of the recall, if expedient actions are not taken. There may also need to be a flowchart of events included as part of the actions.

D. Procedures

The procedures spell out the disposition of the product in the customer's possession and the arrangements for any upgrades planned. Specifically, the procedure section involves one of the following elements: exchange of product, replacement of product, repair of product, substitution of product, or upgrade of product. One of these should be selected, and the procedure for the selection should be spelled out completely.

E. Acknowledgment

This is a very important part of the recall process. The acknowledgment provides evidence of the company contacting the customer, the customer responding to the contact, and the product disposition being executed. Any arrangement of bounce-back cards, telephone, or some other means, is demonstrable proof of the recall engagement showing the responsibility and the fulfillment on each party's part.

F. Compensation

Depending on circumstances, there may be compensation involved. This compensation may also be included in the explanation portion of the recall. It outlines the compensatory steps the company will take in securing customer satisfaction.

G. Closure

This is a formal record to close out the recall. It provides demonstrable proof of the company diligence, recall engagement, resolution, and proof of mitigation. Depending on product circumstances, it may be included in other sections also.

6. MITIGATING THE DAMAGE

From a marketing perspective, there is a critical need to contain or mitigate the damage in a recall situation. Whatever steps can be taken to preserve the company and product reputation must be done. The company reputation is the first priority and the product line reputation is second priority. In either case the idea is to pave the way for tomorrow's business from the same customers that are having problems with your product today.

There is one element for certain: The marketplace and customers' expectations of newly introduced products are ever-increasing. Customers have very little tolerance for product problems, so the degree to which you can reduce the impact on them the better.

THE MYTH OF THE HOCKEY STICK FORECAST

1. THE AGGRESSIVE SALES FORECAST

Remember when the new product was being justified for investment? At that time it was easy to be very aggressive in the forecast. This aggressiveness and optimism helped justify the program. Now, however, it is time to deliver the sales, and the tendency toward an aggressive forecast will give way to something that is more achievable. The traditional hockey stick forecast is something that is generally very difficult to achieve. This is because of the human dynamics of a sales campaign and the mathematics of a geometric progression.

2. WHY VERY FEW PRODUCTS HAVE CHARACTERISTIC GROWTH

Figure 10-6 represents a comparison of planned and typical actual progressions. The format is based on several assumptions. The first assumption is that the planned forecast has the company sales personnel leverage their activities through the sales channel by successfully engaging 10 salespeople from the channel. This corresponds to a 100% acceptance rate of the sales channel. Each of these salespeople has 10 customers available to them for selling the product to (i.e., the available market). The other assumption is that it takes five periods to achieve a success rate, indicated in the success rate column. In this ideal, planned example, the expectation is for a 100% success rate.

Figure 10-6 shows that the first channel salesperson achieves a sale to 2 out of their 10 customers in the first period. Therefore the volume of the first sales person is 2. In subsequent periods more customers are added at a rate of two per period. Additional salespersons generate sales in the same manner.

GROWTH PROJECTIONS

PLANNED SALES	X	CHANNEL SALES	CUST.	SUCCESS RATE	CUSTOMER				
					PERIOD 1	PERIOD 2	PERIOD 3	PERIOD 4	PERIOD 5
SLSMN 1	10	1	10	100%	2	4	6	8	10
		1			2	4	6	8	10
		1			2	4	6	8	10
		1			2	4	6	8	10
		1			2	4	6	8	10
		1			2	4	6	8	10
		1			2	4	6	8	10
		1			2	4	6	8	10
		1			2	4	6	8	10
		1			2	4	6	8	10
				TOTAL	**20**	**40**	**60**	**80**	**100**
ACTUAL					PERIOD 1	PERIOD 2	PERIOD 3	PERIOD 4	PERIOD 5
SLSMN 1	61	1	10	0.50	1	2	3	4	5
		1			1	2	3	4	5
		1				1	2	3	4
		1				1	2	3	4
		1					1	2	3
		1					1	2	3
		1			*0*	*0*	*0*	*0*	*0*
		1			*0*	*0*	*0*	*0*	*0*
		1			*0*	*0*	*0*	*0*	*0*
		1			*0*	*0*	*0*	*0*	*0*
				TOTAL	**2**	**6**	**12**	**18**	**24**
				% ACTUAL/PLANNED	10	15	20	22.5	24

Figure 10-6. Sales Projection

In addition, all sales personnel in the channel start selling to their 10 customers, each at the same time. The progression shows a volume building from 20 to the total available of 100 within five periods.

The actual projection is somewhat different and probably represents a projection that is more typical. In this part of the example, the "hit rate" of success with the customer is 50%. In addition, there are only 60% of the 10 salespeople in the channel that will support the program. This may be indefinitely or just at the beginning. In either case the effect is quite damaging to the program. Another real-life effect is that not all channel sales personnel will immediately embrace the program. More often that not, they will have a tendency to delay their entry in the promotion as they observe others' success or failure. They often do this to protect their customers from newly launched products that could have reliability problems or other product problems.

The 40% not participating are represented by the italicized 0 entries. The delayed participation is represented by the shift in time. The 50% hit rate is represented by the value 1 versus the value 2 in the planned portion of the example (e.g., for salesperson 1 in period 1), 2 versus 4 in period 2, 3 versus 6 in period 3, and so on.

The impact of these adjustments is significant, reviewing the bottom of the illustration where the percent actual to plan shows results are poor.

3. PLANNING FOR REALISTIC RESULTS

The operative lesson is to forecast and justify the program with achievable forecast numbers. Eventually the product team will have to deliver these numbers, so be realistic in the forecast, and allow no complacency in the execution to obtain the numbers. At the onset ensure the forecast can be met. At the launch phase of the new product program, reexamine the initial assumptions made, and operate the new business to a reasonable forecast that can be achieved in the time frame outlined.

FORECASTING AND BUILDING INVENTORY

1. REVISE FORECAST AT LAUNCH, IF REQUIRED

Depending on the type of product line, the revised forecast at launch will need to be accurate enough to drive material procurement and manpower loading in the manufacturing group. Often, an understated forecast will result in too few materials, preventing the company to capitalize on the market opportunities as they become available. An overstated forecast will cause inventories to swell beyond tolerance. Incorrect inventory results in the inability to ship, partially completed inventory, and unbalance in the manufacturing environment.

2. USE LOW/MEDIUM/HIGH ASSESSMENT

Given the project is in the launch phase, there is a good idea on what forecast is reasonable. The product team may want to generate this number as most likely, along with a low- and a high-usage figure. The high-usage figure can drive the materials, and the low figure better be enough to carry the business and amortize the development in a reasonable amount of time.

3. MONITOR RAMP UP AND BUY MATERIALS FOR HIGH VOLUME STAFF MEDIUM, THEN FLEX STAFF

The product team's job, with respect to volume, is to closely monitor the volume and ramp up the materials and the scheduling for manpower and facilities to capture market share. The staff assignment should be targeted at the low figure initially, and then expanded by manpower, flexing to the high volume should it become necessary.

4. CHANGING OVER FROM PLANNING TO RUNNING A NEW PRODUCT BUSINESS

One of the elements of a product introduction is to test the market's response to the product. This can be used to scale the future levels of business. There is a shift in building of inventory between a newly launched product and one as it matures. The newly launched product has a great amount of uncertainty, which needs absorption and reactive planning to match the company's manufacturing capability to the marketplace's appetite for the product. Consequently, there is a lot of management expertise associated with balancing out the initial product quantities.

After the product has settled out and the marketplace response is somewhat more predictable, the forecasting systems and manpower scheduling systems can begin to take over the logistics of the product line.

5. BUILD PLANS

As the transition from initial launch to managing the product occurs, there needs to be a plan to build inventory and have it available for sale in a timely fashion. The initial experience in the launch will generate a trend to base the business levels on. In the manufacturing arena a company must have inventory available for shipment, or be able to transform raw inventory to finished shippable product rapidly. Consequently, a build plan should be developed and implemented to ensure timely shipments.

6. REFER BACK TO OPTIONS AND PRODUCT CONFIGURATION

The job of raw inventory selection for assembly, and the product stocked for sale and shipment, is made easier by the initial product configuration. Referring back to when the product was configured with all of the various versions, future shortages can be inadvertently built into the product line whereby, the wrong raw inventory is in stock at any given market opportunity. This being the case, there will always be difficulty in securing market share through timely shipments. The best defense against this is to prevent too many configurations from being designed into the product line. The degrees of success in this endeavor will be measured at the product launch.

7. COPING WITH SHORTAGES

Coping with shortages is an exercise in loss mitigation. Given that the vendor links are established and the part specifications are in order, expediting parts into the company for manufacture will cost additional funds thereby contributing to a lower gross margin. Depending on manpower loading, there may be overtime charges levied as well, further degrading the margin.

Long-lead items of low value should be placed in inventory as safety stock. Common parts between all products are also a big help. Purchasing queue time at the vendor can yield some flexibility also. Fundamentally, however, the best way to mitigate the damage of lower gross margin and lost opportunity is to secure the most accurate forecast possible.

PRODUCT PROMOTION AND CUSTOMER VISITS

1. THE NEED AND VALUE OF CUSTOMER VISITS

Very few products enjoy the position of being able to sell themselves. Unfortunately, products do not engage customers, establish business relationships, resolve problems, and apply solutions to meet customer needs. Salespeople and application people need to perform this function. Consequently, placing company personnel in front of customers is a healthy activity. It gives the product team primary feedback on the product offering, it removes incorrect assumptions in development by observing the customer reactions directly, and it removes development interpretation of customer issues in the vacuum of customer interface.

From the sales promotion perspective, it allows the sales personnel and the channel to reinforce the promotion objectives and the promotional success factors.

If third-party channel members sell the product line, it affords the direct-company sales personnel to observe the promotion in action and effect improvements or changes in their approach.

2. PARTICIPATING IN INITIAL SALES TRANSACTIONS

The customer visit creates bidirectional awareness between the company and the customer. The company learns about the issues facing the customer within the framework of the new product offering, and the customer learns about the benefits afforded by the company and its products.

There is another, more selfish reason the customer visit is necessary. The initial marketing story is developed in somewhat of a vacuum of the widespread customer base. It is developed with a select few customers, and then a comprehensive promotional campaign is prepared. The customer visit helps validate this marketing story and refine it for optimum performance.

By directly interfacing with the customer, the product group can learn which issues are (and are not) important to the customer, regardless of the original assumption. In this way the promotion can be made more effective.

3. FOCUS ON THE SALES TRANSACTION AND THE CUSTOMER INTERFACE

This is where the new product success is born. By focusing on the sales transaction, a great deal can be learned about the new business and how to grow it. The idea here is to determine what elements of the promotional campaign work in certain customer situations and leverage them into a wide demographic area.

The sales and marketing function needs to learn the "pressure points" of the transaction and translate them into sales. Studying the transaction as it occurs, for both a successful and an unsuccessful sale, can do this. Subsequent refinements or segmentation of the presentation can improve efficiency of the sales activity. The segmentation may need to be done because one presentation may not serve the needs of all customers. Each segment may be interested in the product for different features; consequently, each presentation may focus on a different set of features and cite different benefits to the customer.

4. CORRECT SALES ACTIVITIES AS REQUIRED

If there is a change required to the marketing campaign, as a result of certain feedback in the customer visit, the sales activities may have to be somewhat modified to improve effectiveness. The area of concern here is that changing the entire sales organization focus, based on limited input, can be quite dangerous to the program. Be sure to evaluate these proposed changes in light of the entire market served.

Make only minor modifications where necessary, or structure the program slightly differently to allow the local sales personnel more freedom to customize the program to their specific needs, without compromising the focus and strategy of the business development. The basic core of the campaign should be on target, if the project has been prosecuted diligently.

5. DETERMINE HUMAN ENERGY REQUIRED TO SELL AND PLAN ACCORDINGLY

One of the imperative, initial feedback items is to determine the human energy investment required to secure an order for the new product. This determination of energy level will reinforce or update the original estimate of the sales personnel's time required to support the launch. If the energy required on average is significantly more than originally planned, the focus of the salespeople's time will have to be realigned to prosecute the launch. The worst scenario is to have a shortfall in the time allocation and cheat the new product launch.

If additional time and resources are needed initially to launch the product, realign immediately to secure the incoming planned order rate. Fundamentally, you never have a better opportunity to grow the business and market share than at the initial launch, with the pricing and feature set at their most favorable position.

6. PREVENTING THE MYTH OF THE CASH COW

Complacency is the most formidable enemy of business development. The term *cash cow* is almost a misnomer, because every business and product line driving the business requires nurturing and attention. Consequently, there is a big difference between the model for a new product and the actual activities to support it. Figure 10-7 shows the model presented earlier for a new product.

The model in Figure 10-7 includes an investment for 2 years and a stream of returns for several years afterward. Traditionally, the term *cash cow* has referred to the steadily growing stream of revenues occurring in subsequent years after the initial investment. This is what is depicted in the macro sense from a financial perspective. Operationally, however, there is a lot more going on. The organization must support the business, and include enhancements or value-added additions to the product line. In this sense there is no such thing as a cash cow. Every business needs constant attention.

The lesson for a new product launch is to overcome the tendency toward complacency that may occur with initial success. Constant diligence and attention, even with success along the way, will drive the larger returns.

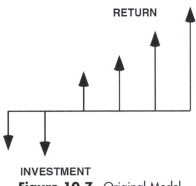

Figure 10-7. Original Model

TOOLS

1. NETWORKING THE SALES FORCE TO YOUR ADVANTAGE

In any organization there is an established chain of command, and a less-obvious set of allegiances that exist between members of the organization. This also occurs in the sales arena. Managers involved in the new product launch should not assume a homogeneous sales department with equal and independent allegiances to sales management.

Given that this is a fact of corporate life, the new product team should use networking to their advantage in launching the product. If this means convincing the most influential manager or salesperson on the merits of successfully "embracing" the program, then this is what must be done for the good of the product launch. This phenomenon should be looked on as a tool to be utilized, rather than as a roadblock.

2. THE MODEL FOR A PROSPECTED SALES TRANSACTION

Every product line has its own characteristic model for a typical sales transaction. In addition, each transaction has its own peculiarities; the transaction involves human beings who interpret data and make judgments. However, it is worth considering all of the identifiable phases of a prospected sale. The term *prospected sale* refers to a sale precipitated with a customer who has not necessarily contacted the company to purchase the new product. It is the model of the sale where the company has prospected the customers for product sales.

A. Prospecting

This portion of the model deals with the decision processes in refining an entire global market and targeting an achievable market share by the company. The prospecting activity, if conducted properly, can reduce a large amount of wasted effort, and can focus resources at the target market segments that will pay off.

B. Identification

The identification portion will further reduce the prospecting list, from the market segmentation activity to the generation of an actual list of accounts for the sales channel to contact.

C. Contact

This is the initial direct contact with the customer to pave the way for the sales activity. It can be done through the sales personnel directly or on behalf of the company by a sales channel.

D. Qualification

The qualification is the interchange between the company and the customer that results in a qualified lead. This means the customer has a need that will be met by some supplier at some point, and that the company is a viable candidate for the business.

E. Engagement

This is where the detective work is done by the sales professional to determine how to go about securing the sale. What tactics, arguments, and features that bestow benefits will be highlighted and focused on in the actual presentation? The engagement can best be characterized by fact-finding and strategizing of the sales presentation so that it results in an order.

F. Presentation

The presentation is the actual constructed result of the engagement. It is the case that is made for the company securing the business.

G. Feedback

After the presentation, the customer will have some type of response. The response may be affirmative, interrogatory, or rejection. In each case there is action that may be required to redirect the interchange to a product sale.

H. Redirection/Confirmation

The redirection and confirmation portion of the interchange is generally where the sales professional will determine if they are going to be successful in selling this prospect. In some cases this may be a reaffirmation of features and benefits, explanation of the benefits, or a discussion of how specific features will satisfy the prospect's needs now and in the future.

I. Close

Simply stated, this is the activity that moves the prospect from conversation to action.

J. Assistance/Logistics

After the sale, there may be assistance required for the new customer. Even as part of the ordering process, certain logistical information and planning may have to take place. The sales professional must consider this as part of the transaction.

K. Reinforcement

After the new customer has had some experience with the product, the sales professional may follow up with the customer to determine level of satisfaction with the product. They may inquire if the product meets the customer's initial expectations and if they are happy with their purchase.

L. Exploration

With a satisfied customer, there is an additional exploration process that should take place to determine what other products or services available from the company can be proposed to the customer. This process enhances the customer's status from that of customer to account.

3. PICK A SALES EXAMPLE AND DRIVE IT TO SUCCESS

One method to generate sales is to select a sales example and drive it to success. This means to target certain sales situations, and certain attainable applications and customers, and take effective action and intervention to ensure that they come to pass. If direct company involvement is required, then direct company involvement will take place. By doing this certain critical success stories are being cultivated by the company. These will be used to leverage other sales as presented in the next section.

4. SUCCESS STORIES: MATRIX DEMOGRAPHICALLY

If the examples have been selected properly, these will formulate the basis for a demographic spread that can be used to leverage sales. In other words, if you don't have a success to refer to in the promotion of the new product, make a success and then refer to it!

With the several success stories in place, demographically leverage them. Use one success in one application to leverage success in similar applications. One example can then be used for repeat applications that are the same *and* for developing new ones in which there is no success story yet.

5. UTILIZING PUBLICITY TO PROMOTE THE NEW PRODUCT

The use of publicity in promoting a product can be helpful in getting the word out. Publicity can be positive or (unfortunately) negative for a product reputation. When it is positive, it leverages the new product story among a wide audience. When it is negative, it becomes a challenging task to overcome. The following describes some of the basic facts about publicity. By understanding the dynamics of it, one can use publicity to his or her advantage while avoiding the pitfalls.

A. Publicity can add credibility

The primary purposes of publicity are to create awareness, raise issues, and convey a message. Each of these elements can be used to promote a new product. For example, awareness can be created about the new product, or specific issues that the new product addresses can be raised via publicity. It can even be used to convey a specific message about the segment of the market or population that the product serves.

B. It must be news-oriented

When attempting to use publicity to promote a new product, it is important to understand the driving forces of publicity. Publicity is about news. It is imbued with a "scoop" mentality that can report positive elements or negative elements of a subject. There is no loyalty to any one thing; the broader appeal of the "new," either good or bad, will get the focus. Consequently, the product team must be very careful in utilizing this medium. A product failure can be higher profile than a product success.

The news can be about change, achievement, events, success, mass impact, or opinions. If you are going to utilize publicity, be sure to frame the story around one or more of these elements. For example, the product could cause change in the marketplace that readers would be interested in. The product line may have won an award or some other achievement. There may be a success associated with the product that may need coverage. There may be impact on a certain mass of people, caused by the new product. Or an expert may render an opinion about the new product in the media. All of these elements can be a positive driving force for the product when considered within the framework of publicity.

C. Characteristics of the media

The media is a very unpredictable means of product promotion, however. Referring back to the positive aspects and negative aspects, the more newsworthy, either good or bad, will carry priority. Consequently, untimely negative news can far overshadow any positive aspects and require significant recovery. In these cases it may have been better not to have any publicity in the first place.

The media have a gossip mentality, and are more acute at listening to the negative aspects. They look for the insight into the events.

D. The dynamics

When dealing with the media, be sure to know what *their* audience is and what their preferences are for reporting, facts, opinions, and style. Become familiar with *their* audience. For example, a medium dealing in innuendo, nonfacts, and sensationalism is not one to consider for promoting and interpreting the benefits of a new product or a new process. Commonsense about the medium will govern your actions.

To maximize exposure, stage the news or stories to have repeated coverage in subsequent presentations.

E. Dealing with the media

There are some accepted protocols used when interfacing with the media.

First: Try to visit the editor personally. This establishes credibility and begins a rapport between the two of you. Be sensitive on deadlines for their projects. Communicate the story to them in a clear, concise format to allow them to explore the story angles easier. Be prepared when issuing comments. Cite reasons for no comments. Be deliberate in comments. Remember they are giving you "air time"; do not waste it. Accept unsolicited calls from reporters. Do not hound or make demands of reporters, or attempt to write the story for them. Keep in mind: *The facts are yours, the perspectives are theirs*.

MEASUREMENTS

1. ESTABLISH LAUNCH GOALS

The business plan generated for the new product establishes the required results of the new product as a business. There is a requirement to evaluate the new product launch separately to determine early in the launch if it will be successful. This may determine the success of the overall new product and give the product team an early warning of problems. In addition to the forecasted sales, it is important to generate a profile for the initial launch. This, along with the actual profile, will indicate degree of success. This may also be segmented by area to see which area is performing and which is lacking. Figure 10-8 is an illustration of the planning and monitoring activity specific to the launch phase.

As shown in Figure 10-8, this comparison can indicate degree of success in obtaining the forecasted volume, as well as give an indication of the time required to meet the volumes. Segmented versions can be used to evaluate the performance of the individual groups.

LAUNCH GOALS

VOLUME	1	2	3	4	5	6
LAUNCH PLAN	111	222	333	444	555	666
LAUNCH ACTUAL	65	75	275	325	435	503

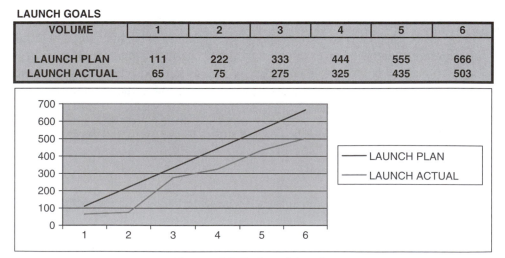

Figure 10-8. Launch Results

2. QUALITATIVE EVALUATION OF PRODUCT SOLD

There is also another factor that must be considered in the launch performance evaluation. The volume estimates yield raw performance; however, there is a qualitative measure needed to determine if the launch is progressing according to the marketing plan, the marketing story, and the new product business strategy.

This qualitative evaluation answers the question of the program being on track from the strategy perspective. Is the positioning of the product correct? Is the customer base responding as planned, or are there unknown circumstances causing the success? These and other questions need to be answered as part of the qualitative evaluation. How is the business plan being executed during the launch? Are there any compromises being made to effect the sales that the product team is unaware of? Are there pricing actions being taken at the local level to stimulate acceptance? Are there products and/or services being bundled at the local level to stimulate sales of the new product? These need clarification so that the product team does not draw the wrong conclusion from the initial results of the launch.

3. PRICE PROTECTION AND MARGIN RETENTION

The basic theme of this chapter is to give an appreciation for the fact that the new product launch is a critical juncture in the product development program. This phase has its own dangers and concerns that must be navigated. One of them is the issue of margin. Pricing concessions made at the local level, without the product team awareness, was discussed earlier in this chapter This is a dangerous practice, because the company must establish pricing in the marketplace for the product.

The establishment of pricing levels is not only important from the perspective of the marketing program; but it also must preserve margin. This margin is what amortizes the development investment. Initially there should be no reason to modify pricing during the launch phase. The product group must understand the dynamics of the marketplace and its response to the product introduction. The issue of pricing actions should not factor into this monitoring as added input.

4. BALANCED SELLING

The sales force and their channel personnel need to focus on making the new product launch a success; however, not at the expense of the existing business. They need to balance their efforts to support the existing business with whatever plans are made to grow the new business.

In terms of measurements, looking at incoming order rates of the new and existing, as well as evaluating the time and energy spent on both, will monitor this balance.

5. PREVENTING PRODUCT CANNIBALISM

All the talk, and all the convincing in the world, will not deter a sales force from leading with the new product and cannibalizing the existing product. It is simply the most natural thing to do as part of the sales effort. Therefore the prevention should be systemic. This means that the product team should have planned out the obsolescence, if there is one, or they should have prevented the cannibalism by some other means. This is done by structuring the product offering in such a way as to protect the existing product, or to phase in its obsolescence to maximize margin while the existing product is in its decline.

Simply stated, if product cannibalism is taking place, it is the fault and responsibility of the product group, not the sales force or any other channel issues.

THE SALES CHANNEL AND LAUNCH OBJECTIVES

1. LAUNCH INCENTIVES

Are there any special incentives or inducements needed to move the sales force into promoting the product? Should compensation be granted for the launch that is different than the compensation granted normally? These and related questions need to be evaluated as part of the launch plans. There is a danger in making compensation launch specific. This may cause disproportionate effort devoted to the more lucrative launch, rather than the balanced selling effort required by the *entire* organization.

2. EASE OF CONDUCTING BUSINESS

One of the responsibilities of the company is to absorb some of the uncertainty for the sales force. The activities associated with the launch should be clearly articulated and communicated. The company should make it easy to do business with the new product, and as such, take steps to reduce any difficulties in selling the product. For example, if the product is technical in nature, make sure the sales personnel have been trained, or at a minimum, have the basic documentation in place to become conversant with the technology. Do not send the sales force out into the marketplace with no information, or assume they can learn the product as they go along. This will delay launch objectives and reduce the sales efficiency. Negative experiences may also cause the sales personnel to drop the product from their efforts, causing further damage to the launch.

3. ACTION ITEMS, FOLLOW UP, COMPLETION DATE, AND RESULTS

These are the watchwords of a successful program. They can be applied to every phase of the program, and should be applied to the launch as well. To be effective, action items as part of the launch must be completed within a time frame to be effective. Consequently, it is important to have the sales personnel execute these programs on time. There is a vocational hazard in sales to delay these action items because of other activities. The product team must instill a sense of urgency in the sales group so they pay timely attention to the launch.

4. CLEARLY IDENTIFY THE DIRECT OBJECTIVE

The "marching orders" for the sales channel must be clear and given within the framework of other company objectives. No single group can place its importance above the balanced requirements of the organization. For example, the objective for the sales personnel when launching a new product should be in a framework as follows:

(Achieve the x unit amount of sales and shipments in the following time frame, with the resulting gross margin, while maintaining the existing product sales base.)

5. TRANSITIONING INCENTIVES

The incentives, if chosen for a product launch, cannot stay in place indefinitely. Eventually, as the product transitions from a newly launched product to the growth phase and product maturity, the incentive must transition from the launch incentive to a standard incentive.

COMMUNICATION, AGREEMENT, AND COMMITMENT TO OBJECTIVES

1. THE ESSENCE OF ANY NEW PRODUCT DEVELOPMENT

The coordination of a new product launch involves many facets of corporate and human behavior. The same characteristics needed for the development are the same for launch. Each part of the organization must contribute to the launch, and as such, require communication, agreement, commitment, and understanding.

This is something that cannot be finessed, sidestepped, worked around, or ignored. The participants, if they want to be players, must effectively and repeatedly demonstrate these basic tenets of group activity.

2. COMMUNICATION TO ALL PARTIES

There must be swift and accurate information disseminated throughout the sales channel as it pertains to the new product launch. This is needed because the organization is evaluating the effectiveness of the product offering in the marketplace, the pricing, the marketing tactics, strategy, and the market's consumer, as well as competitive response to the product introduction.

3. AGREEMENT ON POLICY, PROCEDURES, PLANS, AND GOALS

At the time of product launch, there should not be any disagreements on product policy, procedures on implementation, marketing plan for the company, and overall corporate goals. Keep in mind that all human and corporate energy must be devoted to the success of the product, not infighting or territorial power plays. The corporate objectives govern the individual actions, and the new product launch is an integral part of the objectives that cannot be ignored.

4. UNDERSTANDING AND TOLERANCE OF PROBLEMS AND TEMPORARY OBSTACLES

The product group must establish a spirit of cooperation between the various departments involved with the new product launch. They need to imbue a pattern of perseverance into the sales personnel, as well as plug the entire product development group into field organization feedback. The product group must work with the sales channel to overcome any obstacles that may arise.

It is management's responsibility to create and foster an environment of tolerance and understanding among all of the players in the launch. This is needed because any uncertainty left in the product development program must now be absorbed by the sales function and the product group. Both must work together to secure incremental market share.

5. COMMITMENT TO OBJECTIVES

In a like manner, all elements of the organization must support the product group during this launch phase. The commitment to the corporate objectives must be real and lasting. The organization and its people must lock onto them and execute them. With the entire organization committed and in lock step, the new product will easily attain orbit!

SUMMARY

This chapter transitions the product development, from planning to action, in the marketplace. It is the phase of the process where initial results are due and senior management watches for "early returns." Consequently, it is a basic tenet of this chapter to guide and ensure a successful new product launch.

The chapter starts with the rollout of the product line and the various means for a successful rollout. As part of the launch process, initial monitoring of results is used to evaluate the execution of the plan, the performance of the channel, and critical success factors. Discussed next, were any product modifications for success, any corrective actions, recall sequences, and mitigation.

Initial sales of the product, within the framework of realistic expectations of unit volume, was also discussed.

Transitioning from a newly launched product to a supportable business in terms of inventory and materials was reviewed, as well.

As an integral part of the launch process, the all-important customer visit, was covered in terms of needs and tools available. Measurements of the sales effort and the results it produces, along with incentives for the sales and channel personnel to ensure the success of the launch phase, was also reviewed.

Finally, the chapter ended with a discussion of the basic required elements in any human endeavor involving a group of people: communication, agreement, understanding, and commitment to the corporate objectives.

Given a successful launch, management of the new product as a business becomes the focus of Chapter 11, "The Pursuit."

THE PURSUIT

ABSTRACT: Now that the product is launched, the business of product management comes into play. This chapter deals with the elements required to manage a product after initial launch. This chapter has two basic thrusts. The first thrust deals with the mechanics of product management to keep the product in sellable condition. These include general product-management issues, cost reductions, quality initiatives, and product maintenance issues. The second thrust is directed more toward the marketing aspect of the new product. This thrust is oriented toward growing the new business—through follow on planning, market development and application development, pricing issues, and growth strategy. In many ways this chapter represents a review of several elements of the other chapters. Those elements were presented in detail, whereas this chapter revisits these elements within the framework of the life cycle of the product.

PRODUCT MANAGEMENT

In this first section, the focus will be on the mechanics of product management. The product manager's role in the organization is unique to almost any other role. A fully loaded product-management job represents the microcosm of a CEO for the organization. Their role crosses operational lines vocationally, yet they must remain a cooperative part of the operation, working in concert with their peers.

1. THE RACE: COST, PROFIT, AND VOLUME

Identifying, developing, manufacturing, and selling a new product are all races. The race is marked by goals and milestones. First, the race is to identify the product opportunity, followed by the race to develop the product on time, within budget, and at the appropriate factory cost. Next, is the race to introduce the product and get it established. Finally, the race is to grow the business and usher the next new product enhancement or totally new opportunity. During all of these races, the product manager must shepherd the product through the various pathways toward success.

2. THE DEFINITION AND ROLE OF PRODUCT MANAGER

The product manager is one of the most powerful positions in a corporation within the scope of product. They define the role of the product, the measure of success, and the pathway toward that success. Higher-level management may effect decisions about the corporation and overall direction, but few possess the power and influence on an individual product as the product manager.

The product manager must take care of the product's well-being in the marketplace, and is also responsible for generating sales and profit. If a competitive situation that threatens the product presents itself, the manager must react effectively. If there is a quality issue, the manager must deal with it swiftly. The product manager becomes the champion of the new business.

3. NURTURING THE NEW PRODUCT BUSINESS

The product manager's role is one of nurturer of the business. The product is a means to this end. In any human endeavor in which multiple people are involved, there is always someone else to blame for failure or setback. The product manager is the focal point of responsibility for the corporation.

4. KEEPING CHANGES IN CHECK (PRODUCT STABILITY)

One of the product manager's key roles in effecting success with the new product is to maintain a certain amount of stability with the product's design, manufacturing, and marketing. All change is not good. Each change proposed must be considered and acted on within the framework of the overall program and its goals. Repetitive changes that are not orchestrated well, and have questionable impact, tend to destabilize the product within the corporation and eventually in the marketplace. The manager must hold the reins of these changes and effect only those that will offer a concrete contribution.

This is a difficult task, since those associated with the product—from vendors to manufacturing people—will have a tendency to change the product for their convenience. For example, the vendor will have a tendency to raise prices. The purchasing people may have a tendency to respond to this pricing by changing parts. Supplier obsolescence generally yields to availability, given added component cost. Manufacturing may add steps to a process thereby increasing the labor. Corporate initiatives, implemented in isolation of product objectives, can also contribute to added costs.

5. COST-REDUCTION PATHWAY: LOW VERSUS HIGH VOLUME

A key element of the manager's job is to effect cost reductions in the product to offset vendor issues and labor cost increases. Simply stated, if the costs naturally increase because of these factors, the manager must introduce cost reductions through materials and labor. The materials come from vendor negotiations, and labor comes from less content applied to the product's manufacturer. As time progresses, the cost of labor will go up, simply because the labor force will want a pay increase every year. This must be offset by a reduction in time applied to each product in manufacturing. In the next section there is a more detailed outline of the cost reduction techniques used in this "race."

6. REDUCING LABOR TIME: THE PATHWAY TO IMPROVED THROUGHPUT

As part of the effort to reduce the product cost, the reduction of labor has a spin-off benefit of allowing more manufacturing capacity. By shaving off minutes of manufacturing time, additional units can be manufactured with the time saved. This can be an effective and low-cost method of improving throughput. It achieves economies of scale without the addition of overhead costs associated with added headcount or equipment.

A diligent focus on methods and manufacturing time reduction will have big payoffs in the long run.

7. IT'S A CONTINUUM: LOOK TO THE NEXT PRODUCT DEVELOPMENT

Finally, the role of the product manager should include the issue of balance. Each product is a link in a chain in developing new business for the corporation. No single product should make or break a company. It should contribute to profit, enhance the customer base, and provide leverage. No product lasts forever and should be treated as such. It has a finite life and a distinct period of funds consumption and funds contribution. It is neither a lone star, nor the gathering place for all the corporation's ills. The product is part of a continuum, which starts with growth, the harvest, and positioning for the next generation of product.

LEARNING CURVE COST REDUCTION

The progression down the learning curve pathway toward cost reduction is a required and necessary one. It is accomplished through several mechanisms, which will be discussed in this section. The issue is multidimensional and has a variety of ways to accomplish it. The product manager must keep a watchful eye on product costs and manufacturing throughput. Left unattended, these factors will deteriorate to the unfavorable side. Consider Figure 11-1a, which illustrates the learning curve cost reduction. It consists of the three basic components of a product's cost and illustrates the deterioration that can occur when volumes are not met. In this case the burden is calculated based on the labor content.

Figure 11-1a shows the actual volumes experienced in the marketplace and planned numbers. Materials increase because of the shortfall in volume and labor increases, either as actual hours or in the cost per hour. In this example increased non-value-added labor crept into the manufacturing equation. Since the burden is levied on

LEARNING CURVE COST REDUCTION

P/N	MATERIAL PLANNED 1000 QTY	MATERIAL ACTUAL 750 QTY	LABOR PLANNED HRS 1000 QTY	LABOR ACTUAL HRS 750 QTY	COST/HR 1000 QTY	LABOR COST 750 QTY	BURDEN PLANNED 1000 QTY 1.2	BURDEN ACTUAL 750 QTY 1.3
A1	10	11.5	0.1	0.15	50	7.5	6	9.75
A2	15	16.5	0.1	0.15	50	7.5	6	9.75
A3	16	16.5	0.1	0.15	50	7.5	6	9.75
A4	17	17	0.1	0.15	50	7.5	6	9.75
A5	19	19	0.1	0.15	50	7.5	6	9.75
B1	22	22.1	0.2	0.25	50	12.5	12	16.25
B2	33	33.2	0.2	0.25	50	12.5	12	16.25
B3	44	45	0.2	0.25	50	12.5	12	16.25
B4	55	55.3	0.2	0.25	50	12.5	12	16.25
C1	9	9.5	0.15	0.2	50	10	9	13
C2	10	10.1	0.15	0.2	50	10	9	13
C3	12	12.3	0.15	0.2	50	10	9	13
C4	14	14.2	0.15	0.2	50	10	9	13
TOTAL	276	282.2	1.9	2.55		127.5	114	165.75
VARIANCE		6.2		0.65		32.5		51.75

A

	YEAR 1 PLANNED	YEAR 2 ACTUAL	YEAR 3 ACTUAL	YEAR 4 ACTUAL	YEAR 5 ACTUAL	YEAR 6 ACTUAL
MATERIAL	276	282.2	292.1	302.3	312.9	323.8
STD COST	276	276	276	276	276	276
LABOR	95	127.5	132.0	136.6	141.4	146.3
STD COST	95	95	95	95	95	95
BURDEN	114	165.75	171.6	177.6	183.8	190.2
STD COST	114	114	114	114	114	114
TOTAL	485	575.5	595.6	616.4	638.0	660.3
STD COST	485	485	485	485	485	485
		90.5	110.6	131.4	153.0	175.3

B

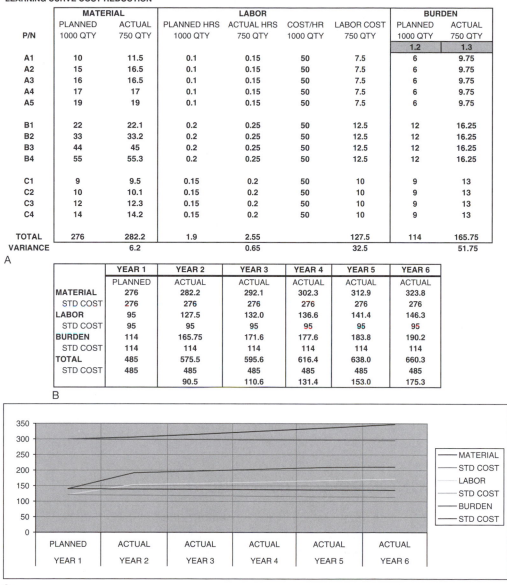

C

Figure 11-1. Learning Curve Cost Reduction

the labor, the creep in labor has a multiple effect when considering the burden and the labor together. This can make cost containment a losing proposition. The data shows the variance numbers. The burden numbers applied to labor change from 1.2 to 1.3 times labor, because of unabsorbed labor to cover manufacturing expenses. Given a fixed set of manufacturing expenses based on 1000 units, a reduction to 750 pieces means that the manufacturing expenses must be applied over the 750 units labor content (i.e., burdening the reduced volume piece part even further).

This describes a cost scenario between the planned and the actual as the product exists in development. Consider the effects after production, where the product is now in the marketplace and has no effective cost containment measures. Figure 11-1b represents this.

In Figure 11-1b the costs of material, labor, and resulting burden increase with each successive year. The table shows the change in total cost as $90.45 from planned Year 1 to actual Year 2. This is the difference of $575.50, shown as the total cost in Year 2, minus the planned standard cost of $485 (shown below it and also as the total in Year 1).

By looking at the previous chart comparing planned and actual, one can see that the $90.45 is comprised of $6.20 in material variance, $32.50 in labor (due to non—value-added time in manufacturing), and $51.75 added because of unabsorbed burden. Figure 11-1c graphically illustrates the deterioration in successive years.

The data in Figure 11-1c show that in year 1 through year 6, the material labor and burden increase dramatically, even though each year represents only a 3.5% increase by itself. This lack of cost containment can cripple a product line quickly and limit the number of available options for the product manager.

The following is a list of methods, along with their summaries, used to achieve cost containment:

A. Procurement-Related Cost Containment
B. Manufacturing Methods and Labor
C. Manufacturing Venue
D. Currency Fluctuations (Foreign Content Related)
E. Other Methods

1. PROCUREMENT-RELATED COST CONTAINMENT

The basic focus of the first method of cost containment is to simply buy better, and to leverage other buying power. A natural tendency for a product left unmanaged is for the manufactured cost to increase. This occurs for several reasons, the primary reasons being that the procurement of materials becomes rote and vendor uncertainty begins to be absorbed by the manufacturer, rather than the vendor. For example, a vendor-obsolete part is generally available for some period at a premium price. Left unmanaged this part remains on the Bill of Material for the product. The design becomes financially compromised because the changeover to the newer, more available part is not implemented. To effectively manage this, the pathway toward containment of materials cost must require active participation.

2. MANUFACTURING METHODS AND LABOR

The focus on labor reduction is to determine ways to produce the product easier, with less labor content. This can be accomplished through the implementation of additional tooling and capital equipment. The capital equipment will require an investment to secure. One of the more proactive methods to the labor issue is to design the product in such a way as to be forward-compatible with more fixtures and automated methods of manufacturer. The less-mechanized means can be employed initially, and as the market volumes dictate, the more mechanized equipment can be phased in. There is a caveat here in that the volume should be well understood with few surprises occurring in the marketplace. Given this, the capital equipment required for mass manufacture should be in place at initial launch. Manufacturing capacity issues should not be a major constraint at this juncture.

By mechanizing the manufacturing content, the labor decreases and, if in balance, can generate additional throughput given no other constraints.

There is a natural tendency to increase the labor cost because of the increasing cost of wages. This is because production personnel desire wage adjustments, just like the rest of the corporation. This must be offset with a decrease in applied labor time to the product. In addition, there may be a tendency to add quality-control checks at various trouble spots in the process, further aggravating the problem by adding labor time.

The material and labor initiatives are designed to widen the cost revenue gap to maximize profit.

3. MANUFACTURING VENUE

One way to attempt to alleviate the pressure of increasing costs is to relocate manufacturing to a foreign venue where labor costs may be significantly less. This may well be the case; however, relocation of the manufacturing venue can create manufacturing problems also. Training in processes and process control and improvement is not an assumed trait of foreign manufacturing capabilities. These people must be trained and evaluated, as do the resident personnel. It may be more difficult to imbue values and process control at these remote locations.

If the materials are procured at these locations without direct control of the corporation, there could be additional performance problems and manufacturing problems introduced by substitute components. Local purchasing without the benefit of corporate resources and technical guidance can lead to a degradation of features and performance. If the tendency was to be subject to the vagaries of a domestic supplier's offering changes, imagine the loss of configuration control at the remote level.

Local foreign practices, laws, and customs may also interfere with the manufacturing of a product with known configuration control. Another consideration is the export compliance laws and transportation of technology to foreign nations.

The decision to use foreign manufacturers should be based on an overall strategic goal and long-term plan, not simply the difference in a labor rate. In many industries recent trends in design and manufacturing have significantly reduced the labor content to a small percentage of the product's cost, thereby making the argument for foreign venue, simply for labor rate reduction's sake, even weaker.

4. CURRENCY FLUCTUATIONS (FOREIGN-CONTENT RELATED)

The issue of currency exchange can significantly affect the product's profitability. Global products are affected by global conditions and currency fluctuations that can work to the company's advantage, or to its disadvantage. If all transactions surrounding a product, from the procurement of materials to the collection of payment, occur instantaneously, then currency fluctuations have small impact. If everything transacted instantaneously, prices and terms would automatically be adjusted to accommodate the fluctuation. However, when time delay is introduced into the equation, either side can get caught short in an unfavorable position or benefit from a favorable swing in currency.

In a typical manufacturing scenario the company deals with this issue in both the buying side and the selling side. It occurs on the buying side if the component parts used in the product are of foreign manufacture. It occurs on the selling side when the product is being sold to a foreign company. This can best be illustrated by the following example: Consider a case where the company purchases a component from an Italian firm. Even though the Euro is used as a normalized currency, we will use the previous currency to illustrate the contrasts from U.S. dollar to Lire (LIT on Figure 11-2) and Swings in valuation. The exchange rate from lire (Italian currency) to U.S. dollars is given at 1650 baseline. At the time the procurement was negotiated, the dollar bought 1650 lire's worth of this Italian company's goods. This is shown in Figure 11-2a.

The currency risk associated in buying materials from the Italian firm is shown first. The upside for the buyer occurs when the strength of the dollar increases with respect to the lire. If the transaction was negotiated at 1650 lire to the dollar, the component cost the company $20.45. If the exchange rate goes favorable to 1750 lire to the dollar, the component only cost the company $19.28, or only 94.29% of the original. If the exchange rate flips the other way, it can be a disadvantage to the company. In this case the dollar weakens with respect to the lire, and at 1500 lire to the dollar, more dollars are required to purchase the 33,743-lire ($21.69) part. This represents a price that is 106% of the original price.

From the seller's perspective, if a part now sourced by an American company costs $20.45 dollars, the Italian company can procure that same part for 33,743 lire. This is at the baseline rate of 1650 lire to 1 U.S. dollar. If the exchange rate goes to 1500 lire to 1 dollar, the buyer's (Italian firm) upside is that after the negotiation, they only need to pay 30,725 lire to procure $20.45 worth of goods. If the exchange rate deteriorates to 1750 lire to the dollar, the $20.45 part will cost the Italian firm 35,787 lire, instead of 33,743 lire, which represents a price that is 106% of the original price.

One of the ways to alleviate the problem of sensitivity to currency is to hedge against these fluctuations to your advantage, depending on the circumstances and your position in

CURRENCY ISSUES

	SELLERS PRICE	EXCHANGE RATE	BUYERS COST	CHANGE IN EXCHANGE RATE	ADJUSTED BUYERS COST	EFFECT %		ACTIONS FOR THE BUYER	ACTIONS FOR THE SELLER
CURRENCY RISK IN BUYING									
BUYERS UPSIDE	33743	1650 LIT/USD	20.45	1750 LIT/USD	19.28	94.29	FAVORABLE	BUY LIT AT 1750/USD	RAISE PRICES IN LIT
BUYERS DOWN SIDE				1500 LIT/USD	21.69	106.06	UNFAVORABLE		LOWER PRICES IN LIT
CURRENCY RISK IN SELLING									
BUYERS UPSIDE	20.45	1650 LIT/USD	33743	1500 LIT/USD	30675	90.91	FAVORABLE	BUY USD AT 1650 LIT/USD	RAISE PRICES IN USD
BUYERS DOWNSIDE				1750 LIT/USD	35788	106.06	UNFAVORABLE		LOWER PRICES IN USD

A

CURRENCY RISK IN BUYING	ACTIONS FOR THE BUYER	ACTIONS FOR THE SELLER
BUYERS UPSIDE	BUY LIRE AT 1750/USD OR ANYTHING OVER	RAISE PRICES IN LIRE
BUYERS DOWN SIDE		LOWER PRICES IN LIRE
CURRENCY RISK IN SELLING	ACTIONS FOR THE BUYER	ACTIONS FOR THE SELLER
BUYERS UPSIDE	BUY USD AT 1650 LIRE/USD OR ANYTHING UNDER	RAISE PRICES IN USD
BUYERS DOWNSIDE		LOWER PRICES IN USD

B

Figure 11-2. Currency Hedge

the transaction. This means buying foreign currency when it is favorable, in the hopes of using it when rates go unfavorable.

Figure 11-2b summarizes actions the buyer and/or seller can take in the buying case and actions the buyer and/or seller can take in the selling case. This chart summarizes the actions that either party can take to minimize cost or preserve profit in response to the fluctuation. In considering the currency risk in buying, the buyer can hedge currency fluctuation by purchasing lire with dollars, when the exchange rate is 1650 or anything over 1650 (e.g., 1750 as listed in the table). The reason to purchase lire at 1650 is to protect against a weakening of the dollar. In this manner, the buyer can buy the part at 33,743 lire and it will cost $20.45 dollars, nothing more. Hedging is designed to protect against the risk of an unfavorable change. In the case where the exchange rate goes to 1750 lire, the seller can raise prices. This is because, the actual dollars needed to purchase the part at 1750 lire would still be $20.45 (outlay of hedged cash in lire, however).

In the downside case, the seller may have to lower prices to induce the company to purchase the part from the Italian firm.

Switching to the other side of the transaction, if the Italian firm was to purchase the part (priced in dollars) with lire, a decrease in the lire-to-dollar ratio will allow the Italian company to procure the $20.45 part for less lire. They may hedge their currency by purchasing dollars at 1650 or anything less (e.g., 1500 lire). This means less lire outlay for them. The option for the U.S. company is to raise the price in dollars to increase profit, because it will mean the same outlay in lire to the Italian company. If the exchange rate goes the other way, and the lire-to-dollar ratio is 1750 or anything over 1650, the U.S. company may have to lower prices to continue to secure the business from the Italian company.

In addition to changes in currency that affect cost and pricing, there is also a domestic factor of inflation that must be accommodated. In periods of low inflation, wage, cost, and price are relatively stable. In periods of high inflation, each of these excursions can severely affect the profitability of a product. Wage changes not only affect the costs, but also affect the split between material and labor, thus affecting burden as well.

Inflation not only affects the cost and pricing of a product, but can also affect the company's ability to invest in new development. If the company's products and transaction speed do not keep pace, there may be a shortfall that may prevent investment. A very high rate of inflation may also make other noncompany, non–development-related investments appear more lucrative. These situations take a company off strategy and are very harmful in the long run.

5. OTHER METHODS

There are a host of other alternatives for lowering costs, including partnerships and subcontracting arrangements. The fundamental issue with these arrangements is to look for the situations in which the arrangement is beneficial for both sides. A one-sided arrangement, will not last very long. Relationships that do not have longevity cause uncertainty and instability in the product line.

Another method for cost reduction is to reevaluate the feature set of the product using a value management approach. This approach is fundamentally an exercise in examining the elements of value transferred with the product to the customer, and attaching value to each of the features by examining the benefits to the customer associated with them. It is possible in some cases to deplete the product of certain nonvalued features, assuming they result in a cost reduction. If so, these depletions will result in certain cost reductions that can be effective margin enhancement means without jeopardizing the customer acceptance. It is important to note that if these are significant, the product team has essentially missed its mark on defining the product.

QUALITY SYSTEM

The previous discussions in this chapter have centered on the various performance issues associated with the new product. We have discussed the technological performance issues, the commercial performance issues, and the financial performance issues. In this section we will discuss quality performance issues. These performance issues are measurements of the overall product line's ability to meet customer expectations.

1. QUALITY SYSTEM OBJECTIVES

The objectives of a quality system are as follows:

A. Data collection

The data collection is the gathering of intermediate and summary data throughout the product's life cycle. It includes design, manufacturing, and field data.

B. Manufacturing monitoring

This is a measurement of the manufacturing performance in terms of efficiencies, speed, and ability to absorb uncertainty and effect conforming product.

C. Field feedback

This is the measure of how the product satisfied the customer's expectations in terms of performance, reliability, and use issues.

D. Corrective action

This is the active participation by members of the product group to improve overall performance and correct mistakes. It is also a measure of the performance of the development/manufacturing value chain in producing the product.

E. Measurement of effectiveness of corrective action

This is the measure of the effectiveness of change decision making of the product group. Are they actually correcting issues, or are they introducing added intertwined issues that will be more difficult to untangle later?

F. Degree of normalcy and stability

Finally, the quality system can indicate a degree of stability of the product line. A product line with many changes, one following the other and even reversing previous changes, is an indication the product configuration is out of control.

2. MANUFACTURING AS AN ARTFORM

The historical and prevailing attitude toward manufacturing is that it is a machine, comprised of human operators that can be turned on, turned off, sped up, slowed down, change course, and produce perfect, zero-defect results consistently. Companies talk themselves into the premise that with enough procedures, documentation, drawings, and "training," they can magically expect this performance. In reality it is more accurate to think of manufacturing as an art form. Humans have a typical response to a manufacturing environment. It usually follows some pattern of training, increasing performance in speed and accuracy, followed by degradation in performance resulting from boredom and other fatigue factors. This being the case, there is a sweet spot of elapsed time in manufacturing where the tasks are new enough to be interesting, and challenging enough for the operators.

A constant introduction of design- and manufacturing–related changes resets the manufacturing cycle to the earlier uncertain time where "training" is a major factor and mistakes are more common. If there is a steady diet of this, the manufacturing group never achieves the second stage where speed and accuracy have large benefit to the company.

3. MECHANICS OF A QUALITY SYSTEM

The basic mechanics of a quality system are best illustrated in Figure 11-3. The specifics of how to obtain the data are system and company specific. The overall mechanism is more universal.

As shown in Figure 11-3, the process begins with the design and development of the product. On completion of the project, the product moves into the manufacturing environment. The quality system is designed to monitor and record data needed for the support of the product. The product is shipped to the customer, and the customer places the product in use. In the unfortunate event of product failure, the customer returns the product to the factory for repair. This can be a warranty or nonwarranty repair.

Alternatively, there may be local repair shops authorized to do the repair. In either case the repair data should be fed back to the factory for analysis. This is very important for the initial return so that the factory may confirm failure modes and evaluate if any corrective action needs to be taken with the design.

The repair analysis is used to effect the repair, but also to feed specific information about the failure modes back to the various functional groups involved in creating the product. For example, the corrective action recommended by the repair department may be safety

PRODUCT FEEDBACK SYSTEM

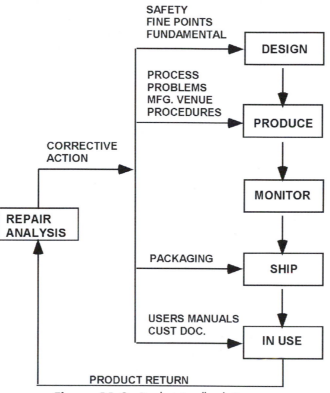

Figure 11-3. Product Feedback System

related, deal with fine points of the design, or highlight fundamental flaws. The feedback for manufacturing may be process related, venue specific, or procedural. Feedback for shipping may include packaging-related issues and shipping guidelines. Finally, in-use issues include clarification issues with the users' manual, and other application information.

4. LOSS MITIGATION

The overall objective (besides financial) behind reducing warranties is to improve customer satisfaction. The financial cost of a warranty, however, can be high. At each stage the product goes through, from component parts through manufacturing, to shipment and in use, the cost of remediation multiplies. Therefore the earlier the problems can be captured, traced, and resolved in a deterministic manner, lowers the exposure to the company.

The quality system therefore should be designed to capture and report these problems at the earliest possible point of occurrence.

5. WARRANTY EXPOSURE MARKET AND DOLLARS SPENT

In every company, from large to small, a warranty is thought of in two perspectives: the "cost of quality" perspective and the customer satisfaction perspective. The cost of quality perspective is oriented toward the financial exposure involved with product failures, returns, repairs, scrap, and labor costs. The customer satisfaction perspective is a measure of the marketplace's acceptance of these failures, and a response to the company's ability to effect speedy reparations.

6. "SHOW ME THE DATA"

Show me the data! This is the battle cry of quality control. All to often, opinions creep into the discussions pertaining to how a product is doing. This happens especially when there may be a recent run of failures. Human-memory patterns seem to focus on the bad and do not correlate very well with the level of shipments, historical rates, and other data. This is why it is important to chart field data on returns, product failures, and shipments. Serialization also will help in pinpointing exposure on certain problems. As will be shown later in the section dealing with recalls, it is important to be able to identify the end of a field problem. Consequently, when it comes to quality levels of a product, converse in terms of hard data.

7. SETTING THE WARRANTY LIMITS; EXTERNAL VERSUS INTERNAL PERSPECTIVES

Whether intentional or not, every organization essentially sets toleration limits on warranty returns. They may be driven by the cost of quality, or the cost of lost market share. The important point to remember is that there are two toleration limits to product warranty, namely: internal limits and external limits. The internal limit is the amount of warranty the corporation will accept before taking aggressive corrective action. This aggressive action may be a selective or total recall of product.

When the level of return is less than this number, little is done in the way of corrective action. When it exceeds this number, the entire organization is aware of the problem and swings into action.

The other limit is the limit of patience the market has for the product problem. Below that limit, the customers' experiences will be scattered, and it's easy to maneuver around in a sales situation. This can happen where the customer base is diverse and unrelated, vocationally or geographically. Above this limit, however, the product begins to get a bad reputation. This is common where members of the customer base are in the same vocational group, or the same industry, where the bad reputation gets passed from one customer to another, even if they personally have not had a bad experience with the product.

8. THE EXPOSURE TO THE ORGANIZATION

There is a significant amount of direct and indirect labor associated with the effort required in supporting a product with problems. With each successive problem, more and

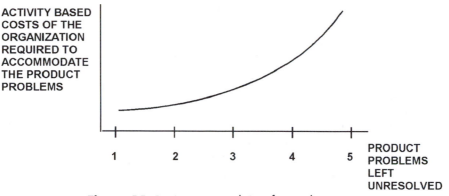

Figure 11-4. Activity Based Cost for Product Issues

more effort will be required to conduct normal operations. This is because each product problem has its own caveats, rules, and issues. As the problems are piled on, more rules (now affecting the other rules), in addition to the problems, are added. Figure 11-4 illustrates the point.

As can be seen in Figure 11-4, the effect of supporting additional problems becomes exponential. Left uncorrected, this requirement for manual effort, outside of the normal operations and system, begins to dominate the quoting, selling, development, and manufacturing operations. This results in nonvalue-added labor to the product and the business. The nonvalue-added labor takes an efficient process and slows it down, while adding uncertainty and the potential for error.

MANAGING PRODUCT CHANGE

1. PERSPECTIVES ON THE ROLE OF PRODUCT MANAGER

In addition to product promoter, the role of product manager is one of product configuration control. There are several forces that tend to drag a product line off of its base configuration, as discussed earlier. The product manager must evaluate the requests for change and determine what changes will be integrated, and what the potential benefit will be. In some cases there will be no choice in the matter. In other cases the choices are discretionary.

2. SAFETY, PERFORMANCE, AND ENHANCEMENT

The product changes fall into the following three broad categories of change:

A. Safety related:

The safety-related issues are further subdivided into three categories as follows:

1. Product

   The product-related issues can be serious enough to issue a stop in production, or less serious such that the changes could be phased in at a future date.

2. Application

   The application issue concerns those cases in which the product is applied in certain circumstances in which there may be a safety concern. In these cases the product should not be applied and the product stakeholders, ranging from the salespeople to the end users, should be effectively notified.

3. Documentation

   This issue pertains to the product documentation, containing a recommendation, practice, or procedure that may render the product unsafe or to incur a safety concern.

B. Performance related

The performance-related issues concern the operative parameters of the product. Information may come to light in which the product performance falls short of the expected or advertised performance. In these cases a certain retraction must be communicated to the people using the product or potentially procuring the product. This is to remain in effect until the correction is implemented. These actions fall into the following three categories:

1. Stop production

   The stop production order is the most aggressive position, in that production of the product is halted until the correction is in effect. A recall of the product may or may not be required.

2. Immediate change

   An immediate change is just as aggressive, with the exception that the product is not halted in production; however, a change is designed and implemented in the same expedient manner.

3. Scheduled change

   The scheduled change is less serious, in which a performance-related issue is identified and then scheduled for implementation without suspending normal operations. The issue may or may not be communicated to all parties, depending on circumstances and seriousness.

C. Enhancement related

The enhancement-related issues are similar to the schedule changes in that they are implemented in a scheduled time frame within the scope of normal operations. They do not necessarily fall under the category of design flaws or corrections. Some manufacturers have used this terminology in the marketplace to describe some of the performance-related issues!

RECALLS: A NEW PRODUCT NIGHTMARE

1. SEVEN STEPS OF PRODUCT RECALL

One of the more distasteful tasks, and the true measure of a product manager, is deciding on and prosecuting a product recall. One of the most disheartening actions to take, a product recall may be the only way to save a product in the long run.

The product recall forces a direct about face in the present pathway of a product. It allows the company to reset the reputation and generates a fresh start for the product. The product recall engagement must be well-considered, decisive, and speedy. Failure of any of these attributes can cause failure of the recall and prompt the marketplace to abandon the new product altogether.

A product recall is a public admission that the company failed to produce a good product. It is a step that can draw criticism and admiration at the same time.

The key to any successful recall is to face up to the product problem, decide on the recall promptly, and prosecute it with diligence and speed. The market has tolerance for some inconvenience. It has no patience with ineptitude on top of product problems. Reviewing from Chapter 10, "The Launch," these are the seven elements of a recall.

1. Contact
2. Explanation
3. Actions required
4. Procedures
5. Acknowledgment
6. Compensation
7. Closure

A framework for a product recall is available for your specific use, should it arise, in The *Tool Box*.

2. DETERMINING IF A RECALL IS NECESSARY

The determination of a product recall is an inexact science. Aside from the protocol necessary in a safety-related recall, or a blatant misrepresentation of the product because of performance, the recall must navigate the perceptions of the customer and the marketplace. The customer perceptions are unpredictable in many cases. The result of this navigation is a corrected product with a reputation for quality intact. In fact, the company issuing the recall exemplifies the reputation for quality during the recall. There is lack of confidence in the customer's eyes in the product at this point. All they can rely on is the confidence in the quality reputation of the company. Only after the product is corrected can the reputation of the product be restored. This is why it is critical to be decisive and correct in the determination of the recall. If not, the company reputation will be damaged also.

There are several factors involved when determining the necessity of a recall. These factors including the following:

A. Known bad component(s)

This is where key components can cause product failure or safety compromises. Once the manufacturer knows of these issues, a recall is most likely in the product's future.

B. Known bad process

This is where manufacturing process used in construction can cause product failure or safety compromises. Once the manufacturer knows of these issues, a recall also is most likely in the product's future.

C. Latent product risks

This is where components will fail at a future point. They are presently functioning, but as time and use accumulate, the latent flaw will surface and cause premature product failure.

D. Safety-related issues

This broad category applies to any safety-related issue of the product that is addressed by design or manufacturing issues, as opposed to application issues. These issues generally cause immediate action, with specific prosecution plans and timelines.

E. Blatant misrepresentation due to a product performance shortfall

This is where an extreme oversight, inadvertent product changes, or a complete specification violation places the company's reputation in extremis.

F. Failure to deliver based on contractual performance

This is the situation where the product, or group of products, was sold to the customer under some contractual performance guaranty. Failure to achieve the performance will necessitate some means of correction or recall.

If your specific situation meets the basic tenets of these criteria, there is a good chance that a product recall is needed.

3. INTERNAL MOVES: HOLD PRODUCTION TILL FIXED

In a recall situation there are several internal moves that must be made to prosecute the recall properly. There are also several external moves to be made. These moves are designed for facilitating the recall, and also to ensure some measure of protection in the case of product liability. One of the internal moves is to designate the recall officer. This position is a product-management position specific to the recall. The person in this position designs all aspects of the implementation and activates these systems. They make critical decisions about the product line to minimize the reputation damage and also to mitigate the financial exposure.

In addition to the logistical implementation, there needs to be an immediate hold on production until the problem is defined, analyzed, and resolved. The production can only resume after the "fix" is in place.

Personnel must be selected to implement the recall. They must be trained. They must be audited. An infrastructure specific to the recall should be set up, and be easily accessible, as well as responsive.

4. DESIGNATING THE RECALL OFFICER

The "recall officer" should be a strong, product manager profile that will act decisively and swiftly. They must go through barriers and have a strong marketing and customer orientation. The product manager for the product line's development is usually good choice unless there is some vocational weakness that prevents them from completing the task.

5. EXTERNAL MOVES TO BE MADE

The external moves to be made depend on the organization's size, route to market, and degree of severity. Field service, customer support, and local sales offices may be involved in the process. Ideally, however, the company should be the focal point of the recall with primary customer contact, and the shortest pathway to and from the customer to provide expedient service to the customer, given the inconvenience of the recall situation.

6. COMMUNICATION

The first instinct in implementing a recall may be to keep it as low a profile as possible. This is true from the reputation perspective; however, it is difficult to keep any product issue silent for any length of time. Rather, the company should be careful in deciding on the recall; but once the decision is made, communicate it effectively to all parties concerned.

The issue of communication can even be the subject of a potential litigation. There is a responsibility associated with getting the information to the right people in the shortest period of time, with minimal misunderstanding. The communication should permeate the company, satellite offices, customer support function development, and manufacturing.

7. PROTECTING FROM LIABILITY DURING A CRISIS

A fiduciary responsibility of the product manager and the product team is to mitigate the financial impact of a recall to the company. This is generally accomplished by diligently assessing the issues and prosecuting them with speed and accuracy, and in the best interests of the customer and the company. With safety as the primary motivator and customer satisfaction as the closing signature, a crisis can have a positive outcome, and the long-term financial impact to the company will be minimized.

Open, well-considered communication is desired; cover up tactics cannot be allowed.

8. PLUGGING IN THE ORGANIZATION EARLY AND COMPLETELY IS KEY

The recall is a time when the organization can assist the product manager in correcting the product. The manager must rely on the organization's infrastructure to support the recall. Although the manager is the architect of the recall, it must be a company-wide effort, not an individual one. To that end, there must be company-wide communication. Action plans must be made known and supporting people must be trained, motivated, and in place. Once the decision is made to recall the product, involve the company immediately. It will serve the recall effort, the product, and the customer base.

9. DESIGN AND PROSECUTE AN END TO THE CRISIS AND MOVE ON QUICKLY

The operational concept behind a recall is to design what constitutes an end to the product crisis and drive the events toward this end. If the recall means to get every unit back to the factory and repair it, then the end of the crisis is when this is accomplished. If the recall means to inform every product owner and schedule a refit at some service location, the end of the crisis is when this is complete. In either case the recall team must prosecute with speed and diligence to this end.

The product team and the company need to monitor for an end to the field problem. This means to look at the quality database and the returns coming back.

The quality database will tell you if the fix is effective by indicating no returns past the production date with the "fix" in place. The warranty returns will indicate the diminishing of labor and materials associated with the recall. This indicates that whatever product is coming back is back, and the financial exposure is minimal, given no latent issues.

10. DRAGGING ON HAS ENORMOUS ACTIVITY-BASED COSTS ASSOCIATED WITH IT

In a similar manner, when living with product problems there are significant costs for the added activity to support a recall effort, as discussed earlier. To that end, it is desirable to end the recall as quickly as possible and get on with the product and the business it generates.

This concludes the basic product-management section of this chapter where general product management issues, cost reductions, quality initiatives, and general product maintenance issues were reviewed. The focus now shifts toward the product initiative this chapter is named after. The pursuit of the successful product is oriented toward growing the new business, follow on planning, market development and application development, pricing issues, and growth strategy. This next section will explore these basic tenets of new product marketing within the framework of product management, to grow the business and pave the way for follow on new product initiatives.

MARKET DEVELOPMENT

1. ONE OF THE MOST ENJOYABLE PARTS OF PRODUCT MANAGEMENT

Taking the new product to market and investigating new applications can be one of the most rewarding aspects associated with product management. It combines military-flavored strategic planning; tactics, and detective work focused on securing additional applications for the product. This process encompasses the identification of market opportunities that are on target with the product launch plan, as well as identifying new opportunities for small changes in the product configuration.

The market development aspect gives true definition to the term *engagement of the customer*. It is most effective when the customer is an active participant in the communication between the sales force and themselves. It takes the detective work to uncover new customers and market segments for the product.

These next sections will discuss the various aspects of the market development process and how they apply to growing the business.

2. MARKET DEVELOPMENT FLOWCHART

There is a generic protocol that can be used in the development of the market. The basic philosophy is similar to those used in other aspects of life: try to sell the product, learn from the interaction, and reapply this new "learned" knowledge to guide in selling in the future. Figure 11-5a represents a simplified version of this flowchart as it relates to market development of new products.

As can be seen in Figure 11-5a, there are two basic paths. The first is where the product fits the marketplace without any modification. The second is where the product may need minor modification to be marketed in an application. This example assumes the product is fixed somewhat, but modifiable in design and construction to accommodate the new application. It requires intimate, but not extensive, development work to accomplish. The term *intimate development* refers to the change in configuration, bills of material, parts, components, and/or assembly methods.

There is another example that we will now consider, in which the product is designed, initially modifiable either in the field or by factory personnel. It is diagrammatically shown in Figure 11-5b.

The product is designed on a basic generic platform, which is set up to serve the target applications. It is also flexible enough to serve other applications which expand the business by product modifications.

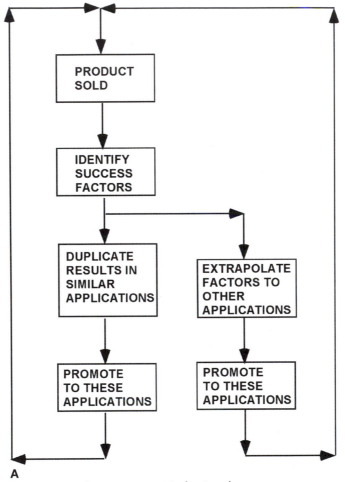

Figure 11-5. Market Development

It can be modified by the alteration of some medium inherent in the product, such as software. The best test to determine which category it belongs to is to determine if it can be modified back to its original state. The first example used in Figure 11-5a cannot. It is engineered and manufactured to be in this state.

As shown in Figure 11-5c, the product has two basic components: namely, a hardware function block and a software function block. The hardware block has generic interfaces from three basic perspectives. They are the human interface (or how the user interfaces with the product), the control system interface (how a hierarchical control system would tie into the product), and a vocational or functional interface (how the product relates to the environment it acts on). The software block serves as a repository for the "personality" of the product and functional modification. The product can serve a wide variety of customers and

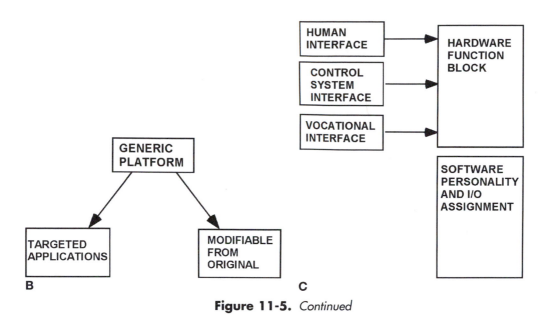

Figure 11-5. *Continued*

needs by modifying the basic function, the human interface, the control system interface, or the vocational interface. Input and output can be reassigned to facilitate this transition also.

In this way a generic design can serve a growing list of applications with a small incremental effort. For example, the hardware of an engine/transmission could be modified for performance, mileage, or comfort by using the same basic platform and altering software in the engine control computer. In a broader product application, standardized generic car platforms can be used to generate a variety of products serving different market segments for automobiles by varying the configuration and accessories.

3. USE CREATIVITY TO DEVELOP NEW CUSTOMERS FOR THE PRODUCT

The marketing professional should use creativity to enhance their product's market share and expand the available market. This is done by engaging the customer again with the standard product offering and broadening the scope of the product to serve the extended needs of the customer base and to uncover new applications for the modified product.

Figure 11-6 illustrates that the product's available market is of a certain size illustrated by the width of the channel on the drawing. By implementing the flowchart and exploring additional applications, the available market is increased dramatically. The activity driving this desired result is shown in the flowchart to the left of the drawing. It starts with the engagement of the customer and the creation of newly served applications generated by the interaction and dialog. To accomplish this, redefine and manage the scope. Then the concept is implemented and refined. Finally, it is locked onto and promoted to a new collection

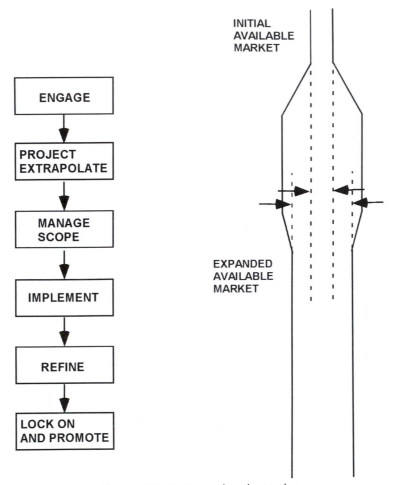

Figure 11-6. Expanding the Market

of customers now available for using the product. The result is a wider pathway to more available markets.

The term *pathway* can best describe the illustration, because as nothing can last forever, the product's sale and growth of market share is a pathway. The width cannot remain fixed forever. The idea is to manage the process faster than the competition and carve out a business segment that is defendable.

4. CLASSES OF CUSTOMERS, TIERED MARKETS, AND TIERED PRODUCTS

By implementing these concepts while widening the available market, it is possible to create classes of products that serve different classes of customers. The terms *tiered market* and *tiered product line*, which serve that market, refer to a fit between the product and the customer. In many cases simple changes can create better identity for the product in the mind of the customer.

There is also a selfish motive for the corporation behind this approach. By creating the tiers, additional margin can be realized in the product line by commanding increased pricing and adding little incremental cost.

5. SEGMENTING THE MARKET AND PRODUCTIZING THE SEGMENTATION

In summary, the basic process is to segment the market during the promotion phase and productize additional offerings to widen the available market and expand the business. As a starting point for this exercise, go back to Chapter 3, "Refinement of the Product Concept into a New Product and Business," seek out all of the ideas for products and versions, and test them in the marketplace. Implement ones that require little change.

APPLICATION DEVELOPMENT

1. TAKE TECHNOLOGY BUILDING BLOCKS AND APPLY

The issue of application development requires technological engagement with the customer, in addition to commercial engagement. This is where the basic building blocks of technology embodied in the product are available for redistribution and reconfiguration into different products altogether, from the factory perspective; however, they are an on-target solution to the customer's problem. An example of this reconfiguration of features is to think of the original part as a "toolbox of technology". Both development and manufacturing that can be used to solve different customer problems in different applications .

Another example of this is a pizza parlor where pizza is made and sold to consumers. The infrastructure for storing raw materials (foodstuffs), assembling them into desirable portions of food, baking them in an oven, and delivering them uses the manufacturing technology and the vendor base to serve new products to new customers. This is only limited by the imagination of the customers and the owner of the pizza parlor!

2. DEGREE OF MODIFICATION CONSTRAINTS

The manager needs to be wary of going too far with modifications and segmenting to serve too many customers. This will dilute any economies of scale and destroy any leverage created. In addition, too many product versions are difficult to track, implement, support, and procure components for, unless each generates significant volume and profit to absorb the overhead needed to implement them.

3. SEARCH FOR LATENT ISSUES SOLVED BY A RENDITION OF THE PRODUCT

As the product is being accepted into the marketplace and is being used, there may be latent needs not met by the product. They may be functional, human-factors related, or ease of usage related. The commercial and promotional effort should also focus on seeking out these issues and producing a version of the product that does meet these needs. The timing of this depends of course on seriousness and degree of change. This is nothing more than attempting to meet all of the customer's needs with the product, something that should have been done throughout the product program.

4. ELIMINATE THE COLLATERAL EQUIPMENT AND ABSORB FUNCTION

Another way to promote the product, if it has the capacity for easy modification, is to engage the customer to determine what other collateral equipment is required with the product to make the application work. Then embody the collateral equipment or its function into the product at a lower total cost and easier use. This will simplify the customer's use of both products and give your product the advantage.

MARKET FEEDBACK AND PRODUCT MODIFICATIONS

1. WATCH THE FEEDBACK

Now that the product is out and has an established base, the product group must watch the market feedback carefully. Earlier in the business development process, we discussed the need to refrain from being all things to all people. It simply cannot be done well and support any initiative for product leverage. The retention of the market feedback must be organized to evaluate the potential return for each added or exchanged feature. These requests can be evaluated like the original opportunity, with the exception that they represent incremental business. The decision needed is to ensure that they do not drain resources to the point of disadvantaging the base product.

The product group may even want to take an active role in soliciting the feedback from the customers and users, to make certain determinations themselves for planning and product maintenance purposes.

2. SELECT CATEGORIES FOR SPECIFIC ITEMS AND GROUP CHANGES

In the same manner that the original product features were qualified and selected for incorporation, the specific market feedback can be organized into the following three basic categories:

A. Minor modifications. These are simple changes; sometimes even cosmetic or user support changes can be made to improve acceptability of the product in the target application. In some instances these minor modifications can be effective enough to open new opportunities for the products.

B. Product feature enhancements. These are more involved changes from a development and manufacturing point of view. They involve product changes and re-qualification. They are scheduled into the development hierarchy of activities.

C. Major platform changes. These are the market feedback, which define the next generation of product. They are significant enough to warrant a complete redesign and selection of possibly a new platform.

All of the market feedback must be evaluated and categorized for decision and implementation purposes. The benefits of each of these implementations can then be evaluated in a financial manner, along with the strategic implications, before action is taken.

3. THE NEXT GENERATION OF PRODUCT

This feedback related to major platform changes, and to some degree the enhancement requests, define the next generation of product. The time to start formulating this next generation is now. Keep in mind that these requests represent the customer's input to the product development process. It is only one component of the process. The strategic objective, and the technological and cost initiatives, also must drive the plans for this next generation of product. In the same manner as the product evolution flowchart was completed initially, the chart must be updated with the customer feedback. In addition, the two elements should be compared to each other. The first element is the market feedback; the second is the original plan for what the follow on generation would look like.

In this way the trends of the marketplace can be cataloged, and the rate of change can be tracked to leapfrog the next versions of the product accurately.

4. MARKET RESPONSE GIVES ONLY PART OF THE PRODUCT DIRECTION

The market will give the overall direction, but the product team must select and qualify the details. It cannot be overstated that the market feedback is what the customer can desire now; it cannot forecast where the competition may be, what the future cost structures will be, and what competitive moves should be taken by the company to be a player in the future.

5. HOW CAN YOU IMPROVE YOUR POSITION?

The gathering of market feedback and the analysis of data are directed toward only one goal. That goal is to improve the company's competitive position with the products available. The staging of the product introductions, the improvements, and the new products are designed to improve the competitive position and gain advantage. They are not for esoteric purity, line completion, or any other singular reason; rather, they are part of an overall plan to secure market share and build a profitable business.

CONTINUING THE PRODUCT EVOLUTION FLOWCHART

1. HOW DOES THE PRESENT MARKET FEEDBACK COMPARE WITH THE ORIGINAL ASSUMPTION?

Most product evolution flowcharts are developed as part of the process of launching into a new product development. Unfortunately, if they are not integral to the process through to product introduction, they soon fall by the wayside and lose their impact and guiding direction. There is a tendency to react to the competitive tactics and the market dynamics in the vacuum of this chart and run the risk of a program without direction.

The comparison to the original product and market projection is a very important one. It serves as a basal temperature reading on the speed at which the market is reacting, the competitive resolve, and the degree of the company performance. If the projected product from the chart falls short of the actual market feedback, the market is moving faster than anticipated. If the projected product is far ahead of the actual requests, the market is more lethargic than anticipated, and leapfrog introductions may be too advanced for the customer base to accept. Figure 11-7, comparing the original plan and the current market feedback, illustrates this point.

As shown in Figure 11-7, the product evolution flowchart indicates an initial product offering. A second is planned for after the end of period 3. A third offering is planned for after the end of period 4. The gray products represent the plan. The height of the products

ACTUAL VERSUS PLANNED EVOLUTION

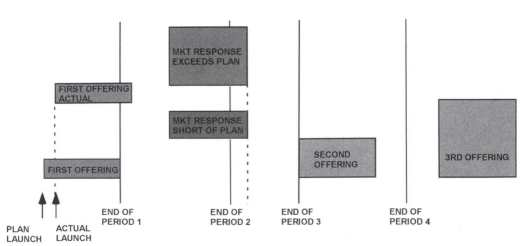

Figure 11-7. Product Evolution Comparison

indicates their feature/benefit/ and/or performance. The higher the product diagrammed, the more features and performance.

The actual product introduction for the first product version is a little late and shown accordingly. The accelerated market response is shown where the market expects more features. The lethargic response is shown underneath it, where the market expects less features and functionality. The dotted line at the point of measurement factors in the late product introduction. If initial market feedback outlines a product like the product with significant features and performance requirements exceeding the planned offering early in the product cycle, you are in a fast-paced market and must act accordingly to keep up.

2. SLOW? FAST? ADJUST THE LEAP

The comparison will detail the response of the next product action. The leapfrog must be enough of an improvement to position the company effectively. Even at this point be sure to factor in ample improvement or enhancement so as to be in a favorable competitive position, even after the product enhancement is developed. With knowledge doubling in shorter amounts of time, it is especially important to target properly. This is especially important in a fast-paced arena, as competition is not sitting by while you launch the product, gather market share, and develop follow on enhancements. In fact, the introduction of the product in certain circumstances may ignite a product-capabilities race to secure market share.

3. ATTACKING YOURSELF AS THE NEXT STEP

Let an introspective philosophy permeate the organization to stay current with the product. Constantly attack yourself before the competition does. If your product-planning personnel begin to think *like* the competition while they are working in *your* best interests, they will guide your moves to always be ahead of competition.

PRICE PRESSURES AND MARKET SHARE

1. MARKET IS NOT STAGNANT, IT RESPONDS

If there has been one shibboleth in the new product arena, it is that new product development occurs in dynamic and changing market conditions, which include performance and pricing. In the field of physics Newton's third law governs: *For every action there is an equal and opposite reaction*. In the field of product development and marketing, for every product action: an introduction, a pricing action, and a feature enhancement, there is a market reaction. It may not be equal and opposite. In certain circumstances it may be a larger response or a backlash. The best that could be hoped for is a competitive response clouded in ignorance.

Practically speaking, however, the market does respond to product actions. The manager must learn to anticipate the reaction and plan for it as part of the initial action.

2. RESPONDING TO THE RESPONSE

This being the case, the manager needs to formulate a response to the market reaction; otherwise the company may be left in a vulnerable position. The battle for market share is almost analogous to a chess game, with each move counteracting another, and an overall strategy is implemented with tactics. The manager should be aware that once the game commences there is no time for complacency. You are in the game and the game must be played or forfeited. The same holds true with the management of a product.

3. EXAMINE PRODUCT MARKETING TACTICS

With each advance and retreat there is a response to market conditions. The response needed may not always be price; it can be some package of values along with price. The expected responses are somewhat peculiar to the individual market players. If we examine the model presented earlier, there are four major classifications of market players, namely:

A. Defensive orientation
B. Direct Pursuit orientation
C. Oblique Pursuit orientation
D. Opportunistic orientation

There is an expected response from each of them, based on their position and outcome of an interchange. Figure 11-8 summarizes the expected responses when one takes on another. This is not to say that under certain conditions and circumstances the response will always follow this form; however, it does represent a typical expected response, given the nature of the players.

As shown in Figure 11-8, the reactions to the market actions vary. At the base of the illustration is a key, which summarizes the various actions possible. *PA*, for example, can represent a pricing action. One can either raise the price or lower the price to effect certain responses. The company can introduce a product enhancement, indicated by *PE*. A new rendition of the product can be a leapfrog of the existing product, Indicated by *LP*. One can choose to abandon the market altogether, indicated by *LM*. If you are an entrenched player and are clearly out maneuvered in the marketplace, you may choose to harvest the residuals only (indicated by HR). If your company is clever enough and has the influence, they can change the playing field to their advantage. This is indicated by *PF*. If your company has the commitment and wherewithal, they may choose to acquire the product, indicated by *AP*. Finally, if your company wants immediate

MARKET DYNAMICS

PRIMARY MARKET PLAYER	DEFENSIVE THIS IS THE ACCEPTED LEADER IN THE MARKETPLACE. THEY HAVE THE LARGEST MARKET SHARE	DIRECT PURSUIT THIS IS USUALLY #2 TRYING TO BECOME #1. THE CONTEST IS THE BATTLE OF THE TITANS	OBLIQUE PURSUIT THIS PLAYER GENERATES PRODUCT OPPORTUNITIES THAT ATTACK MARKET OPPORTUNITIES OBLIQUELY AND DIFFICULT FOR THE ENTRENCHED COMPETITOR TO COUNTER	OPPORTUISTIC PURSUIT THIS PLAYER DISCOVERS OPPORTUNITIES AND ACTS TO SECURE THEM. WILL LEAVE MARKET AT A MOMENTS NOTICE TO PRESERVE CAPITAL IN A FIGHT FOR MARKET SHARE
DEFENSIVE	AC, AP	AC, AP, PE, PA	AC, AP, PA, HR	AC, AP, PA
DIRECT PURSUIT	LP, PE	AP, PE	PE, PA	PA
OBLIQUE PURSUIT	PF	PF	PF	PA
OPPORTUISTIC PURSUIT	LP, LM	LP, LM	LP, LM	LM, HR

PA	PRICE ACTION +/–	PE	PRODUCT ENHANCEMENT
LP	LEAP FROG PRODUCT	LM	LEAVE MARKET
HR	HARVEST RESIDUALS	PF	CHANGE PLAYING FIELD
AP	ACQUIRE PRODUCT LINE	AC	ACQUIRE COMPANY

Figure 11-8. Market Dynamics

results, they may simply acquire the company and integrate it into their operation, indicated by *AC*.

The traditional company roles would indicate that an opportunistic player normally would not take on a defensive player. There are enough market dynamics, with ownership changes and acquisitions, that it could be acquired by an offensive player and funded to take on a leader with the backing of the offensive player.

The chart is organized such that the left side of the chart takes action against any of the players along the top of the chart. Their tactics are listed at the intersection of the horizontal and vertical rows and columns. For example, an opportunistic pursuing a defensive player has the options of leapfrogging the product introduction to attempt to displace unsatisfied customers, or to leave the market altogether.

4. IS THE PRICE RIGHT?

"How's my price?" is an often-used phrase in the selling process. Is it worth cutting price to obtain business? How low should the company set prices to introduce the new product? All of these questions may not have simple answers, depending on the circumstances; however, the company must take a position on the questions and act accordingly.

There are extenuating circumstances in which a company may want to lower prices to get business, even at a slight loss for a short period. However, there is one thing for certain:

Once the company lowers the price in the marketplace, there is little chance of it ever being raised. This is because the expectation has been set in the marketplace and competitors will react to the price change in some manner. If they reduce their price, you are now in the position of responding to the reduction. Good luck trying to increase the price under these circumstances.

Aside from the strategic and market-driven reasons for preserving pricing, there is a pure financial perspective. Consider Figure 11-9a, which shows what happens when prices are reduced to drive product volume and increase profit.

FINANCIAL IMPACT OF PRICE REDUCTIONS WITH ELASTIC DEMAND

UNIT VOLUME	0	100	200	300	400	500	600	700	800	900	1000	1100
BASE REVENUE	0	50000	100000	150000	200000	250000	300000	350000	400000	450000	500000	550000
PROFIT NORM	−120000	−97500	−75000	−52500	−30000	-7500	15000	37500	60000	82500	105000	127500
PROFIT+10%	−120000	−92500	−65000	−37500	−10000	17500	45000	72500	100000	127500	155000	182500
PROFIT −10%	−120000	−102500	−85000	−67500	−50000	-32500	−15000	2500	20000	37500	55000	72500

PR. NORMAL		$ 500
PRI. INC. 10%	0.1	$ 550
PR. DEC.– 10%	−0.1	$ 450
VAR.CST/UNIT		$ 275
FIXED COST		120000

	UNIT PROFIT	B/E	VOLUME EFFECT	
GP NORMAL	$ 225	533		
GP + 10%	$ 275	436	18%	LESS
GP – 10%	$ 175	686	29%	MORE

A

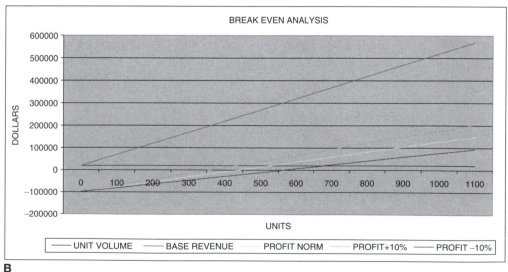

B

Figure 11-9. Price Reductions

As shown in Figure 11-9a, the product generates revenue and profit. This profit offsets the fixed cost of production. A reduction in the price lowers the profit, forcing additional unit volume to generate the same profit dollars. Conversely, an increase in price increases the profit margin dollars per unit and requires less unit volume to generate the same profit. Break even is where the profit generated is equal to the fixed costs.

In this example shown in Figure 11-9a, the product sells for $500 normally. The per unit cost is $275. Under normal conditions, the profit for a single unit is $225. The revenue and profit is shown for unit volumes, from 0 through 1100 units. The fixed costs are $120,000.

Figure 11-9a shows what happens if the price is lowered and raised by 10%. As can be seen the break even point shifts from the normalized value of 533 units to a low of 436 units, given a price increase and a high of 686 units, given a 10% reduction in price. A price reduction requires a 29% increase in unit volume to generate the fixed cost offset. Lowering prices to generate volume can be an entry into a race that cannot be won.

Figure 11-9b displays this point on a graph. It shows the change in break even points based on the pricing action.

5. HOLDING PRICES AND ADDING VALUE TO PRESERVE MARGIN

If a pricing action carries this liability, how can the product manager compete in the face of changing market conditions? The response to this question lies in the bundling of product values that support the existing price. In other words, give more for the same price. The challenge is to do this without incurring too much additional cost, which will deteriorate the gross margin and effectively cause the same problem as a price reduction. This is because it is, in fact, a price reduction.

6. CHANGING THE PLAYING FIELD TO YOUR ADVANTAGE

If the circumstances allow it, and if the idea is clever enough, change the playing field of competition to your advantage somehow. Create some offering that the competition cannot respond to easily or with equivalency. As stated before, it is a chess game that is won or lost based on these moves. Make better, more informed moves than the other players and you will be successful.

GROWTH STRATEGY

1. HOW WILL THE BUSINESS GROW WITH THE NEW PRODUCTS?

The responsibility of product management includes a certain degree of continuity of the business planning cycle. In the same way that the product features and technology have an evolution, the business must have and evolution too. At the onset of the program, a comprehensive business plan was developed to guide the development and the marketing of the product. The product, however, represents only a small part of the process of growing the business. The follow on activity must now be directed toward growing the business and satisfying the original business plans with market modification. Go back to the overall plan that was developed and start to work on the follow on programs and products. This follow through is what will create the anticipated success.

Take the market feedback from the first product introduction and develop the next generation of business plan.

2. REMEDIATION

If the original assumptions of the business plan are incorrect or need updating, do so and begin to execute them. The saying, "Do not argue with success," is true enough; however, it is utterly imperative to understand the reasons for success. It is the basis for future actions. Conversely, if a midcourse correction is needed, do not hesitate to make it—decisively and completely. Take lessons from those things that did not work out and incorporate the corrections in the next version's plan. In summary, learn from the success as well as the failures. Build on the success and remedy the contributions to failure.

LIFE CYCLE MANAGEMENT OF A PRODUCT

1. EVALUATING NEW PRODUCT PROGRAMS

How has the new product development program performed against plan? Is it a success? Is it marginally successful? Did the product team deliver the results promised? There are several ways of answering these questions; indeed, several factors can be used to evaluate the degree of success or failure. Fundamentally, outside of the strategic initiatives, a new product development is supposed to generate funds to the corporation at a rate that far exceeds the rate of funds consumption. This profit is used to fund new programs. The dynamics represent the corporate covenant between the product group and corporate management.

There is a simplified method of evaluating the degree of success of a new product program from a purely financial perspective. It consists of a comparison between the original expectations and the actual results. Figure 11-10 outlines an example.

As shown, Figure 11-10 summarizes the various scenarios in the planning stage of the program, and compares them to the actual results. The various case scenarios allow the actual results to be placed in perspective of the original business plan. The chart shows an investment of $1.5 million, a product price of $500, and a cost of $250 in the planning stage. A total of five cases with different volumes were generated to assess the internal rates of return and the net present values. The actual results indicate an increase in costs, as the product life cycle continued, with the best-case volume. Actual net present value and

NEW PRODUCT PLAN		YEAR 1	YEAR 2	YEAR 3	YEAR 4	YEAR 5		
PRICE EACH		500	500	500	500	500		
COST EACH		250	250	250	250	250		
CASE 1 VOLUME		1000	2000	3000	4000	5000		
CASE 2 VOLUME		800	1600	2400	3200	4000		
CASE 3 VOLUME		600	1200	1800	2400	3000		
CASE 4 VOLUME		400	800	1200	1600	2000		
CASE 5 VOLUME		200	400	600	800	1000		
	INVESTMENT						NPV BASED ON 5% INFLATION	IRR
GROSS PROFIT CASE 1	−1500000	250000	500000	750000	1000000	1250000	$1,563,427	30%
GROSS PROFIT CASE 2	−1500000	200000	400000	600000	800000	1000000	$965,027	22%
GROSS PROFIT CASE 3	−1500000	150000	300000	450000	600000	750000	$366,628	12%
GROSS PROFIT CASE 4	−1500000	100000	200000	300000	400000	500000	($231,772)	0%
GROSS PROFIT CASE 5	−1500000	50000	100000	150000	200000	250000	($830,172)	−17%

ACTUAL RESULTS	INVESTMENT	YEAR 1	YEAR 2	YEAR 3	YEAR 4	YEAR 5	NPV BASED ON 5% INFLATION	IRR
PRICE EACH		500	500	500	500	500		
COST EACH		250	290	330	370	410		
CASE 1 VOLUME		1000	2000	3000	4000	5000		
GROSS PROFIT CASE 1	−1500000	250000	420000	510000	520000	450000	$323,807	12%

Figure 11-10. New Product Evaluation

internal rate of return is much less than planned. There is a sample chart available for your use in the *Tool Box*.

2. OPPORTUNITIES AND THREATS DURING THE CYCLE

Previously we discussed the several threats that endanger a product program. There are also several opportunities that are made available during the business cycle. If the company has difficulty in focusing on goals, and demonstrating the resolve to attain them, the opportunities that "befall" the company can also become threats. These are the worst kind of threats because they foster complacency and demand the most costly extravagance, namely time and energy.

As important as the vocational aspects of new product development are, the corporate characteristic traits are equally as important. If the program is on target and the product management is diligent, the real competitive threats will be handled effectively in most cases. If, however, the corporation has a "weakness" for the prettier opportunity, it is virtually impossible to carry a program through to completion, let alone success.

Focus and execute throughout the business and product cycle. Do not be driven off course by seemingly better opportunities. They detract from your primary objective.

3. CONTINUOUS EVOLUTION IS INTEGRAL TO THE LIFE CYCLE OF THE PRODUCT

The product life cycle demonstrates the need for vigilance and continuous improvement in the product to nourish the business. When the product volume is building up, the costs are reducing, and the profits are beginning to be realized, it is easy to get lulled into a sense of complacency. However, the time to embark on the new version or the next product is when the business is riding high. The cash position is healthy, the profitability is peak, the market presence is good, and there is only one way to go if nothing is done: down! Therefore the best time to question your market strength and do some business introspection is when the natural tendency is to be complacent. The fact of the matter is, *that* at that time, the competition is making their next move. This can be illustrated in the chart in Figure 11-11a.

The graph in Figure 11-11a shows the build up of volume under the five cases outlined in the business plan. Since year 5 is the peak of the product line in volume, product management should launch new product efforts sometime before the end of year 5.

Figure 11-11b shows the gross profit, under the plan scenarios, which will be used for the investment.

Figure 11-11c shows the actual gross profit of the product run, based on volumes and updated cost data. It would indicate that the peak of the product's contribution to corporate profits occurs earlier in the cycle, and reinvestment should probably occur nearer to that time rather than in year 5.

As shown in the various ways, there is a point on the product life cycle's characteristic curve where reinvestment must be made. The shape of the curve may change from product to product; however, the operative lesson is to recognize the peak of the product cycle and take effective action to preserve it in the long run. The spreadsheet and the accompanying graphs are available for your use in the *Tool Box*.

4. MAXIMIZING PROFIT

The product manager's responsibility is to maximize profit for the corporation, if possible. This directive is comprised of three essential elements, all things being equal:

A. The volume of product must be pushed. Meet or exceed the target volumes, but do not fall short!

B. Maintain the pricing of the product in the marketplace. Take every possible step, including value enhancement to reduce the need for rolling back pricing.

C. Finally, make a conscious attempt to contain cost increases and work to reduce costs. By reducing costs, the company is in a better position to offer enhanced value or meet competitive pricing, should it be required.

SUMMARY

This chapter represents the vocational capstone of the new product development process. The chapter is presented in two sections. The first section is oriented internal to

NEW PRODUCT PLAN		YEAR 1	YEAR 2	YEAR 3	YEAR 4	YEAR 5	YEAR 6	YEAR 7	YEAR 8
PRICE EACH	PRICE EACH	500	500	500	500	500	500	500	500
COST EACH	COST EACH	250	250	250	250	250	250	250	250
CASE 1 VOLUME	CASE 1 VOLUME	1000	2000	3000	4000	5000	3500	2000	500
CASE 2 VOLUME	CASE 2 VOLUME	800	1600	2400	3200	4000	2200	1000	200
CASE 3 VOLUME	CASE 3 VOLUME	600	1200	1800	2400	3000	1500	750	175
CASE 4 VOLUME	CASE 4 VOLUME	400	800	1200	1600	2000	900	400	125
CASE 5 VOLUME	CASE 5 VOLUME	200	400	600	800	1000	400	200	75
	INVESTMENT								
GR. PROFIT CASE 1	−1500000	250000	500000	750000	1000000	1250000	875000	500000	125000
GR. PROFIT CASE 1	−1500000	200000	400000	600000	800000	1000000	550000	250000	50000
GR. PROFIT CASE 1	−1500000	150000	300000	450000	600000	750000	375000	187500	43750
GR. PROFIT CASE 1	−1500000	100000	200000	300000	400000	500000	225000	100000	31250
GR. PROFIT CASE 1	−1500000	50000	100000	150000	200000	250000	100000	50000	18750

ACTUAL RESULTS	INVESTMENT	YEAR 1	YEAR 2	YEAR 3	YEAR 4	YEAR 5	YEAR 6	YEAR 7	YEAR 8
PRICE EACH		500	500	500	500	500	500	500	500
COST EACH		250	290	330	370	410	410	410	410
CASE 1 VOLUME		1000	2000	3000	4000	5000	3500	2000	500
ACTUAL GROSS PROFIT CASE 1	−1500000	250000	420000	510000	520000	450000	315000	180000	45000

Figure 11-11. Life Cycle Management

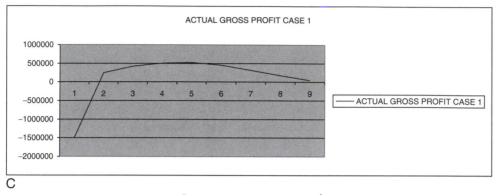

C

Figure 11-11. *Continued*

the organization and focused on a discussion of the mechanics of product management. The second section centers on the external focus of promoting the new product in the marketplace.

The chapter starts out with the discussion of the role of product manager as champion of the product, within and external to the organization. Next is a discussion of the learning curve and the resulting cost reduction that should be achieved. Without the learning curve cost reduction, very few products will have a long life cycle in the marketplace. With cost reductions come venue changes in manufacturing and outsourcing. The international focus concerning currency changes and how they affect the product profitability was presented next.

The quality system is integral to the product, and a basic requirement for success. It cannot be an overlay to the company. Since change is inevitable, the management of change is crucial to success. The manager must learn to evaluate the field returns incidences and react accordingly to correct issues. The reaction must be practical and effective. A framework for decision-making is presented. The final entry in this first section is a discussion on the product recall. A framework for deciding if a recall is required is discussed, as well as mechanics needed to prosecute one.

The next section focuses on the external marketing orientation of the product. In this section we discussed the market development of the product, and also application development. Both approaches are designed to increase volume of the product and market presence. The market feedback is used to determine the next version of the product and the means for deciding which feedback to accept, and how to structure the implementations. Next is a discussion on price pressure in the marketplace, and how it can affect the profitability of the program. A model for tactical responses to market actions is also presented.

To complete the business and product planning cycle, a growth strategy is presented within the framework of the original business plan. In addition, a model for evaluating the program's overall financial performance is presented within the framework of the life cycle management of the product line.

With the new product development basics in place, you are now ready to embark on the new product journey of your own! To assist you in this endeavor, Chapter 12, "New Product Development Records Format," presents perspectives on the product development process and outlines a framework for you to organize your own project. Good luck in developing your new product!

NEW PRODUCT DEVELOPMENT RECORDS FORMAT

ABSTRACT: This chapter consists of two parts: a strictly mechanics-oriented section, and a section outlining final perspectives on new product development. The mechanics section is designed to show the reader how to structure their files so that information and data are easily stored, presentable, and retrievable in an organized manner. It can also serve as a checklist for the various aspects of new product development process. The second section reviews operational perspectives on new product development as a vocation and a career pathway.

ORGANIZATIONAL FORMAT

BACKGROUND

1. General Usage Patterns of Product Types

2. Trends in Industry and the Marketplace

3. Driving Forces in the Functionality

4. Driving Forces in the Technology Employed

5. Product Opportunity Summary

6. The Niche/Advantage/Uniqueness Of Position

7. Statement of Fit with the Enterprise

8. Brand Label, Joint Ventures, Acquisitions

   A. *Miscellaneous Topics to Agree on*

B. *Brand-Label Checklist*

C. *Diligence Checklist*

MARKET DATA

1. Primary Market Data

2. Secondary Market Data

3. Product Mix-Analysis

   A. *Product/Services Mix*

   B. *Weighted Evaluation of Opportunities*

   C. *Value Pricing*

4. Pricing Levels and Trends

   A. *Creep Impact on Cost*

5. Estimated Costs

6. Volume

7. Demographics

   A. *Value Engineering*

8. Classification of Customer Base

CORRESPONDENCE

1. Internal Correspondence

A. Sales Department

B. Development

C. Finance

D. Manufacturing

E. Quality

2. External Correspondence

A. Customers

1. Account Development Model

2. Action Plans

3. New Items/Issues

4. Problems/Solutions

B. Agencies

1. Qualifications

2. Testing

3. Approvals

C. Reps/Distributors/Agents

1. Contract

2. Apr/Territory

3. Sales Goals

4. Forecasts

5. Issues

FINANCIALS

1. Product Life Cycle Forecasts

   A. *Pressure Cycle Forecast*

2. Revenue with Mix

   A. *Profit in Backlog*

3. Factory Cost with Learning Curve

4. Development Cost

   A. *Project Cost Analysis*

   B. *Project Evaluation Worksheet*

5. Tooling Costs

6. Capital Equipment

7. Sales General and Administrative Costs

8. Return On Investment Calculation

A. ROI Calculation Worksheet

B. Comparative ROI Worksheet

9. Return on Net Assets Calculation

10. Income Statement

A. Breakeven Analysis

11. Balance Sheet

12. Funds Flow

ROUTE TO MARKET

1. Flowchart of Route to Market

2. Funds Flow for Transaction

3. Percent Cost Added through the Channel

4. Sensitivity Analysis through the Channel

5. Analysis of Transactional Pressure Point through the Channel

6. Alternate Routes to Market

7. Impact of Alternative Routes

MARKETING REQUIREMENTS SPECIFICATION

1. Format and Outline

A. Background

B. Industry Trends

C. Market Opportunity

D. Tie to Strategic Plan

E. Scope of Product Line

F. Component Parts and Product Configuration

G. Functional Sequence of Operation

H. Performance Requirements

I. The Operating Envelope

J. Standards

K. Cost Target

L. Timing to Introduction

M. Human Factors Engineering

N. Safety

O. Longevity, Product Life

P. Service Plan

Q. Field Replacement Parts

R. Corporate Standards

DESIGN SPECIFICATION

1. Outline Each and Establish Format

   A. Scope of Product Line

   B. Component Parts and Product Configuration

   C. Functional Sequence of Operation

   D. Performance Requirements

   E. The Operating Envelope

   F. Standards

   G. Cost Target

   H. Timing to Introduction

   I. Human Factors Engineering

   J. Safety

   K. Longevity, Product Life

   L. Service Plan

   M. Field Replacement Parts

   N. Corporate Standards

2. The Responsibilities of the Design Engineer

3. Matrix the Marketing Spec and Design Spec: Resolve Changes

4. Commitment to What Will Be Achieved

COST DATA

1. Bill of Material

2. Cost Roll-Up

   A. ***Design to Cost Versus Cost Plus***

   B. ***Parts Cost Roll-Up***

3. Cost Basis and Assumptions

   A. ***Cost Systems***

   B. ***Product Cost Analysis***

   C. ***Profit Sensitivity***

4. Vendor Links

5. Agreements with Vendors

6. Labor Analysis

7. Vendor Plan

8. Labor Plan

9. Methods Plan

DRAWINGS AND DOCUMENTATION

1. Design Documentation

2. Manufacturing Documentation

3. Repair Documentation

4. Electronic Medium Plan

5. Retrieval and Modifications

6. Portability

7. Structure and Format

8. Methods Documentation

9. Improvement Plans

CUSTOMER VISITS

1. Outline of Marketing Plans

2. Customer, Agent, Reseller Matrix Profile

   A. Agent Reseller Management

3. Time Line for Customer Visits

4. Status Report for Each

5. Link to Communications Section

6. Site Visit Report

LITERATURE DOCUMENTATION

1. *Type A Info: Effecting the Purchasing Decision*

2. *Type B Info: Product Usage Information*

3. *Type C Info: Training on the Product*

4. *Type D Info: Application of the Product*

5. *Type E Info: Effecting the Repair of the Product*

6. *Type F Info: Prosecuting the Warranty*

SIGNIFICANT PERFORMANCE AND APPLICATIONS FEATURES DOCUMENT

1. Operating Envelope

2. Degree of Derating

3. Margin to Advertised Spec

4. Corporate Derating Guideline

5. Spider Web Chart of Parameters

6. Spreadsheet of Market Requirements/Design Objectives/Actual Parameters and Percentages of Each

COMPETITION

1. Competitive Comparison Chart

    A. *Competitive Comparison Chart*

2. Spider Web Chart for Each Competitor

3. Pricing Analysis

4. Route to Market

5. Sensitivity Analysis

6. Technology

7. Trends

8. Strategy Against Each Competitor

FIELD FEEDBACK

1. Problems

2. Complaints

3. Suggestions

4. What's Right/Needed

5. Value Analysis

    A. *Beta Test Feedback*

    B. *Beta Test Form*

    C. *Beta Test Summary*

6. Next Generation or Product Iteration

MANUFACTURING

1. Manufacturing Plan

    A. *Lead-Time Summary*

2. Manufacturing Processes/Procedures

3. Layout of Line/Flow

4. Formula for Logistics

5. Manufacturing Documentation

    A. *Vendor Profile*

    B. *Bill of Material*

    C. *Part Card*

6. Training Plan

    A. *Employee Certification*

7. Failure/Fault Repair System for Production

8. Testing Plan

9. Record Keeping

    A. Serialization

    B. Histogram of Modules

    C. Test Data

    D. Failure Data

       1) *Manufacturing Yield Data*

    E. Critical Components

10. Quality Plan

QUALITY REPORT

1. Product Birth Record

2. Serial Number Database

3. Spares Repairs Tie-in

4. Field Incidences

    A. *Incident Information*

5. Quality Database System

6. Procedures

7. Reports Data Entry

8. System Platform—MIS Tie-in

9. Product Life Cycle Database

CATALOG NUMBER BREAKDOWN

1. Product Versions: Scope of Product Range

2. Options

3. Product Evolution Flowchart

4. Catalog Number Breakdown

ENGINEERING CHANGES

1. Product Configuration System

2. ECR/ECN System

   A. *Sample Engineering Change Procedure*

3. ECR/ECN Database

4. Master Log and History

5. Cost-Impact Analysis, Factory Cost, and Activity-Based Costs

   A. *Engineering Time Log*

PRODUCT EVOLUTION FLOWCHART

1. Conservative Development of Product Line

2. Aggressive Development

3. Most-Likely Scenario

4. Detail of Present Product Plans and Disposition

5. Specs of Future Units

PRODUCT LIABILITY

1. Scope Analysis

2. Scope of Product Line/Exposure

3. Financial Exposure

4. Risk Analysis

5. Recoverability Index

6. Problem Identification

7. Problem Engagement

8. Solution

9. Communication

10. Mitigation/Containment

11. Recall

   A. ***Product Recall Format***

12. Definition of Completeness

PROMOTION

1. Marketing Manual

   A. ***Matrix***

   B. ***Quotation Database***

   C. ***Sales Order Entry Analysis***

   D. ***Sales Representative Management***

   E. ***Sales Report***

2. Promotional Literature

3. Direct Mail Pieces

4. Telemarketing

5. Powerpoint Presentation

6. Computer Multimedia/Video

CUSTOMER SUPPORT

1. ***Customer Support Database***

2. ***Customer Service Evaluation Tool***

3. ***Format for Sales Compensation System***

PROGRAM MANAGEMENT

1. Business Plan

2. ***List of Deliverables***

3. *Narrative*

4. *Risk Analysis*

5. *New Product Evaluation*

6. *Life Cycle Management*

PERSPECTIVES ON THE CONTINUITY OF NEW PRODUCT DEVELOPMENT

1. CONTINUUM (NOT ISOLATED, DISCREET ACTIONS)

To effectively win the competitive war, the process of new product development and business development should be thought of as a continuum, rather than a series of discreet, reactionary actions. The manager must weave the actions into a holistic theme and communicate it to the marketplace. These actions must also be effective in countering competitive thrusts. The worst scenario is to react to every competitive action and burn out the development personnel. The "art" of management is to determine what items to address and what items to ignore.

2. MYTH OF THE ONE-PRODUCT BUSINESS

When evaluating the prospects of a new business based on a single product, be careful to understand the sensitivities to external actions involved. It is much easier for a one-product business to fail, rather than a multiple-product business to fail. Given this, no one-product success should make a company successful, nor should one product cause a company to fail.

There have been many businesses that have "changed ownership" because of placing too much credence and hopes in the success of a new product. True business success is comprised of many little successes. The ones that chase the "Holy Grail" of opportunities are the ones who accomplish little in the long run and have little sustainability in the marketplace.

It can even be said that the opportunistic tactics used by the guerilla players can allow a certain amount of success; however, the guerilla knows when to cut losses and chase a more-achievable opportunity. They also tend not to invest substantially in any one opportunity. The larger company may not always be "streetwise," or may have politics that delay exiting a loser prospect. These companies waste time and resources and remain small-market players with a long list of stories and disappointments as part of their folklore.

3. SET UP THE TEAM AND PROSECUTE OPPORTUNITIES

If your company is serious about growing the business through balanced internal and external product development, establish procedures and assemble a group of motivated people to prosecute real opportunities on time and budget. Become proficient at selecting and completing these programs and harvesting what you can from them. Learn from the success, and also from the failures. Keep the momentum of selection, implementing, and prosecuting the opportunities at pace. Do not let adversity deter the group from their objectives. Reinforce that the programs must be small in scope and numerous, rather than large and singular. Each should contribute to profit and be completed in a market-timely fashion.

4. GET EVERYONE ON BOARD

Make sure there are no loners or prima donnas in the group. Every member of the group succeeds or fails based on the performance of the group. Tolerate no personal agenda that may detract from the energy of the group, but use the programs as a means for personnel to experience personal satisfaction and professional growth. Be fair, deliberate, and swift in personnel matters because the team members are wasting multiples of the time you waste in deciding to take action.

Foster a success orientation within the group and the company. Anticipate areas of conflict and navigate around them before they affect the performance of the group.

5. CORRECT OR ELIMINATE NONBELIEVERS EARLY

Ensure the total commitment of the group. Make an attempt to convince nonbelievers; they may become your strongest supporters. If they cannot be persuaded, remove them from the group. This includes everyone in the value chain, from the development person, to the sales channel personnel. Do not utilize the energy of the group, either in part or in total, to overcome negative energy broadcast by these people. It is not fair to your people, and is analogous to running an eight-cylinder engine with two dead cylinders. Not only do the remaining six cylinders have to produce the required horsepower, they must overcome the "dead weight" of the nonproducers. It is better to have a six–engine, or better yet, a full-functioning eight-cylinder engine.

6. COMMUNICATION, COMMITMENT, AGREEMENT, AND ORGANIZATION ARE KEY

Like any endeavor involving people, there are four requirements for success. They are:

A. Communication. This is a basic requirement among all of the group members and the customers offering product insight and input. The need for accurate, timely, and unbiased communication is a fundamental requirement. Do not accept anything less.

B. Commitment. This requirement ensures some degree of momentum and sustainability in the program. A project involving people without commitment will languish and fail. The program with committed people will succeed. Do not assemble a group that lacks this trait. You cannot add it in the midst of the program.

C. Agreement. It is difficult for any two people to agree on everything all of the time. It is even more difficult to get a group of people to agree with every decision in a program, especially when judgment and more than one solution are involved. The important thing to remember, however, is to allow everyone to express their opinion and make a suggestion; but in the end, the manager must take all the input and make the best decision possible, which might not necessarily be the most popular decision. They must have enough leadership skills to make the decision and see it through. Once the decision is made the entire product group must support it without exception or reservation. Consider the facts, consider the recommendations, and make the decision and prosecute it with everyone supporting it.

D. Organization. Every project has a multitude of details. The challenge is to organize them for usage and recall. Take the required time to organize the program in a format that is usable. Do not put the task off till later; you will never organize the program, and in the meantime, you will suffer from lack of organization. Lack of organization will degrade the decision-making process. (It is human nature to make an attempt to look for information, but soon forgo the attempt when it is inconvenient, and make a decision from memory or loose facts.) A well-organized information system will aid in decision making by improving access when it is needed.

7. NEW PRODUCT DEVELOPMENT: IT'S THE MOST FUN PART OF THE BUSINESS

The development of a new product can be the most fun part of the business. It is usually on the cutting edge of technology, deals with new elements, and allows for the engagement of new customers.

This can represent an environment for challenges, both technologically and vocationally. New product development will challenge the limits of a manager's skills and allow for the opportunity to develop even more refined skills.

8. IT'S THE FASTEST WAY TO LEARN AND INFLUENCE THE BUSINESS

Developing a new product and integrating it into production and moving it through the sales channel to the marketplace requires touching most, if not all, aspects of an operation. It is the fastest way to learn about the business because it forces you to use the resources to accomplish your goals. By learning all aspects of the business, you are placed in the position of being the most capable of influencing the business and its future. If the desire for a person is to control their corporate destiny, the product development arena is an excellent means to this end. You will create and implement products that will affect the organization and effect certain changes. As such, the manager can become the "architect" of growth for the organization through the new product development process.

9. IT'S ONE OF THE MOST REWARDING FACETS OF THE BUSINESS

Finally, the process of new product development can be one of the most-rewarding activities in the corporate world. It will challenge you and allow you to succeed in ways that would otherwise take many years to accomplish. It will "season" an executive with real-life experiences, rather than him or her digesting a selection of articles. It serves as a litmus test for peers and upper level managers to see how they confront challenges. It allows an individual to excel in an organization and gain almost immediate access to a variety of career pathways. It will yield lessons learned and give unique insight as to how the business changes, grows, and becomes profitable.

APPENDIXES

SELECTED NEW PRODUCT DEVELOPMENT EFFORTS

FAMOUS PRODUCT DEVELOPMENT EFFORTS

This section is a review of some of the new product developments that took place within the last 100 years. This list is by no means conclusive or representative of all the developments that significantly affected our lives during that time period. They have been selected to illustrate several points about new product acceptance in the market place and growth patterns from inception to market penetration. The following are the programs in review. Each will have a summary about the development, some statistics on its acceptance, pervasive use, and a measure of the impact on society.

COMPUTERS: 1834

Background

The origins of computing machines date as far back as 1642, when Blaise Pascal built the first digital calculating machine. Subsequent endeavors included 1671, when Gottfried Wilhelm von Leibniz invented a stepped gear arrangement, and a century later when Thomas Colmar developed the first commercially successful mechanical calculator.

However, the conceptual application of a general purpose computer date back to approximately 1834 when Charles Babbage designed the first plans for a calculating machine. The memory for the computer was stored on punched cards. Improvements that were added by Herman Hollerith and James Powers, working for the U.S. Census Bureau. Cards with mechanical feelers were used to analyze the results of the 1890 census. The lack of any material progress in the next 40 to 50 years gave way to the use of thermionic valves to allow or restrict the flow of currents, which were then related to numerical values. The 1940s brought the development of the binary code, which was easier to mechanize and allowed numerical manipulation.

Technology Development

The first electronic digital computer was completed at the University of Pennsylvania in 1945. The ENIAC (electronic numerical integrator and calculator) was a massive machine employing over 18000 vacuum tubes. It could execute 5000 calculations per second and occupied over 1800 square feet. It consumed 180000 watts of electrical power. This technology was successfully used between 1946 and 1955.

In 1947 the origins of algorithm development and software started to root. A newer generation of computers with RAM (random access memory) and physical changes required less space and power.

In the 1950s, progress was made in advanced programming techniques, and the computer's use was brought into the business environment, as payroll calculators; however, it was the development of the transistor and magnetic core memory that spawned the next generation of computers. The transistor was then integrated into massive memories (considered so, at the time) to improve performance and thus make its usage even more pervasive.

The 1960s brought enhanced performance and speed, as well as a host of peripherals such as input/output devices and magnetic storage means. The commercially viable machines were used in applications such as accounting, payroll, inventory control, vendor links, and billing. By 1965, the next generation of computers employing silicon chips with this medium-scale integration performed millions of calculations per second and reduction in size was dramatic.

The 1970s saw movement away from the large centralized computational centers employing time sharing and batch processing toward more broader applications and less costly smaller systems. This decade also saw the use of photo printing of conductive circuit boards to eliminate wiring.

The 1980s saw the advent of the personal computer with very large-scale integration, with hundreds of thousands of transistors on a single chip. These personal computers were small enough and affordable enough for personal use and ushered in the era of personal and portable computers. The late 1980s saw a focus on application and software development to facilitate the pervasive use. The 1990s even more so.

Time to Market Acceptance

The time from initial germ of an idea through the development stages and requisite technology development to commercialize the idea took 116 years! Add to this the time period from 1950s to the 1980s when larger scale integration and software development allowed wider acceptance and usage, the real growth spurt for the mass market didn't occur for almost 150 years!

Pervasive Use

The pervasive use of the computer by the mass market began in the 1980s. The developments in this industry are astounding. As of August 2000, 51% of all American households or 53,716,000 households used a computer. If the automobile industry had followed the computer industry down the cost learning curve, a Rolls Royce automobile would cost today approximately $2.75 and get 3 million miles to the gallon! Virtually one out of every two households has a computer. The computer's performance has out stripped most people's ability to comprehend. But, as a figure of merit, the portable laptop on which this book was written has over 160,000 times the memory capacity of the computer used to land the first Americans on the moon!

Impact on Society

The impact on humanity is equally impressive. Computer usage is so ingrained as part of our lives that they seem virtually invisible to most items they are controlling. For instance, it is not readily obvious to the driver of an automobile that a computer is controlling a multiplicity of factors affecting engine performance every fraction of a second. A computer is used in appliances and in entertainment systems. It is the preferred medium of choice in communication systems. It has created a new benchmark for obtaining information, gathering data, and communicating. As fast as the computer ran up the product life cycle, that's as fast as it has become a commodity from a hardware point of view. The difference now is in the operating systems and software.

Computers are used in virtually every aspect of our lives. For instance, they are used in the design of drugs to assess the body's reaction and assimilation without the extensive testing before the development phase begins. They are used to design and test products before the prototyping stage to optimize the design.

The computer has also increased the power of legitimacy of the written word. If it's from a computer, printed on a laser printer with professional-looking graphics, almost any content carries a certain amount of legitimacy.

THE PHOTOCOPY: 1903

Background

The photocopy, a process of rapid reproduction of a document by instantaneous development of a negative, was invented in 1903 by G.C. Beidler. Beidler, a clerk in an attorney's office, developed a machine to easily make copies, which eliminated the need for manual or typed copies. The device was patented in 1906. A company known as Recigraph introduced the first commercially available machine in 1907.

Thirty years later, Chester Carlson, a New York law student, made the first dry image by photocopying on dry untreated paper. His process was inspired by the previous work of a Hungarian, Paul Selenyi. The process, originally called electrophotography, was then patented.

Technology Development

Between the years of 1939 and 1944, twenty companies failed to show interest in the technology, and the patent wasn't developed. In that year, however, the Memorial Battelle Institute signed an agreement with Carlson to develop the technology. In 1947, a small photo company named Haloid signed an agreement to commercialize the technology. In 1959, the first Xerox machine employing Chester Carson's concept of xerography was introduced.

Time to Marketplace

As you can see, the technology development preceded the commercialization by over 50 years. It is also the nature of the process that the optimum solution is not obtained immediately. It takes several attempts to perfect the processes, before an idea becomes a product.

Pervasive Use

The Xerox machine is one of the most widely used machines today. Most offices cannot operate without one, even though the world would like to go to "paperless" business operations.

Impact on Society

The impact of the Xerox machine has revolutionized certain productivity measurements in the business and office world. It sped up communication and accelerated understanding by making knowledge and facts available to several people at the same time, with minimal time and energy required. It improved efficiency in documentation and record keeping.

It would be an interesting exercise to go through a normal business day without the Xerox and observe impact on productivity. The technology has developed in concert with computerization and imaging enhancement techniques, so that there is a gradient in machine performance that spans from simple residential home office units through complete document production systems that will print books, manuals, and special effects on standard paper. Some of these machines are sophisticated enough to even diagnose themselves and automatically call in service people to correct problems.

We can look forward to the computer and the Xerox machine merging even more.

THE VIDEO CAMERA (TELEVISION): 1923

Background

The video camera and subsequent development of television was a development that also spanned a long time. In fact, many of the examples were not an overnight success but did go on to revolutionize certain areas of our lives. The video age of transmission and reception however underwent a constant evolution during the first 20 plus years.

The video camera was initially invented by a Russian-American, Vladimir Kosma Zworykin in 1923. He is often credited with the title of "Father of Television." In December of that year, he demonstrated the technology to RCA. By 1933, RCA held the first experimental transmission of electronic television from the top of the Empire State Building in New York City. Earlier attempts at television, both color and black and white, preceded this 1933 date; however, they employed mechanical rather than electronic means to image processing.

Technology Development

Subsequent developments in formats and implementations occurred through the years 1954, 1956, 1962, and 1968 that elevated performance, integrated the color cited in this section, functionality, and improved quality. This technology is somewhat different than others in that many people were involved with the developments, each building on the work or failure of the previous.

Time to Marketplace

The time from initial experimentation to widespread use was rather short by comparison with other technologies, some 30 years. This may be in part driven by the fact that as we get further into the 20th century, the general acceptance and application of technology occurs faster. This parallels the increasing rate of information doubling occurring at this time. Also, the US was growing faster, and the need for mass communications and entertainment drove the need.

Pervasive Use

There is little doubt on the pervasive use of the technology. At current estimates, there are over 235 million television sets in use in the US alone. This represents 98.3% of the households with at least 1 set. Most households have multiple units, with different family members watching different programming simultaneously. The average number of sets per households is 2.4. Even with the effect of inflation, which would drive prices higher, the cost of the product has drastically decreased. For instance, a basic feature 25-inch color unit sold for approximately $500 USD in 1970, when the median household income was $10250 USD per year. In the 1990s a 25-inch unit with a standard full-feature mix can be bought for less than half of that amount at certain stores. This is at a time when the median income exceeds $44000 USD annually. Clearly, the need for entertainment and information available in this medium coupled with the drastic reduction in cost to income levels has driven unit volume to an incredibly number.

By most estimates, the use of television as a medium is so pervasive that one-half of the earth's population has ready access to a television set.

Impact on Society

Never before in our recent history has a product had the impact in our society as television has had in the time frame of its growth cycle. The television medium is our information, our training, our baby-sitter, our security, and for some, the blueprint of life. Countless children have become electronic moles, sitting in front of the tube. However, it has dramatically altered the vehicle and speed of information dissemination. We have watched men walk on the moon, presidents being attacked by gunfire, war declared and prosecuted. The television's new product development can be liked, admired, disliked, or hated, but one cannot overlook its significance.

MANHATTAN PROJECT: 1940 TO 1945

Background

It seems rather odd to evaluate the development of the atomic bomb as a new product development effort. There are numerous polarized views on the development and the deployment of the atomic bomb. There were two devices used at the closing of World War II; the objective was to force the Japanese, who were prepared to fight to the bitter end, to capitulate. The strategy was to demonstrate overwhelming force of the Allied forces through the deadly deployment of atomic ordnance. Right or wrong, the assumed need for this product existed early in the war, and a new product development effort was initiated to produce a weapon of such mass destruction that any war would be foreshortened.

Before we examine this as a product development effort, however, it is only fitting to recognize and acknowledge the massive pain, suffering, destruction, and loss of life this bitter fruit of research and development caused on Japanese citizens.

The development of atomic technology started with concept development by Albert Einstein early in the 20th century. Ernest Walton experimentally proved the concept of converting mass to energy in 1932. By 1939, the work of Otto Hahn and Fritz Strassmann showed that a net gain of energy could be achieved in the reaction. In 1939, several physicists from Hungary persuaded Albert Einstein to write to President Roosevelt to develop atomic technology into a weapon before Germany did. Roosevelt was driven to act by the letter, however, the desire was not accompanied by any significant assistance until Roosevelt named Vandevar Bush a few years later. By 1942, the technology development and Roosevelt's desire to productize it in the form of a weapon came under US army control. The program was called the Manhattan Project.

Technology Development

The development of atomic technology into the atomic bomb took several parallel steps, as the uncertainties of each step were daunting. There were essentially two approaches taken with experimental development taking place all over the country. The Uranium (U) 235 bomb was a fission device, engineered to create energy from mass by a gun type mechanism. Two pieces of uranium (enriched) not large enough to sustain a chain reaction separately were brought together in the barrel of a gun to form a super critical mass that exploded instantly. The simplicity of this straightforward design prompted its use without testing. This bomb was named Little Boy and was detonated over Hiroshima, Japan on August 6, 1945.

Prior to the projected success of the U235 unit, another approach was also being developed. A plutonium device, more desirable because it was easier to fabricate the basic elements for operation, could deliver more power. The plutonium could fission spontaneously or so rapidly that to achieve the required effect; the separate pieces had to be brought to critical mass faster than conventional ballistics. A new technique known as implosion was used to rapidly combine the plutonium into a super critical mass. This approach was tested at Alamogordo, New Mexico on July 16, 1945. This device was detonated over Nagasaki, Japan on August 9, 1945.

Time to Marketplace

The time to marketplace is somewhat of a misnomer for the atomic bomb; however, it can be concluded that this product development occurred lightning fast: A short 13 years from early experimentation to deployment!

It should be noted that this development and one other high profile development that is discussed later happened very rapidly without core technology available at the time of initiation, although both share US government financing with objectives clearly defined.

The US government spent a total of $2 billion between 1932 and 1945 in the development of the atomic bomb. Two billion dollars in those years as compared to a gross national product of less than $250 billion. By comparison, the Apollo moon landing, (the other fast track program) cost approximately $25 billion against a 1970 GNP of $1,039.7 billion. The atomic bomb development costs were less than one-half to one-quarter (adjusted) of the entire moon landing. What is even more amazing is that the US government kept the expenditure a secret from the American people!

Pervasive Use

The use of atomic bombs has been limited to the first two wartime uses, and the testing that occurred up until the Nuclear Non-Proliferation Treaty of 1968. In 1945, the United States was the only country to possess these devices. By 1949, the USSR possessed them, Great Britain in 1952, France in 1960, and the Peoples Republic of China in 1964. India possessed the atomic bomb by 1974. Despite the treaty, several countries are believed to possess the technology, including Israel, South Africa, North Korea, Iran, and Pakistan. Unfortunately, the Soviet superpower had weapons deployed in their republics, which are not under Moscow's tight rule anymore, and the republics of Russia, Ukraine, Belarus, and Kazakhstan are further uncertainties in the nuclear chess game.

Impact on Society

The real benefit of the atomic bomb is that it is so powerful that it was considered a doomsday machine and never meant to be used, only to be used as a threat. The nuclear nightmare essentially paralyzed the world for over 40 years. The real tragedy of the atomic bomb was its use on two occasions and the tremendous expense of all the countries in developing the technology and the cost of its proliferation throughout four decades.

FREEZE DRIED FOOD 1946

Background

Another smaller, less visible invention was the process of freeze drying. Invented by Arsene d'Arsonval and F. Bordas in Paris in 1906, it was similarly discovered by an American, Shackwell, 3 years later. A Swiss laboratory had developed coffee based on the freeze-dried process in 1934, although it was discovered early in the century, it wasn't until 1955 that freeze drying entered the food industry.

Technology Development

In 1946, E.W. Flosdorff demonstrated the process could be applied to the preservation of food, such as coffee, orange juice, and meat.

Time to Marketplace

Evaluating the direct invention time for this technology is not straightforward, since the time from original process of freeze drying in 1906 and the beginning of widespread usage in 1955 would represent 49 years to the marketplace.

Pervasive Use

The use of freeze-dried foods is in relatively widespread use.

Impact on Society

The impact on society may go unnoticed; however, the technology offered preservation techniques, which were not in widespread use at the time.

MICROWAVE OVEN: 1947

Background

The microwave oven was a spin-off of the US development of radar. The basic technology was discovered in 1946 and reduced to product in 1947. The discovery was made by accident when Percy Le Barron Spencer was experimenting with the emission of short wave electromagnetic energy. Working for the Raytheon company, he observed that a candy bar he had in his pocket melted in the presence of this type of energy.

Technology Development

Microwave oven technology developed rather rapidly, and the Raytheon company developed and marketed the first microwave oven in 1947. The "Radarange" was initially designed for commercial use.

Time to Marketplace

The time to get to the marketplace was almost instantaneous. One year from observance of the phenomenon to introduction of a product is incredibly fast for the timeframe in question. The domestic use of the microwave oven didn't start in the US for another 18 years. In 1965 domestic units were available and in 1972 units went on sale in Great Britain.

Pervasive Use

The microwave oven has changed the way we cook in our society. Speed and convenience outweigh quality. New types of food, packaged specifically for microwaves, have surfaced.

Impact on Society

The microwave oven not only changed the way we cook foods, it has also changed the way foods are prepared so that they can be cooked in a microwave oven. These units have also allowed certain foods to retain minerals and vitamins during the cooking process, where traditional methods used to rob foods of these nutrients. Microwave ovens speed up the time spent in preparing meals and consolidate the pots and pans and serving dishes into one item.

THE TRANSISTOR: 1947

Background

The transistor is probably one of the most pervasive and high impact inventions of the 20th century. By itself, it was a nebulous invention. At the time, its invention probably went unnoticed by the general public. However, it allowed the development and implementation of technology on a scale so grand that it is not definable in numbers.

This new era started in 1947 at the AT&T Bell Laboratories. Three American research scientists, John Bardeen, Walter Brattain, and William Shockley, demonstrated the transistor. The new device was designed to replace the vacuum tube used in radio electronics at that time. Today, the transistor is the fundamental electronic building block in virtually every piece of electronic gear; furthermore, its development made products possible that could not exist without transistor technology. In 1956, these three scientists were awarded the Nobel prize in physics for their discovery.

Technology Development

The time from proof of concept to implementation was less than 10 years. The first transistor radio appeared in 1955. Through the use of transistor technology, the battery life of the portable radio increased dramatically over vacuum tube technology. The real growth of the transistor occurred in the 1960s to the present day. The devices were improved dramatically and the beginnings of integration started to occur. The silicon wafer on which these devices were fabricated became the means for interconnection and therefore the beginnings of small-scale integration. In the 1970s, medium-scale integration occurred, and in the 1980s large-scale integration was possible. (Millions of devices fabricated and interconnected on a single chip!)

By 2012, it is expected that 1 billion transistors can be integrated on a single production die. The use of transistors and their interconnections made microprocessors possible and today are commonplace. And the integration continues. Since the invention of the transistor, there have been over 30 doublings of the density within the same chip area. Few technologies can parallel this progress.

The transistor grew in complexity, power ranges and base materials employed. There is a device and technology available for almost any application. Computer technology and power electronics development drove the development of smart motor control devices, making machine control ever more accurate and higher performing.

Time to Marketplace

The transistor came to the marketplace in 8 years from discovery to implementation, as previously stated; however, the devices grew in popularity and usage. This drove further development and fueled the demand for performance and integration. The performance and integration opened up new markets, which fueled more research and development and fueled more demand. The results are staggering as examined in the impact section of this summary. This technology is unlike others examined in that unparalleled speed of development and improvement in performance drove demand and usage that accelerated the development and performance.

Pervasive Use

The use of the transistor continues to accelerate. The transistor is not only on the logic side of electronics but also on the power side. There are an uncounted number of units in use, and an uncounted number being produced each day. Hundreds of manufacturing facilities with multiple processes each producing thousands of integrated circuits each day with thousands of devices on each I/C. In 1997, it was estimated that approximately one-half billion transistors would be manufactured each second! The cost has plummeted also with its use. Since its invention in 1947 the costs of a transistor fell from an initial $45 to $2 dollars through the 1950s. As the applications became more and more pervasive, the cost for a transistor today is .0005 cents!

Impact on Society

The impact on society is almost immeasurable. To illustrate the point of pervasive usage, the following scenario outlines the average American's daily encounter with the transistor.

At 6:00 AM, your alarm clock sounds to wake you up. You turn on the television via a remote control device. You fire up your computer to retrieve any fax messages that came from overseas last night. You read them while your coffee maker automatically brews your coffee. You get your voice mail messages off the telephone and answering machine; you clip on your pager to get any incoming calls.

You start your car with electronic ignition and fuel injection, back out of the garage, and close the automatic garage door with the remote. You stop at the gas station to fuel up and read the fuel pumped into the car on an electronic read out. Hesitant to walk to the service station's counter, you pay by the credit card at the pump. You now arrive at your office and insert your code to enter the premises. Your cup of coffee still in hand, you check the manual fax machine and your telephone messages. You need a refill so you go to the company vending machine to get a cup of brewed coffee.

You are now ready for your first meeting of the day. Not content with the bright lights this early in the morning; you tone down the brilliance with the dimmer control. You check the order-entry computer and send an electronic message to all your agents. The piece of equipment you ordered last week arrives, and you sign the deliveryman's electronic clipboard that he uses to keep track of the shipments and receipts.

You are getting hungry as lunch approaches and order pizza from the local vendor. As you call, the pizza shop's caller identification device identifies you as a good customer. After the pizza arrives, you sit down in the conference room to enjoy your meal and turn on the television. You see an advertisement for a sale at the local department store. On your way home you get the car washed and stop by the sale at the department store. Worried that you may be a little late, you call your spouse on your cellular telephone to let your family know you are on your way home.

These are only a fraction of the devices we touch each day that employ transistor technology. Transistors have reduced the size of the equipment we use and made equipment not before thought possible in years gone by.

THE VIDEO CASSETTE RECORDER: 1948

Background

The video cassette recorder dates back to the early 1950s. Previous work on formats and recordings done in 1948 served as the basis; however, Mincom, a subsidiary of 3M company and RCA and Bing Crosby Enterprises tried through several demonstrations to play these recorded images. All had similar drawbacks in the quality of the reproduced images.

It was in the later part of 1951 that Alexander M. Poniatoff, the founder and president of the American-based Ampex Corporation, put a team in place to resolve these problems. Again, the results were mediocre. Another strategy was attempted in 1954. With better results, the team presented the refined concept to Ampex management in 1955. Product launch of the first video tape recorder was in April 1956. The first color unit was developed 2 years later, with a transistorized version in 1963. Today, this machine has worldwide acceptance and usage.

Technology Development

The technology was developed as time went on. With a good technological start and early marketing, Ampex had established a market for the machine by 1958. But as a classic lesson in the importance of the pursuit of the business, the Japanese embraced the technology to capitalize on it. By 1958, Toshiba had announced the first single head recorder, in 1959, JVC developed the two-head recorder. By 1962, Shiba Electric (now Hitachi) introduced their first transistorized version. By 1964, Sony marketed their first video recorder to the general public. By 1965, Hitachi went small and portable with their first size-reduced version.

The 1970s saw many improvements with German, Japanese and American manufacturers participating in the market by 1980 the first camscope (the generic term for video camera) combined the camera and videocassette recorder in one unit. By 1982, the top five producers agreed on a standardized design. It would have a built-in videocassette recorder and an 8-mm camera-scope with a tape length of 1 hour.

In the 1980s, continued evolution resulted in the cost reduction of the video camera. Also, the video recorder came down in price several-fold.

Time to Market

If we use the year 1980 as the year that the market for this technology really started to move, then the time to mass market was some 30 years. This technology was somewhat different than others in that the Japanese were really the ones to capitalize on it. Americans lost the expertise for producing the electronics, even though we invented the technology. It was the shape of things to come for many products. The lesson learned was the importance of the pursuit of a market and that you only need abdicate the market opportunity for a short while before others are there waiting to take it away. In this example, 30 years of preliminary work could have and should have been harvested.

Pervasive Use

The pervasive use of the embodied technology is widespread all over the world. It is the basis for an entire new entertainment industry. Prerecorded movies and home movies have become a mainstay in our lives.

Impact on Society

The impact on society has been enormous. An estimated amount of time spent by adults 18 years old and older was 57 hours per year using this technology. These devices have enhanced our ability to entertain and be entertained. They allow the average person to capture moments of family life and easily edit or erase unwanted parts. Capturing these images is affordable in comparison to the standard film-based movie camera. A roll or cassette in the movie camera costs $3 plus several dollars to develop. The standard VCR cassette cost $2-3 and that is the only cost. The elapsed running time is a twenty-fold difference. The videocassette is 2 hours, whereas the movie roll of film was only 3 minutes.

MERCURY, GEMINI, APOLLO PROGRAMS: 1959 TO 1969

Background

Without much discussion, the 1969 landing on the moon was probably one of humankind's greatest achievements. It rivals the Egyptian pyramids in grandeur, the Hoover dam in immensity, and the development of the atomic bomb in speed, from concept to practical application. It represented the best in terms of amassing personnel, technology, funds, support, and execution in order to meet the objective set forth by President Kennedy. Originally conceived as a competitive venue between Soviet and US superpowers, the real benefit to mankind came in the spin-off technologies employed through the years. From a pure product development effort consideration, it remains unchallenged as the epitome of achievement.

One of the parallels that can be drawn from this program relating to any product development effort is that the mission was eminently clear. "Land a man on the moon and return him safely to earth by the end of the decade." A mission clearly defined becomes all that more achievable. It is the mission statement against which all else is measured and any uncertainty is evaluated.

The programs represented the finest example of personal sacrifice on behalf of the participants, including the three astronauts that gave their lives in the Apollo 1 fire.

From a purist marketing/product development perspective, it also represented probably the first time that a government initiated and prosecuted a program with no positive direct impact on society attributable at the onset. As we shall examine later, a monumental impact did and is still now occurring.

Embodied in three sub-programs the landing on the moon started with the Mercury program, progressed to the Gemini program, and proceeded to the Apollo program.

Technology Development

The program was structured into three elements that were concentric in nature. The first element was Mercury. The objective of the Mercury program was to basically launch a man into orbit and retrieve him from space. The program was embodied in 6 flights of increasing duration and scientific and medical tests and observations. Starting with a sub-orbital flight in May 1961 and culminating with a 22-orbit flight in May 1963, the Mercury program resulted in 53 hours, 55 minutes, and 27 seconds of flight experience toward the goal of a lunar landing.

The Gemini program was a 2-man vehicle designed to test and qualify the fuel cells, maneuvering procedures, docking procedures, extravehicular activity, and long duration flight, in addition to other elements required for the landing. A total of 10 flights completed this program and served as a test bed for many systems to be used on Apollo. The Gemini program added momentum to the lunar landing program. With 969 hours, 52 minutes, and 4 seconds, the stage was set for Apollo. The facts now in evidence do not properly underscore the effort to prosecute the results. There was monumental number of tasks executed to obtain these results.

With the Apollo program early in 1967, NASA and the contractors had established a culture of "go" fever within the entire organization. This resulted in design compromises and safety oversights. The unfortunate loss of three astronauts refocused the program onto a safety orientation, in addition to achieving the goal.

The goal was clearly stated to land a man on the moon and safely return him to earth. To examine the Apollo program, we will review total hours to achieve the goal and subsequently review the additional hours in the Apollo program.

The accumulated time in space to prosecute the lunar landing was 776 hours, 8 minutes, and 37 seconds. The total space-time after the first landing was 1466 hours, 28 minutes, and 0 seconds.

At the peak, between June 1964 and June 1966, there were over 400,000 personnel employed to prosecute the lunar landing!

The Mercury program cost approximately $392 million, while the Gemini program cost $1.281 billion. The Apollo program cost approximately $21.349 billion, for a total expenditure of $23 billion.

Ask anyone involved with the program, and you will find that the primary driving force behind every move in the program was time. Did it save time? If it did, it was a worthwhile expenditure. With short duration market opportunities, we could learn a lesson from this philosophy; time is of the essence in new product development.

Time to Marketplace

The time to marketplace can be evaluated from two different perspectives. One is to examine the time from commitment to lunar landing. The other is to examine the time from development of the various technologies to infiltrate the mass market.

The time required to prosecute the lunar landing was established by President Kennedy. Essentially, it took 10 years from the germ of the idea to the actual landing. The pathway was not without adversity; there was an 18-month moratorium on space travel during this time in order to redesign the Apollo command module after the Apollo 1 fire in 1967. An introspective and cautious NASA launched the Apollo 7 craft into orbit in October 1968. So successful was the flight that NASA again planned a bold move. Apollo 8 would leave earth orbit and orbit the moon and then return to earth. From that point on, every mission until the Apollo 11 landing mission was stretching the envelope to achieve the goal. From the recovery mission of Apollo 7 to test the new spacecraft in October, 1968 to a July, 1969 lunar landing, 4 missions were executed in a short 9 months!

Unfortunately the next mission seemed anticlimactic. Complacency set in and again a near tragedy of Apollo 13 shook the core of the effort back to reality. We were sending three human beings some 200,000 miles from the safety of their environment to walk on the moon with artificial life support and very sensitive and complex equipment operating in a very hostile environment! Once again, technical introspection was required; however, there was only a short delay this time to the next mission.

Pervasive Use

In terms of pervasive use, we really don't use the 1969 lunar landing per se. What the program generated was spin-off technology that humankind's marketplace took and developed products to improve life here on earth. This was the collateral benefit that was envisioned only by a select few at the time. I doubt if any of the political leaders making the financial decisions at the time realized the immense impact of the program in future years. Some would say that the program generated up to 400,000 jobs during the height of the effort. But this was not the legacy of Apollo; the legacy was the jobs created by the new products developed from the spin-off technology.

If we examine the use of the technology developed, we see it touches many aspects of our daily lives. Integrated circuits for virtually every electronic device, photographic imaging for medical use, insulation and thermal management products, improved television cameras, tools, fire protection systems, construction materials, and many more all have their origins rooted in the Apollo program. Since it was not a direct product development for the mass market, it is somewhat difficult to measure the pervasive use nonetheless it is overwhelming. In addition, it also facilitated and improved the methodologies for translating the raw technology into usable products for the mass market.

Impact on Society

At the time, the government was being asked to justify the expenditure. Arthur C. Clarke, a noted science fiction writer, stated that the opportunity and financial reward to our society in the next 20 years will return 10-fold. How right he was! A later section outlines the benefits of the Apollo space program to humankind and what our world would be like without Apollo.

In terms of the financial benefits, few investments have been better. The spin-off technology embodied in products may possibly generate enough revenue on corporate taxes to pay for the entire program annually. Point in fact, the revenue and tax base generated by the integrated circuit/electronics industry alone would probably provide the offset by itself.

The program also generated practical methods for managing high technology and multiple team projects. This becomes more valuable as the technology base expands and the usage by multiple teams increases.

The Mercury, Gemini, and Apollo programs set a new benchmark for achievement. Over 30 years later, it still is a means for comparison against many things. It still represents the epitome of goal setting and success. More importantly, it united the earth for a brief moment, as we saw a human being walk on the moon. As the amoeba slithered on to the land to venture into new territory, so too was the promise of Apollo. It was about venturing beyond our comfort zone to achieve greatness, to take risks, to manage risks, to confront adversity, to rise above, and to achieve another day. The same principles apply to new product development, no matter how large or small.

The lessons of Apollo are numerous, including its relation to new product development, strategic planning, and the resolve to carry out plans. It is also an example of an event-driven program. When the event was over, the program lost momentum. This can occur with single-product business development and the lack of momentum, continued development, and enhancements. These are bound to contribute to eventual failure of the business enterprise. With regard to this product development effort, the original plan called for a manned landing on Mars by the end of the century.

Each of the product development programs discussed are unique, by the markets they served, the driving forces involved, the circumstances behind the development and the critical factors in their success. They only represent a sampling of several programs, and are not meant to lend disproportionate importance to one over the other, nor over any not listed, They are simply cited for illustration purposes in discussing principles and precepts in the new product development field.

As each of these programs was unique, so are any future programs you may encounter. Each circumstance must be evaluated, facts weighed, plans made and program guided to success factoring the uniqueness that applies to it. For each of these programs, there was no magic formula for success. Each had to champion the program, navigate the uncertainties and the setbacks and execute the desired result. The same is true for any new endeavor.

NEW PRODUCT FLOPS

This section details a few new product failures that have occurred in the marketplace. These are presented for the sole purpose of illustration in examining the causes from a historical perspective. As discussed in this chapter, new products fail for a variety of reasons, and new product successes can be achieved by addressing the reasons and accounting for them up front in the product definition and development execution phases. It is also important to discuss outright failure versus lack of staying power in the marketplace.

FORD MOTOR COMPANY, 1959: THE EDSEL[*]

Original Objectives

The original marketing objective for the Ford Edsel was to introduce a futuristic car with many innovations. It had horizontal coolant flow in the radiator, as well as relocation of the exhaust away from the passenger compartment.

Driving Forces

The driving forces behind the program were to uplift the look and perception of the fleet. It was to be "the car of the future."

Capitalizing on the Opportunity

The Ford Motor Company tried to capitalize off of the prosperity of the 1950s by introducing an exciting futuristic automobile that the public would clamor for.

Causes for Failure: Historical Perspective

The causes for this product failure were many. The car was beset with quality problems. The hoods stuck, and the power steering had problems. It was overpowered, had a lot of gadgetry, and had more expensive accessories than any other car in its class. The result to the Ford Motor Company was a $250 million loss.

Lessons to be Learned

The lesson to be learned from this example is to take care in assumptions, especially when it comes to buyer's tastes. The buyer is the ultimate decision-maker. Just as Mother Nature has the final say on engineering issues, the consumer has final say on marketing issues. Initial poor quality on a new product is very difficult to overcome; many, such as the Edsel, never do overcome the consumer's perception.

DUPONT'S CORFAM[†]

Original Objectives

The original objective behind DuPont's Corfam was to create synthetic leather for shoes. DuPont predicted in 1964 that by 1984, over 255 of America's shoes would be made of Corfam.

---

[*]Edsel is a Trademark of the Ford Motor Company.
[†]CORFAM is a trademark of the Dupont Company.

Driving Forces

Material cost and increasing labor costs for natural leather drove the search for a substitute.

Capitalizing on the Opportunity

DuPont wanted to capitalize on the success of what nylon did for stockings. By inventing synthetic leather, shoes could be made cheaper and faster, thereby driving volume and profits.

Causes for Failure: Historical Perspective

Leather was simply much better. The time on the market for the material was 7 years.

Lessons to be Learned

The lessons to be learned in this example are that one success doesn't necessarily produce another. Just because a synthetic substitute for stockings was a success, didn't mean that all synthetic substitutes will have market acceptance. The estimated loss to the company was $250 million.

POLAROID POLAVISION*

Original Objectives

The objective behind the Polaroid Polavision was to extend the wet chemistry used in still photographs to home movies. In this way, users could take movies and play them back instantly.

Driving Forces

It seemed a natural extension of the instant pictures that could be used in the home movie market. The driving force was the inconvenience of having to have the delay in developing the movie film and then playing back the movie on a conventional projector.

Capitalizing on the Opportunity

The Polaroid Company wanted to capitalize on the portable wet chemistry process that had been used in still photography for years.

Causes for Failure: Historical Perspective

The primary cause for failure was that the introduction of videotape in the form of a video camera was easy to use, used standard cartridges from a home VCR, could be played back on a VCR, and was better quality. It's a classic example of technological displacement. The better mousetrap won out.

---

*POLAVISION is a trademark of the Polaroid Company.

Lessons to be Learned

The lesson to be learned in this example is to be cognizant of technological displacement when developing a new product. While you are developing something, your competitor is not standing still. Moreover, an unforeseen competitor may be in the wings eyeing up your targeted segment with technology that may obsolete yours.

RCA's VIDEODISC

Original Objectives

The original objective behind the RCA videodisc was to capture a portion of the growing video recorder market. Movies prerecorded on videodisc could be played at home with good quality.

Driving Forces

The driving force behind this development was the growing home entertainment industry in the 1980's.

Capitalizing on the Opportunity

RCA was planning on capitalizing in this growing market segment by offering a new and alternative technology to the videocassette medium.

Causes for Failure: Historical Perspective

The primary cause for the failure of the program was the fact that the video cassette recorder could record television shows off of the television and the videodisc could not. The mass market was using the VCR for two purposes: the first was to time shift television shows around individual schedules. The second was to play prerecorded movies. The videodisc could only do one of these things where the VCR could do both.

Lessons to be Learned

The lesson to be learned is that new technology or even better quality does not always get the market acceptance one desires. The benefits of the new technology must significantly outweigh the risk and additional expense of going to the new technology. The financial impact: 500 million-dollar loss.

IBM's PC Jr

Original Objectives

The objective behind the IBM PC Jr was to increase portability of a personal computer.

Driving Forces

The driving forces were changes in the demographic makeup and how people worked and their needs for portable information.

Capitalizing on the Opportunity

IBM tried to capitalize on their eminent position in the personal computer market. Their strategy was to leverage off of their strength in this area to launch a more portable solution to the mass market.

Causes for Failure: Historical Perspective

There were several causes for the failure of this product. The keypad was objectionable to the user, the price was unattractive and the microprocessor used to platform the machine was slow and lacked performance. In addition it was launched late. All these factors together contributed to a $40 million loss for IBM.

Lessons to be Learned

The lesson to be learned is that an established company with an established reputation needs to keep up that reputation. This could be considered IBM at its worst. Late launch, poor performance, and non-user-friendly focus were antithetical to the IBM credo.

NEW COKE*

Original Objectives

The original objective behind the New Coke was to chase the Pepsi leader with the sweeter formula to shore up declining market share.

Driving Forces

Coke was loosing sales to the sweeter Pepsi formula and the fresher advertising.

Capitalizing on the Opportunity

Their strategy was to adapt their existing formula to reach Pepsi drinkers and thereby stop the declining market share.

Causes for Failure: Historical Perspective

Their strategy was flawed in that although it had some chance of success to reach Pepsi drinkers as well as other cola drinkers with the altered formula, Coca-Cola failed to consider the impact the change would have on their loyal drinkers. How would they feel about the age-old formula being altered? Needless to say the plan backfired.

---

*COKE is a trade mark of the Coca Cola Company.

Lessons to be Learned

The lessons to be learned were quite simple: don't forget your loyal customers when anticipating growing your market. You must please your existing customer base and at the same time, either through positioning or other means, gather new market share without loosing or alienating the existing customers.

The financial impact to Coke Cola was $35 million loss. In their defense, however, Coke quickly rectified the situation by withdrawing the mistake after 77 days.

R. J. REYNOLDS' PREMIER*

Original Objectives

The original objective behind the Premier was the smokeless cigarette. It was the result of seven years of research to come up with a cigarette, which would satisfy the needs of the smoker but not have the secondhand smoke.

Driving Forces

The diving forces behind the development included a growing antismoking movement as well as concerns about secondhand smoke for non-smokers.

Capitalizing on the Opportunity

R.J.Reynolds wanted to capitalize on its existing customer base as well as please non-smokers. It also could plow profits from the existing cigarette sales into the development of the smokeless cigarette.

Causes for Failure: Historical Perspective

The historical causes for the failure of Premier included that the company created something that didn't burn, didn't emit smoke and tasted bad.

Lessons to be Learned

The lesson to be learned is never to fail to recognize the needs and wants of your purchasing customer base. The smokers wanted to smoke with taste and flavor in their cigarettes. Don't expect them to buy a product that lacks these attributes voluntarily.

These are only a few of the new product failures that have occurred recently. Some were true innovations like the DuPont Corfam and RCA videodisc. Others were technological developments that missed the evolution timeline like the Polaroid Polavision. Still others were outright marketing mistakes by all classical definitions like New Coke and the R. J. Reynolds Premier cigarette

---

*Premier is a trademark of the R. J. Reynolds Company.

MAJOR CAUSES OF NEW PRODUCT FAILURE

After reviewing some of the failures, the following discusses the major causes of new product failure. This list can be used as a checklist for product development programs to verify approaches to a market and business and execution of the programs.

New Product/Business is Off Strategy

This is one of the most important reasons for failure of a new product effort. The company may find itself off strategy with the new product and no way to effectively tell the story in the marketplace or effectively promote or sell the product. This generally happens when management takes a hands-off approach to planning and does not do an effective job of screening new product opportunities and lay them out within a strategic framework. The company then finds itself with a poorly positioned product within the sales force or channel, an ill defined means or set of executable tactics to sell in the marketplace, and a huge expenditure that needs to be amortized. Consequently the sales channel cannot effectively create a sales volume and the product fails to succeed.

The issue of getting a product on strategy needs to be addressed up front and early in the product development cycle. Programs that are not defined or positioned well generally drag on, gather expense and can never be amortized. The organization just expends energy to no profit. This starts with management and later manifests itself as financial problem that must be eventually dealt with by senior management.

Inconsistent Organizational System for Evaluating New Product Ideas

The ineffective or inconsistent systems for evaluating new product ideas consists of three basic causes as follows:

1. Arbitrary Criteria

This cause generally is due to lack of strategy against which the idea is evaluated. Each idea is evaluated against the comfort level and interest level of an individual or group of individuals. There is no objectivity or financial evaluation necessarily and many time's good ideas are not pursued and poor matches for product opportunities and companies are pursued.

2. Poor Procedures

This is another sign of poor management. It generally indicates a personalization of the decision rather than some objective means. When this personalization takes place, the company is then limited to the strengths and biases and weaknesses of the individual making the decision. Worse yet is the case where poor procedures result in inconsistent decision making, second-guessing, false starts and stops and loss of momentum in a program.

3. Lack of Coordination of Departments

This cause is a very dangerous one. It can indicate a fractured organization, where departments act as individual islands with little regard for their internal customers. Eventually this manifests itself with the customer base in lost sales. The task of galvanizing the organization into a working group to execute the development is the primary role of the product manager or product champion. In the absence of senior management providing this environment, the program leader must. If the group of departments cannot be galvanized, severe problems result, and the program should be halted and the organization straightened out, otherwise, a select few individuals will be expending a tremendous amount of energy without progress or leverage.

Poor Market Planning

1. Wrong Positioning

A failure can occur when the product line is not positioned properly. This happens when the fit between the product and the demand for the product features is inexact. There is always a fine line between the over featuring and under featuring the product. If it is off, the fit is not correct. In the case of over featuring, the marketing effort pushes the product into the hands of the customer base and drives more features than are required. In the case of under featuring, the goods delivered are not sufficient to satisfy the need.

2. Poor Segmentation

Poor segmentation is another cause where, the market segment to which the product is targeted is not defined properly. Is this case, the match between the versions or models of product do not effectively match up with the marketplace. An example of this would be designing a line of automobiles: a entry level segment for young people, a roomy practical segment for mid aged people, and a luxury segment for older people. The mistake in segmentation would be to unilaterally use the same power platform for each. Designed for the worst case load of the luxury vehicle, the entry-level car would be overpowered. Although it would be exciting for that 16-year-old new driver, most parents would opt for a tamer performance vehicle, making that small entry-level vehicle with the overpowered drive train poorly positioned in the marketplace.

3. Underbudgeting (Capitalization)

Under capitalization is a frequent cause in companies where there are a lot of groups, divisions and programs. In this case, there is a management problem in selecting the best program. Consequently, several programs are initiated, in the hope that one of them will rise to the top and be successful. Unfortunately, all this type of activity does is diffuse the development effort and expend energy. They generally drag on in time because the appropriate horsepower has not been brought to bear on the program. This delays introduction and causes missed opportunities. If a program is good and right for the company, it is the management's fiduciary responsibility to fund it properly.

4. Overpricing

Overpricing is generally the result of another problem. It generally occurs when the cost of the product gets out of control. At the end of the program when final pricing is set, a squeeze play occurs between the cost, required margin, the price and tolerable pricing level in the marketplace. The manufacturer must try to recover the cost of development, and the fixed and variable costs associated with the product itself. This sets the selling price high and volume is reduced. Unfortunately, the situation is a non-correctable one. It simply must be avoided.

5. Slow Development, Missed Opportunity

This is self-explanatory. Every program must have a certain momentum to be successful. Lack of sufficient momentum results in delayed product introduction and lost market share. In new product development, we aim at a moving target, a window of opportunity; therefore, it is incumbent to be timely in the response. Lack of the ability to deliver the goods at a fair market price in a timely manner compared to competition, will result in missed opportunities.

Lack of Product Distinctiveness or Customer Benefit

The lack of product distinctiveness is an indication of a dysfunctional market analysis and product planning activity. The "me too" product rarely succeeds because it has plenty of competition and a severe loss in market timing. It simply is too little too late in the market and has little to offer the customer base. It also is at a disadvantage from a cost viewpoint.

If the marketer of a "me too" product tries a low price strategy, they will be met with an entrenched competitor that has the benefit of progressing down the learning curve with volume generated. This will allow a pricing action by the entrenched competitor immediately after the "me too" is introduced. In product planning, the new product development effort must lead the market target. The development takes time, and as such, the market does not stand still.

The market and your competitors will be dealing with advanced expectations and advanced features when your product is introduced; therefore, you must lead the target.

The specification outlines the features and benefits to the customers. Either the specification is off target or the development team missed the target.

Top Management Push in Spite of Supporting Evidence

This is another example of a dysfunctional company. Here, a top manager or senior manager uses rank and position in the company to bypass the normal evaluation criteria.

Generally classified as pet projects, these programs can become successful if the senior manager is correct in their original assumption. If they are not correct, these programs generally fail. Unfortunately there are two problem with this approach. One is that if the program was that good, it generally will pass muster in the original evaluation. The second danger is the lost opportunity that the organization spends on the program instead of working on the programs they are supposed to. The activity diffuses the effort required to prosecute the qualified programs.

Poor Market Size Assessment, Forecasting, and Research

This can be a fatal flaw in the product planning process. Product development must be based on facts, data, and somewhat concrete numbers. Wishful thinking or "leading an unsuspecting market" will not result in the volume required justifying all the expense to develop the product. Many times senior management lulls themselves into a false sense of security by rationalizing this step, however, fiduciary responsibility demands that you select the best product opportunity with the most payoff that makes strategic and tactical sense for what your long term goals are. If this is compromised, the future of the company is compromised.

Incomplete or Inaccurate Product Definition

"We'll get started now, and we'll work out the details later." How many times has management used this line to cover for lack of due diligence in product definition? Unfortunately, the key to product success is to create understanding and agreement throughout the process, from beginning to end. This includes understanding of the marketplace, the technology, and timing and execution. Lack of understanding in segmenting the product or outlining the offering contributes to eventual failure.

Lack of sufficient product definition at the onset results in creeping functionalism and cost pressures later in the program. If a definition consensus cannot be reached, perhaps test the market or customer base to confirm the approach.

It is permissible to get started with near term product definition step, however it is imperative that this step is taken. All to often, this step is glossed over, or worse, engineering locks on to an approach which may not be the best means to a marketing end for the business.

Product Design (Reliability Problems)

Reliability problems are often the result of rushed schedules, changing specs, and cost pressure. The lack of product reliability is simply inexcusable in today's marketplace. There exists virtually any number of tools in the development arena to enhance and ensure product reliability. There is a constant fight to keep product cost low and this can contribute to reliability problems, however, this issue is not a subjective criteria that gets evaluated amongst the customers. Product reliability is a specification item that is reduced to numerics. Failure of a product in the marketplace due to reliability problems is simply missing a critical specification item. It cannot be added on, it is very difficult to recover from.

Unexpected Product Development Costs

Most product development programs are somewhat predictable. In addition, with an established team that has a few programs under their belt, the costs and times are fairly well known at the onset. In these types of programs, an additional 20 or 30 or 40 percent of development cost will not affect the return significantly. The items that affect the return are volume, factory cost, and timing. This will become more obvious as we examine the math of a typical return calculation

In true innovations, where the technology is unknown and much of the program depends on sequential successes of this unknown technology development, there can be severe cost overruns as far as pure development costs. In most cases, these cost overruns manifest themselves as timing problems that force the company to miss opportunities in the marketplace. The costly development is a corporate cash and expense issue, not a direct cause of product failure.

This common cause should be renamed "unexpected technical difficulty," which may result in cost overruns to surmount, and result in dangerous delays.

Missed Factory Cost Target and Subsequent Price Impact

This common cause of new product failure is a result of carelessness in the planning and execution stage. It is not an easily recoverable step other than starting over. A products cost structure forms the basis for the feature set and the position in the marketplace. If these elements are less than optimized, there is little chance for success long term.

The poor cost structure removes any flexibility in the future and limits the manufacturer's options when doing marketing battle. In order to recover the initial development cost, the pricing is adjusted prior to introduction. This puts undo pressure on the marketing effort as the pricing strategy now becomes a dominant element of the initiative rather than trying to gain market share.

The high price strategy then contributes to the lowered volume, which in turn reduce market share. The entire effort is then out of balance and contributes to product line failure.

Unexpected Intense Competitive Response

Never underestimate the entrenched competitor's desire to hold on to their business. Many new products fail because a competitor digs in their heels and decides to retain their market share at all cost with all else being equal. In these cases, the entrenched competitor has a box of marketing tools to use to protect their turf.

Most importantly, they have traversed down the learning curve while you are trying to enter it. They have a trained sales channel with inventory and can place goods at the customer while you are trying to gain momentum. Finally, they are more familiar with the risks and dangers in the marketplace where your may have to "feel your way".

LESSONS FOR FUTURE PROGRAMS

In this section, we have examined several new product failures and have reviewed their objectives, causes for failure, and made observations on errors made. If there was a summary for the "lessons to be learned" in new product failures and their causes, it would include the following caveats.

- Carefully execute a marketing assessment of the opportunity and plan out a well-considered development plan.
- Continuously monitor the progress of the development effort checking original assumptions against real time data.
- Effectively react to changing market conditions in order to protect your development investment in effort, time, cost, and momentum.
- Start each development program with enough of an advantage such that as conditions change or actual results fall short of initial expectations, there is some contingency room in the overall effort.
- Never make the mistake of believing that you're the only company working on a development and that all others stand still while you're executing your program.
- Do your part in galvanizing the organization to embrace and execute your new product initiative. Have the organization contribute energy to the effort rather than use only your energy to move the organization to action.
- Make sure the program delivers the goods at a price that the marketplace is willing to pay and in a time frame that allows success.
- Don't intentionally or inadvertently manufacture sub-standard goods.
- Watch your competition, they are watching you!
- Target your customer base, and cater to it. Do not cater to collateral pressure. The collateral elements are not paying the bills or contributing to your profits.

THE FASTER CHANGING WORLD

The world is an ever-changing place, and the changes that we are currently undergoing are occurring at an ever-increasing pace with wider swings and larger differences. The process of new product development must factor in this uncertainty. Product iterations are coming faster and demand for changes is more prevalent than before. Not all of these changes are necessarily good nor do they necessarily affect a program; however they must be recognized, evaluated and factored into a program and possible modifications of that program, for it to be successful in the marketplace. A presentation of man's trends in information and knowledge will be examined as well as futurism, information availability, connectivity, and speed.

CHRONOLOGICAL TIMELINE OF INFORMATION MULTIPLES

In terms of evaluating mankind's knowledge base, each discipline needs to be evaluated for their progress and accumulated knowledge. Many disciplines have a half-life of only three to four years. Since this accelerates with the passing of time, the amount of change in certain disciplines is enormous. Consider for a moment some of these fast changing technologies like computer science where the level of available knowledge doubles every couple of years. Not only does it set the pace for the manufacturer and designers of these products, it challenges the users to keep up also.

The following represents an attempt to look at data that is representative of the growth of knowledge, so as to flavor the discussion of how new product development must operate in these fast changing times.

US Patent System

The US patent system can be considered a measure of knowledge by certain standards. All else being equal, in terms of novelty, technological innovation, and application, the number of patent applications and issues are a measure of a level of knowledge.

Figure B1, A and B, are graphs showing the number of US patent activities from the period from 1963 to 2003. This 38-year period in our most recent history shows a steady increase of the number of applications and issues.

Population and Time

Another measure of humankind's knowledge level is attributable to the sheer growth in the population. More people, more studies, more ideas, more creativity, and more time ultimately generate more knowledge base from which to draw. In addition to these, the advent of computers and Internet communication schemes accelerate information transfer at lightening

TRENDS OF PATENT ACTIVITY IN THE UNITED STATES FROM 1963 TO 2001						
YEAR	PATENT APPLICATIONS	PATENTS ISSUED	YEAR	APPL	ISSUE	% FOREIGN
				% Increase	% Increase	
1963	90982	48971	1963	0	0	18
1964	92971	50389	1964	2.2	2.9	18
1965	100150	66647	1965	7.7	32.3	19
1966	93482	71886	1966	−6.7	7.9	20
1967	90544	69098	1967	−3.1	−3.9	21
1968	98737	62713	1968	9.0	−9.2	22
1969	104357	71230	1969	5.7	13.6	25
1970	109359	67964	1970	4.8	−4.6	26
1971	111095	81790	1971	1.6	20.3	28
1972	105300	78185	1972	−5.2	−4.4	30
1973	109622	78622	1973	4.1	0.6	30
1974	108011	81278	1974	−1.5	3.4	33
1975	107456	76810	1975	−0.5	-5.5	N/A
1976	109580	75388	1976	2.0	−1.9	N/A
1977	108377	69778	1977	−1.1	−7.4	36
1978	108648	70513	1978	0.3	1.1	37
1979	108209	52412	1979	−0.4	−25.7	37
1980	112379	66170	1980	3.9	26.2	38
1981	113966	71063	1981	1.4	7.4	39
1982	117987	63276	1982	3.5	−11.0	40
1983	112040	61982	1983	−5.0	−2.0	41
1984	120276	72650	1984	7.4	17.2	42
1985	126788	77245	1985	5.4	6.3	44
1986	132665	76862	1986	4.6	-0.5	45
1987	139455	89385	1987	5.1	16.3	47
1988	151491	84272	1988	8.6	−5.7	47
1989	165748	102533	1989	9.4	21.7	47
1990	176264	99076	1990	6.3	−3.4	47
1991	177830	106698	1991	0.9	7.7	46
1992	186507	107394	1992	4.9	0.7	45
1993	188739	109747	1993	1.2	2.2	44
1994	206090	113587	1994	9.2	3.5	43
1995	228238	113834	1995	10.7	0.2	43
1996	211013	121696	1996	−7.5	6.9	43
1997	232424	124068	1997	10.1	1.9	44
1998	260889	163147	1998	12.2	31.5	44
1999	288811	169086	1999	10.7	3.6	44
2000	315015	175980	2000	9.1	4.1	45
2001	345732	183975	2001	9.8	4.5	46

Figure B1, A

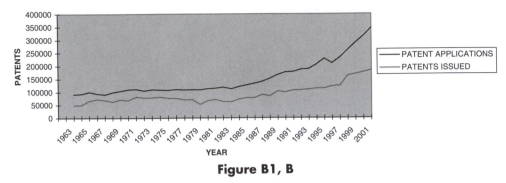

Figure B1, B

speed. Figure B2, A and B, are graphs showing the increase in population over the period from 1963 to 1996.

EFFECTS OF FASTER DOUBLING OF MAN'S KNOWLEDGE

Faster Pace of Life

The effects of this increase in mankind's knowledge are creating a faster pace of life. Information flows faster, with more sources of input. There is more inaccuracy in the information causing false starts and changes in direction. Many aspects of life in modern times have become competitive from a quantity rather than a quality point of view. People are racing toward an ever more mobile target. The result is a faster pace with questionable value added in many of the transactions.

Increasing Pressure

This faster pace is causing increased pressure on our society. We can see the pressure manifest itself in loss of patience, wider swings in violence and crime and loss of reflection and relaxation in people. In many cases this pressure, unnatural by its extreme, is causing anger in many. The pace is also accelerating as time progresses. There is a higher level of learning among people, with requirements and demands that are ever increasing.

Disconnection Index

With all this excessive movement forward, the danger of becoming disconnected is ever more present. This "disconnection index" is a measure of a person's lack of keeping pace with the changes around them. Fifty years ago, a person could have conceivably left the earth for 10 years, returned, and essentially picked up where they left off. For example, in 1947, compared to 1957, the amount of change in life was rather small and not easily discernible, compared to the changes from 1987 to 1997. A person leaving the earth for 10 years in 1987 and returning in 1997 would find sweeping changes in foreign policy, security of the European continent, vast changes in communications, computer

WORLD POPULATION DATA			
YEAR	POPULATION	DIFFERENCE	%GROWTH
	3,136,082,730		
1963	3,205,488,224	69,405,494	2.21
1964	3,276,491,201	71,002,977	2.22
1965	3,345,409,879	68,918,678	2.10
1966	3,415,545,874	70,135,995	2.10
1967	3,485,194,887	69,649,013	2.04
1968	3,556,946,153	71,751,266	2.06
1969	3,631,478,167	74,532,014	2.10
1970	3,706,601,448	75,123,281	2.07
1971	3,783,996,830	77,395,382	2.09
1972	3,860,789,944	76,793,114	2.03
1973	3,937,179,217	76,389,273	1.98
1974	4,012,904,290	75,725,073	1.92
1975	4,086,387,665	73,483,375	1.83
1976	4,158,437,090	72,049,425	1.76
1977	4,230,768,221	72,331,131	1.74
1978	4,303,037,183	72,268,962	1.71
1979	4,378,225,681	75,188,498	1.75
1980	4,453,863,820	75,638,139	1.73
1981	4,529,899,224	76,035,404	1.71
1982	4,610,062,597	80,163,373	1.77
1983	4,690,307,356	80,244,759	1.74
1984	4,769,630,537	79,323,181	1.69
1985	4,850,224,998	80,594,461	1.69
1986	4,932,580,072	82,355,074	1.70
1987	5,018,293,296	85,713,224	1.74
1988	5,104,636,805	86,343,509	1.72
1989	5,190,697,978	86,061,173	1.69
1990	5,277,725,410	87,027,432	1.68
1991	5,360,628,665	82,903,255	1.57
1992	5,443,740,826	83,112,161	1.55
1993	5,525,753,998	82,013,172	1.51
1994	5,606,338,688	80,584,690	1.46
1995	5,687,011,326	80,672,638	1.44
1996	5,766,435,620	79,424,294	1.40

Figure B2, A

power, access to flow of and availability of information. There are also sweeping change: in the database.

Aging Population

Unfortunately, these changes are being thrust onto the populous as if it was homogeneous and equal in terms of ability to assimilate, absorb, digest, and accept these "improvements." For example, how many people of the "baby boom generation" can easily program a VCR?

WORLD POPULATION DATA

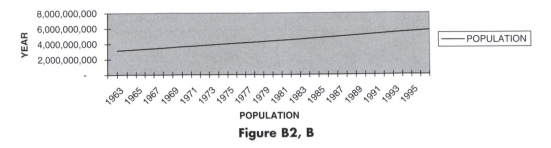

Figure B2, B

How many elderly people feel comfortable with the seven-segment readout for numerics? How many people enjoy reprogramming all of the clocks in the house after a momentary power outage?

As more products get more sophisticated and the population ages, there will be an existence of this disconnection index that will grow naturally. From a product development perspective, it will have to be addressed in order to appeal to the mass market with a single product platform.

Technical Focus Versus Value Focus

Another change that has been taking place is that young people, in their quest for keeping up with these changes, have become technology focused rather than value focused. This is probably a little-observed fact as we lead our busy lives, however, it is a crucial fact to recognize in understanding the demographic changes, the needs, the preferences and the general attitudes of the different segments of our population. It could also be said that this is only occurring in the US, and agreeably it is more noticeable here, however, these changes are occurring all over the world, just at different rates.

Improving the Human Condition

It is important for the new product developer to be cognizant of these trends not only from a planning point of view but also from a social responsibility perspective. The objective of new product development is to make money, foster progress, and improve the human condition. Technology is good, it's required, and it must be accompanied with the wisdom to effectively use it.

EFFECTS OF THE SHORTFALL IN WISDOM ACCOMPANYING THE KNOWLEDGE BASE

Reduced Ability to Cope with Changes

For technology to have a long-lasting benefit to humankind, it must be accompanied by the wisdom to use it. Failure of the wisdom element in the equation results in abuses, and misapplication, and eventual nonacceptance. With many of the recent technological advances, we have seen the accompanied moral dilemma associated with them. These seem to be especially prevalent in the medical and health-related fields. Failure to garner the wisdom to utilize the technology results on an inability to cope with the changes that it brings.

Effects on the Younger Generation

The use of technology without wisdom has a deleterious effect on our younger generation. In the formative years when value systems and responsibilities are being learned, our society has been busy creating technological advances that are virtually limitless. The freedom of speech amendment applied to certain items, generally considered inappropriate for children on the Internet, results in dilemmas for their parents: for example. At the present time, technology is thrust on society. Society then plays with it, and both good and evil may result from its use. Unfortunately, the younger element of our society may be confused in terms of a value system, or they learn their lessons through hard knocks.

Technology Development and the Ability to Manage It

There is always the danger of developing and introducing technology at a rate that outpaces society's ability to manage and integrate it effectively. Rampant development of technology is not progress, as it must relate to the betterment of mankind. The manufacturer must bear some of the brunt of social responsibility in cultivating and introducing new products. For example: a company developing a recreational drug which has the lowest cost, highest margin, no barrier to market entry, no competition, and instant acceptance in the marketplace. However, it is ultimately destructive to the user, is not progressive, and is not a worthwhile endeavor, even though it may satisfy the classical measurements for product success.

Shifting Value System

We have even seen a reduction in the value of human life as a result of the shift in value systems. The rapid increase in knowledge not accompanied by the wisdom to manage it, has refocused our young on the technological and material aspects of life. Hence, we see examples of bizarre behavior, like the theft of designer jackets and shoes and murder of the owner.

Danger of Losing Civilized Status

With this type of behavioral activity, the natural tendency is to try to become desensitize to it and it soon becomes a way of life. Acceptance settles in and we run the risk of loosing civilized status. This is not to say by any means that evil results from technology. It is to recognize that things have a cause and effect and that there is a humanistic element that must be factored into new product development.

Loss of Continuity in Passing on the Value System from Generation to Generation

We must accept that the continuity of a value system is critical passing from one generation to the next. New product development must support that value system. In this way a planned, responsible growth in technology and knowledge will be accompanied with the wisdom to manage it.

Upward Mobility Results

Advanced technology and knowledge will result in upward mobility for the society as a whole. It will discourage fragmentation and splinter groups and allow synergy in improvement in people's lives.

HISTORICAL PERSPECTIVES

Civilization Development

There have been other times in our history when there were periods of rapid growth of knowledge. Depending on which theories one may subscribe to in knowledge development, generally there is agreement in principle of the periods and the driving forces.

Society's progression was made from tribal, fragmented groups driven by symbols and ritual to some of the early empires. Growth and empire development continues through the early empires, ancient Greece and through classical Greece and Rome. Tribal Europe gave way to the feudal Europe and then Medieval Europe. The Renaissance, the Enlightenment, the Victorian era, and the 20th Century brought information and knowledge development of their own.

There were several periods in humankind's history like the Renaissance, and the period following 1859 where rapid development of knowledge occurred. Each brought with it its own developments. The residuals left as the civilization emerged, evolved, and eventually morphed into the next generation civilization.

Parallel to Product Development

It would seem that each era spawns a new generation of knowledge that emerges, grows, flourishes, declines, and dies off to give way to the next generation that evolves. This holds a close parallel to the product life cycle, in that the product emerges, gains acceptance, is copied by competitors (indicating, oddly enough, acceptance in the plurality), flourishes, and then eventually dies off, making way for the next generation of product.

Caveats for the Future

It's important to understand how knowledge levels and development affect our society and its development. Just as the steam engine, a technological revolution that drove far reaching changes in the industrial society, the technological revolution that made knowledge transmission will revolutionize and accelerate drastic changes in the post industrial society.

The caveats for the future especially for product development is that one must understand that change and growth are an inevitable part of the cycle of life and also are inevitable as part of the new product development process. Hence there are only temporary cash cows that generate profit with no additional effort, no products that can enjoy a free ride up the acceptance curve indefinitely and no products that can be sold year after year without investment in their improvement. This again parallels society in the macro sense and the individual sense, that change is inevitable and actually the change must be improvement in order to survive and flourish.

THE PATTERNS OF CIVILIZATION AND EFFECTS ON PRODUCT DEVELOPMENT

There exist patterns in our civilization as time progresses. This affects the marketplace drastically and the market's acceptance for new products. One perspective is that it appears that one pattern has a time base of approximately 80 years or a single human being's long life span. The pattern seems to be broken up into 4 distinct segments as follows:

1. High
2. Awakening
3. Unraveling
4. Crisis

These four segments can be likened to the seasons of change, namely spring, summer, fall and winter respectively. In the high (Spring) a new beginning emerges. This usually follows a crisis of some sort and as society emerges, there is unbridled optimism, and growth. Society expands, prosperity increases civic responsibility increases, and generally, advancement occurs in most aspects.

This phase is followed by an awakening phase, in which society begins to question itself, and its direction. Spiritual agendas begin to dominate the social landscape. Scientific agendas begin to loose influence, public order begins to reduce, and society's general level of enthusiasm begins to wane. War and conflict, driven by newfound idealisms, break out.

The next phase is the unraveling phase. This phase emulates the tone of its title. Society begins to acquiesce, public trust declines, guilt begins to dominate the moral landscape, and general attitudes of pessimism abound.

People "feel" and no longer do. Wars and other conflicts are fought with moral fervor and debates on idealism and morality abound. There are "liberating" forces at work, fragmenting society as they have had their fill of "rebirth and soul searching". It has low resistance to external forces, placing it in extremis.

As can be expected, this phase leads into the crisis phase. This phase separates the conversation and soul searching of the previous phases into actual conflict. The threats become real, there is a sense of public urgency, and the family unit strengthens. War and conflict are prosecuted furiously and with purpose. The government "governs" and extraneous laws are shunted aside to make way to meet the crisis head on. Crime diminishes and optimism increases as the society shifts from individualism to national or civic purpose. This phase then wanes and ushers in the next high phase. Each phase seems to average 20 years in length.

For a historical perspective, World War I and II represented an international crisis where national and international purpose drove many changes into society. The government governed, and created a Promethean war machine. Society shifted to a more unified entity with purpose.

Following this was a tranquil period of growth or "high" in the 1950s followed by the beginning of the awakening in the late 1960s and early 1970s. The awakening evolved into the unraveling of the 1980s and that brings us to the eve of the next crisis.

The 1950s and early 1960s were a tranquil time where growth and prosperity abounded. This high spawned the moon venture, the consumerism, the freedom, and enthusiasm. The late 1960s and early 1970s ushered in the awakening, a period of searching, questioning, and shift to individualism. This gave way to the 1980s when unraveling started to take place. This unraveling started in Poland in 1981 with the Solidarity movement and ended with the unraveling of the Soviet Empire. During these years, personal freedoms liberated people from oppressive, complacent empires. The unraveling continues until the next crisis.

The history of cycles and trends needs to be understood and considered within the framework of new product development. Some products, while they may be achievable, are not right for the "times." For example, would instant access and connectivity and the freedom of the Internet be successful in the forties and fifties with the Red scares and security breaches? Would products promoting individualism and personal freedom be successful in an era where national and civic purposes dominate the society? Or more basically, would fuel efficiency and environment issues embodied in automobiles be accepted in an era where performance and styling dominate the taste of the consumer? These are just a few examples of where product planners need to understand the periods we are in, and factor this into the product and company strategy.

Figure B3 is a chart showing the various time periods and the keywords that tend to underscore the market preference or focus.

In a crisis period, products need to rapidly get practical. The market is looking for products that are functional, affordable, able to be manufactured with available materials, and are reliable. Technology generally needs to be proven because the market will have little tolerance for ushering in new technology. After all, they are prosecuting a crisis.

In a period following the crisis or high, the market is much more accepting of new things, implementations of new technology, and taking products and the people they serve in new

TIME PERIOD	PRODUCT STRATEGY AND THRUST
CRISIS	FUNCTIONAL, PRACTICAL, AVAILABLE, SECURITY, TECHNOLOGY, RELIABILITY
HIGH	MARKET EXPLORATORY, NEW DIRECTIONS
AWAKENING	LESS MATERIALISM, THOUGHT ORIENTED, REJECTION OF THE PREVIOUS
UNRAVELING	NICHES, RESPOND TO FADS, SHORT CYCLE WITH NO RESIDUALS

Figure B3

directions. Product planners have more freedom in implementations and platforms, as well as pricing. This is when the market, which is becoming more prosperous, can afford higher prices that allow the introduction of new technologies and start the pathway down the learning curve.

The awakening period, an era of questioning soul searching and learning must be addressed with new products aimed at a philosophy of less materialism, cerebral orientation and alternatives to those espoused by the previous generation. These are good times for flanking product maneuvers, as they present nontraditional approaches to satisfy the customer need.

Finally, the unraveling period uses products oriented toward market niches, and the company must be able to respond to fads. Product life cycles are incredibly short with no residual sales, as the market locks on to a new product for a short time and then moves on to something else.

These represent some perspectives in product planning as it relates to the various periods society goes through. It is by no means complete; however, it can serve as a starting point for your program to focus in on the approach that will have the best outcome for your company.

US NEWS AND WORLD REPORT'S 50-YEAR PREDICTION FROM 1983

As an integral part of the discussion of product development and product planning, it is imperative to understand patterns of history and the driving forces that affect change. Consequently a review of some examples of predictions made in terms of society and technology from a previous point in time and comparing them with the actual results to date will provide some understanding. The examples listed come from the *US News and World Report's* news magazine, May 9, 1983 issue. The issue deals with forecasting elements of our society in the year 2033, 50 years from the date of issue. Table B4 summarizes the original projection, as can be seen by the forecast made. Many things have come to pass.

Table B4
Trends from 1983 to 2033: Predictions by
***US NEWS AND WORLD REPORT* magazine**

AREA	PREDICTIONS
Education	60% of Americans will have attended college vs 30% in 1983.
	Industry will become more involved in education.
	Gap between the educated haves and have nots will widen.
Values	Family unit will strengthen.
	Family unit will lead and dictate policies.
Cities	Population will be more spread out.
	Underground delivery systems will move goods and people.
	Focus will shift from industry to entertainment and pleasure.
Professions	Education is king, continuing education is queen.
	More changes in careers.
	Dual career families—80-90% of women working.
Cultural	Differentiation in our culture, cliques based on interests.
	Technology will allow people to limit news received; fragmentation of society.
	Home will remain focal point of leisure/electronic moles.
	Home computer access to vast amounts of updated information.
Aging	1983: 26.8% of Americans are 65 or older. In 2033: 65.8% will be 65 or older.
	Intergenerational conflict because of shift in the federal budget.
Resources	Global resources will be less costly and more plentiful.
Energy sources	Autos: 100 miles per gal; family sedan: two seater electric runabout.
Politics	Politics will focus more on minority issues.
	Hispanics will overtake African-Americans as largest minority.
	US will remain #1 as USSR falls behind.
Economy	Economic strength through know how.
	Manufacturing supremacy will be a must.

As a starting point, let's examine what type of world we lived in 1983. The year 1983 was an eventful year both domestically and internationally. The time was market by new and exciting products, changes in the political complexion of the world, and last bastions of outdated systems trying to hold on. If history has taught us anything, it's that change does occur, and that very little is exempt from change. The 1980s started out with the US still reeling from the Iranian hostage crisis.

The US was loosing international ground fast, both in stature and market share. International freedom among all peoples was still a germ of an idea, and a small fire smoldering in Eastern Europe.

In 1983, Lech Walesa, who was a Polish freedom symbol, was awarded the Nobel Peace Prize for work on an underground movement named Solidarity. Ronald Reagan was the President of the United States.

The USSR was a superpower that was a threat to peace worldwide. A Star Wars initiative was introduced by the US designed to enter the US and the Soviet nation into a research and development program so immense and an ante so large that would eventually break the financial back of the Soviet Union.

The Soviet Union, in a display of belligerence, fired a missile at a Korean passenger jet-liner with military fighters, killing all aboard, over Soviet airspace, to international outrage. Sally Ride was the first US woman astronaut in a flawless shuttle program.

The US invaded Grenada in order to secure our visiting students and stabilize the government.

A Central American, Manual Noreiga was named commander of Panama defense forces, later to be removed from Panama by a US-led invasion. Benito Aquino, husband of Corazon Aquino (leader of the Philippines) was assassinated, prompting her entry into leadership to continue the work of reform.

Yasser Arafat, the head of the Palestine Liberation Organization (PLO), and 4000 PLO members were expelled from Lebanon to Tunisia. A car bomb was detonated by a terrorist in Beruit, killing over 200 US marines.

A small plastic record album like disc was introduced to music audiophiles as a means for playing high quality sound from this new medium. The year before a computer company known as Apple would break the paradigm barrier for computers with a new product known as Macintosh.

Now, reevaluate Table B4's prediction with this as a starting point.

WHAT THE FUTURE COULD BRING

Impact of these Driving Forces on Product Development

Demand

The impact on product development in these driving forces is widespread. New product development managers must now think in terms of product platforms that are modifiable rather than fixed designs. The fixed-design approach is simply too costly in terms of time and money over the life of the business. The platform approach has a higher initial cost and higher barrier to entry but offers lower costs to change and modify. Market conditions have positioned this platform to force managers to forecast future product features and create platforms that will accommodate the future improvements as the market develops. The product platforms must now be modifiable, iterative, flexible, and still remain reliable, and be consistent with quality programs.

Speed

The speed at which these products must be developed in order to gain market share is getting faster. Since the product features must be implemented in an iterative manner to meet the timely demand for these features, it is imperative that the platform on which the product is developed accommodates the improvements. These features must be thought out and provided for. Even though the speed is getting faster, the safety concern, reliability, and accuracy of product configuration must all be maintained. In addition, many companies are now competing by improving their collateral services.

Changing Pace

This changing pace in product development not only affects the manufacturer; it also affects the customer base. As fast as the manufacturer develops the features, the customer infrastructure must absorb the change. This means training, cataloging, documenting and factoring of any spare parts if appropriate. The manufacturer must make the feature palatable to the customer to facilitate acceptance. The degree to which they can make this seamless is the degree to which acceptance is made easier.

Functionality and Complexity

As time marches on and the products developed gain functionality, and complexity, the manufacturer must appropriately match the functionality to the customer need without undo complexity. We are all familiar with the VCR scenario, all the functionality of delayed starts and stops, and other features that is often considered too complex for the average patience level of the adult user. Similarly, other products can embody too much functionality in trying to meet all market needs with a single version. Undo complexity can be a large deterrent to sales, and must be factored into the equation.

Cost Pressure

All product management roles must concern themselves with cost pressures. With most products having any labor content there is built in potential for costs to increment. In addition, materials play a major factor in contribution to cost increases. Vendors who have been pressured into initial lower prices eventually try to recover margin lost. Obsolescence of certain materials necessitates replacement with many times a costlier component, thus reducing margin. The ability to increase functionalism without additional cost can be key to holding on to market share.

Spin-Off Developments

The increase in the speed of development of products also many times results in potential spin-off products being ignored. While these technologies may make very valuable products, often times they are left behind in favor of the immediate marketing effort.

Response of Manufacturer

The manufacturer should be planning for future not just immediate market response This point is not to be underestimated. With the product planning referred to earlier in this section, the manufacturer must plan the product evolution, and factor in all of the future embodiments. Failure to incorporate this will result in undo development costs and time. The platform must be designed so as to lay in the improvements over time in a seamless manner, with little or no incremental expense. A diligent process in the beginning will result in faster platform improvements and better competitive positioning.

As a final note to this section, Maslow's hierarchy of human needs identifies several levels of need, starting with basic survival requirements and progressing through the levels of social interaction to the pinnacle level of self actualization. The important thing to factor into the new product planning process is to position the product along with the hierarchy and visualize where it fits into the customer's needs. As part of this exercise, the strengths and weaknesses of the marketing story and the transactional analysis must be reviewed against this hierarchy. Understanding the driving forces behind the acceptance or rejection of the product is important. A successful product marketing strategy targets the right level and factors in the customer's motivation in light of the hierarchy of needs.

INFORMATION FLOW, CONNECTIVITY SPEED

As evidenced by recent Internet activity and other electronic communications schemes, the world is connected by a medium allowing information to flow in both directions. Information, facts, figures, data, and opinion are available at the touch of a button. This connectivity and speed are driving the pace of our society faster. Information is currently available at one's fingertips in seconds, and what was once only available in a library after hours of research is available immediately.

Assessing the Stretch in Product Planning

How much of a stretch should be taken in planning the future products? Once the product concept is defined, the effort must be directed to determine how much of a stretch in terms of features that the product should have. How much is enough, but not too much to limit manufacturability or acceptance? New product planning is like aiming at a moving target. One must "lead" the target so that the ordnance meets the target at a point in space when the two meet.

The analogy holds true for development where the velocity of the ordnance is the efficiency and speed of product development and the target's speed is a measure of the market's preference for functionality of the product.

Select the point in time the product meets the market and make sure they connect. With new product development, you have an advantage to make mid course corrections to ensure you hit the target; however, the target (market) keeps changing its course as well. One can increase their chances of hitting the target by using a rapid fire approach of rapidly introducing products with small incremental improvements; however, that is cost-and time-prohibitive in most cases.

STRETCHING VERSUS INCREMENTALISM IN PRODUCT DEVELOPMENT EFFORTS

The Necessity of the Stretch

As discussed earlier, the competition does not remain motionless while your company develops a new product. It is therefore necessary to provide enough of a stretch in the planning stages to position the product favorably *at* introduction. It is also incumbent on the product manager to see to it that enough resources are available to execute the required development in a timely manner. Otherwise, the opportunity will be missed.

The Faster Pace

The faster pace makes increments only enough to keep up (me, too). This is the challenge for the product planning function; whether it is a committee or a product champion. The delicate balance must be struck between stretching too far or not enough. Negotiating this balance successfully, however, will pay off big returns at product introduction and during the initial growth stages of the product life cycle.

INDEX